Geometries, Groups and Algebras in the Nineteenth Century
–
A History

by Professor Isaak Moiseevich (I. M.) Yaglom

Translated into English by Sergei Sossinsky

with a new foreword by

Richard Bozulich
Emeritus Lecturer of Mathematics
Shonan Institute of Technology
Fujisawa, Japan

ISHI PRESS
INTERNATIONAL

Geometries, Groups and Algebras in the Nineteenth Century – A History

by Professor Isaak Moiseevich (I. M.) Yaglom

Translated into English by Sergei Sossinsky

with a new foreword by
Richard Bozulich, Emeritus Lecturer of Mathematics, Shonan Institute of Technology, Fujisawa, Japan

Cover design by Sam Sloan

First published in Moscow, Soviet Union in 1977 as Novoe v zhizni, nauke, tekhnike : Seriiâ Matematika, kibernetika; No 11, 1977

Current Printing in October, 2009
by Ishi Press in New York and Tokyo

ISBN 4-87187-836-8
978-4-87187-836-4

Ishi Press International
1664 Davidson Avenue, Suite 1B
Bronx NY 10453-7877
1-917-507-7226

Printed in the United States of America

Foreword to this New Edition

I am honored to write a foreword to this new edition of an English translation of a very important, masterful work of exposition by Professor Isaak Moiseevich (I. M.) Yaglom. For those like me who were taught advanced mathematics dogmatically, without any indication of the origins of those abstract concepts, this historical, biographical book is a delightful revelation.

Yaglom's treatise is a history primarily of geometries, groups and algebras in the nineteenth century, culminating in the works of Felix Klein and Sophus Lie in the latter part, spilling over into the early part of the 20th century. The less advanced mathematics is clearly explained, while the more advanced material is described, examples are given, and the reader is provided references in the notes for further study (I have listed many more recent books at the end of this foreword). Yaglom includes much biographical information that makes his presentation quite colorful.

A theme that occurs throughout the book, beginning in Chapter 2, is the rather rough classification of certain mathematicians as left-brained "logicians" vs. right-brained "physicists," who have approached their subjects differently and were sometimes antagonistic to one another. Thus Leibniz vs. Newton, Weierstrass vs. Riemann and Klein, Bolyai vs. Lobachevsky, Hilbert vs. Poincaré. In the twentieth century, Jean Dieudonné vs. René Thom. Most great mathematicians, such as Barry Mazur or Jean-Pierre Serre in our time, do not fit comfortably into either pigeon hole.

Yaglom described this book as follows:

"The book is meant for different groups of readers ... This explains its unusual structure: the main text is followed by numerous and often extensive notes which contain, among other things, most of the numerous references to the literature. When reading the book for the first time, the reader is advised to ignore the notes or to make selective use of them...

The possible variety of interests and backgrounds of the book's prospective readers has resulted in the diversity of its contents. Some parts may interest the mathematician, while other parts are addressed to people who are more interested in history in general and in the history of science in particular. This is why the levels of difficulty of the references vary greatly.

The book is based on the author's lectures to graduate-level students majoring in "pure" mathematics at Yaroslavl University, most of whom subsequently go on to teach in secondary schools; this probably influenced the choice of material and the nature of the narrative.

The book was significantly influenced by discussions on the subject with the author's friends and colleagues, in particular S. C. Gindinkin, B. A. Rosenfeld and A. M. Yaglom. However, the author assumes full responsibility for the book's content, in particular for the thoughts which may occasionally seem controversial.

The author would like to express his gratitude to the translator Sergei Sossinsky, for his precise rendering of the Russian original into English, to my colleague Alexei Sossinsky for his assistance, and to Hardy Grant and Abe Shenitzer for editing the English text."

I summarize highlights of this book:

Chapter 1

Begins with the tragic story of the very short life of Evariste Galois, killed in a duel, balanced by the everlasting new ideas he introduced to reveal the group and field-theoretic structures that underlie the problem of solvability by radicals of polynomial equations. (Below in my list of new references readers will find that various other Galois-type theories remain a vital area of research in current mathematics.)

Niels Henrik Abel first showed the unsolvability by radicals of the quintic equation. He did other work in analysis of lasting importance, not recognized by Gauss but eventually published in Crelle's new journal and finally recognized by Jacobi, who named the new mathematical gadgets after Abel — too late for poor Abel, who died very young from tuberculosis.

It was Camille Jordan who belatedly discovered Galois' letters to Cauchy, understood them, and, in his *Treatise on Permutations*, published the first systematic exposition of Galois theory, elaborating on the basic group-theoretic ideas needed (such as normal series of subgroups).

Yaglom explains these basic notions in some detail. He discusses the much earlier work by Tartaglia, Cardano et. al. in solving cubic and quartic equations by radicals. Galois' theorem is that an equation is solvable in radicals if and only if its Galois group is a solvable group.

The chapter and its notes also mention (1) the contributions of Lagrange and Euler that preceded the work of Galois, plus biographical information about those two great mathematicians; (2) the work of Paolo Ruffini on algebra and unsolvability by radicals, unappreciated in his lifetime; (3) some elementary group theory results found in Cauchy's papers after his death.

Chapter 2

Introduces Klein and Lie, who both studied with Jordan (and with Darboux, a leading differential geometer) in Paris.

Hermann Weyl's distinction between mathematicians who are primarily logician-algebraist-types like Leibniz, Weierstrass and Landau, vs. those, such as Newton, Klein, Riemann and Helmholtz, whose minds work more like physicists.

Chapter 3

Reviews and explains the varied works in projective and inversive geometry throughout the nineteenth century, a "golden age." Discusses and explains work by the most prominent contributors: Monge, Poncelet, Gergonne, Chasles, Möbius, Steiner, von Staudt and Plücker. Earlier work by Desargues, Pascal and Euler. Controversy between the synthetic and algebraic approaches. Colorful biographical information about all these men.

Chapter 4

Reviews and explains the discovery and development of non-Euclidean geometries in the 19th century, starting earlier with sowing work by Saccheri and Lambert, budding work by Schweikart and Taurinus, independent flowering bursts by Gauss, J. Bolyai and Lobachevsky, finally reaping with models and applications by Beltrami, Cayley-Klein and Poincaré. Klein's unification of Euclidean and non-Euclidean geometries within projective geometry.

Then the grand new conception of geometries on manifolds of all dimensions propounded by Riemann. The rigorous foundations by Hilbert for elementary Euclidean and hyperbolic geometries at the turn of the twentieth century, his proofs of Dirichlet's principle (its use by Riemann had been sharply criticized by Weierstrass) and of the impossibility of smoothly embedding the hyperbolic plane in Euclidean three-space.

As a mathematician in Russia, Yaglom knew more about aspects of Lobachevsky's work than we usually hear about in the West and he presents some of that information to us. Lobachevsky had in his work the tools needed to prove that his non-Euclidean geometry was consistent relative to the Euclidean, but thinking like a physicist about the nature of space, he did not fret about that question the way the "logician" Bolyai did.

The notes to Chapter 4 mention Minkowskian geometry (the setting for Einstein's special relativity), Yaglom's own work on other Cayley-Klein geometries, Clifford, the Italian school of Riemannian geometry, plus other fascinating historical details.

Chapter 5

Presents the new algebras that appeared in the nineteenth century and their applications to geometries: (1) Grassman's n-dimensional vector spaces, his algebra (called "exterior" by Bourbaki, applied to differential forms), the initial rejection of his work, fully recognized after his death by Klein and Lie; (2) Hamilton's quaternions; (3) Cayley's and Graves' octaves/octonions; (4) studies of "hypercomplex" numbers by the German schools (e.g., Hankel, Lie, Frobenius, F. Schur); (5) Clifford algebras.

The notes to Chapter 5 mine biographical nuggets about French encyclopedist d'Alembert, abandoned as a baby; Englishman Sylvester, considered the founder of American mathematics when he was at Johns Hopkins; German Grassman, who could not obtain a professorship and had to teach in secondary schools; Italian Peano's arithmetic influenced by Grassman; Swiss Schläfli's work on the n-dimensional Euler formula and its application to regular polytopes; applications to physics by Brits Maxwell and Dirac; American Gibbs' vector calculus; Ukrainian Yaglom's own applications to geometries.

Chapter 6

Jordan taught the importance of transformation groups, such as the continuous group of automorphisms of a geometry or the discrete group of symmetries of a geometric figure. Sophus Lie developed a general theory of continuous groups with a smooth underlying manifold structure, now named after him. His main tool was the algebra (also named after him) that he associated to each of his groups, whose underlying vector space is the tangent space at the identity element and whose multiplication is the Lie bracket induced by differentiating the commutator function of two group elements near the identity. (Example: for the group of rotations of Euclidean 3-space, the Lie bracket is the familiar vector cross product.)

Lie proved that each abstract Lie algebra comes from a local Lie group (which Élie Cartan later proved can be extended to a simply connected global Lie group).

The solution to the problem of classifying simple Lie groups was given with incomplete proofs by Lie's colleague Killing; a complete proof was later provided by Élie Cartan, and his proof was improved by both van der Waerden and Dynkin. The four infinite series of simple Lie groups are the groups of direct isometries of real Euclidean spaces (one series for odd dimensions and one series for even), and the groups of automorphisms of complex and quaternionic Hermitian spaces of all dimensions.

There are also five exceptional simple Lie groups, which Freudenthal and Tits in the 20th century showed are derived from isometries of various octave planes. We thus see the connection of the work on algebras discussed in Chapter 5 with Lie group theory.

Lie applied his theory to differential equations, obtaining results analogous to those of Galois for polynomial equations — the analogue of solution by radicals is solution of differential equations by quadratures.

Lie's contact transformations and their relation to Möbius circular inversions and Laguerre's transformations.

Lie's early joint work with Klein on "W-curves" (curves which admit a transitive group of projective transformations).

The notes to Chapter 6 also mention (1) the determination of all crystallographic groups by Fedorov, Schönflies and Barlow; (2) Hilbert's Fifth Problem, to determine which topological groups are Lie groups, solved in the 20th century (by Gleason, Montgomery and Zippin); (3) work by B. and C. Pierce on linear associative algebras; (4) Dynkin diagrams applied also to classifying singularities of smooth maps, etc.; (5) classification of simple finite groups, including Chevalley groups of Lie type over finite fields and the 26 sporadic groups.

Chapter 7

High school Euclidean geometry breaks into a mixture of two subjects: similarity geometry, where there is no measurement of lengths and areas, and metric geometry, where such measurement is derived from an arbitrarily chosen unit segment. The properties studied in each geometry are those that are preserved by the corresponding groups of automorphisms — the group of similitudes for similarity geometry, and the smaller group of isometries for metric geometry.

Felix Klein generalized this example in his Erlangen Program: he initially defined "geometry" as the study of the invariants of some group of transformations of some domain. Yaglom's examples: (1) in metric plane Euclidean geometry, the only homogeneous "curves" are lines and circles, whereas in similarity geometry there are also the logarithmic spirals; (2) the projective group, its subgroup the affine group, its subgroup of similitudes, and its subgroup of Euclidean isometries; (3) the subgroups of the projective group consisting of hyperbolic or of elliptic isometries.

Klein's definition had to be refined to include a "generating element" which is the building block for all the figures to be considered; thus for Plücker's line geometry, the generating element is a line, not a point, though the group and the domain are still those of projective geometry. The generating element has a subgroup leaving it invariant, its stabilizer. Klein's ultimate approach was simply to start with a group and one of its subgroups; the cosets with respect to that subgroup become the possible generating elements for his alternative approach to geometry, the figures in that geometry being the sets of cosets that are invariant under transformations from the ambient group.

Application to the Galilean principle of relativity, and more generally, to Einstein's principle of relativity, based on the Lorentz group of transformations depending on a constant c (the Galilean case has $c = \infty$).

The notes mention Chern finishing off his teacher Blaschke's study of "integral geometry" defined within Klein's framework, so that Gelfand felt free to steal that name for his different work.

Chapter 8

Further biographical notes about the careers of Lie and Klein. While they were both very successful and famous for their highly original work, nevertheless they both suffered from bouts of depression and had "nervous breakdowns."

Klein did pioneering work on discrete groups of automorphic functions. He failed to notice that they could be understood as subgroups of the group of isometries of the hyperbolic plane, a relationship first discovered by Poincaré (in a famous incident while he was on vacation, boarding a bus). Klein's subsequent interaction and rivalry with Poincaré in these investigations.

Klein's valuable later work in reforming mathematics education.

The notes mention Ehresmann's unification of Klein's geometry with Riemann's under the rubric of a smooth manifold having a fibre bundle endowed with a connection. Further discussion of Poincaré and Hilbert.

A Criticism and a Recommendation

Yaglom's treatment of geometries in the nineteenth century omits a more detailed discussion of the algebraic topology founded by Poincaré, whose work should have been given as much prominence as that of Klein and Lie. Riemann's surface theory also included topological ideas.

A wonderful supplement to this book is Michael Atiyah's lecture *Mathematics in the 20th Century*, American Mathematical Monthly 108 (Aug-Sep 2001), 654–666. Atiyah distinguishes between algebraists and geometers instead of logicians and physicists ("Algebra is concerned with manipulation in time, and geometry is concerned with space.").

Biographical Material about I. M. Yaglom

Isaak Moiseevich Yaglom was born on March 6, 1921, together with his twin brother Akiva Moiseevich, in Kharkov, Ukraine. At age 17 they shared first prize in a Moscow mathematics competition for schoolchildren. Both boys grew up to be famous mathematicians and they wrote several books together.

Akiva (nickname Kika) was known for his outstanding research in both physics and mathematics. His most substantial achievements were related to turbulence theory and applications in fluid mechanics, and, in mathematics, to random processes theory and statistical estimation. He wrote joint papers with the great Russian mathematicians Kolmogorov and Gelfand. He outlived his twin brother by 19 years, dying in Boston in 2007 when he had been visiting M. I. T.

Isaak (nickname Isya) did research in various geometries, particularly in the area of degenerate projective metrics and the application of algebras with degenerate norms to geometries. He was known primarily as a great teacher and prolific author. He was the leader in improving high school mathematics education in the USSR for many years. He published over 40 books and many articles in Russian. His books that have English translations are (besides this one):

Complex Numbers in Geometry
Geometric Transformations (4 volumes)
A Simple Non-Euclidean Geometry and its Physical Basis
Mathematical Structures and Mathematical Models
Challenging Mathematical Problems with Elementary Solutions (with A. M. Yaglom)
Probability and Information (with A. M. Yaglom)
Convex Figures (with V. G. Boltianskii)
The USSR Olympiad Problem Book: Selected Problems and Theorems of Elementary Mathematics (with N. N. Chentzov and D. O. Shklarsky)
Higher Mathematics for Beginning Physicists and Engineers (with Y. B. Zeldovich)
Induction in Geometry (with L. I. Golovina)

According to the seven prominent Russian authors of Isaak's obituary ("Russian Mathematical Surveys" 44 (1989), pp. 225-227):

"...the breadth of his interests was truly extraordinary: he was seriously interested in history and philosophy, passionately loved and had a good knowledge of literature and art, often came forward with reports and lectures on the most diverse topics (for example, on Blok, Akhmatova, and the Dutch painter Escher), actively took part in the work of the cinema club in Yaroslavl and the music club at the House of Composers in Moscow, and was a continual participant of conferences on mathematical linguistics and on semiotics."

In a reminiscence by B. A. Rosenfeld (p. 98 of the second edition of *Golden Years of Moscow Mathematics*, Smilka Zdravkovska and Peter Duren eds., AMS 2007), who collaborated with Isaak in making Russian translations of Western mathematical works and in doing geometric research:

"... Isya became a graduate student of V. F. Kagan and Kika became a student of A. N. Kolmogorov. The brothers were as alike as 'two drops of water' and they were often confused for each other. When Isya defended his candidate's dissertation they had one suit coat between them; at the time of his inaugural speech Isya wore this coat, which he then gave to Kika. After the Academic Council decided to award the degree and everyone went to congratulate Isya, Kagan embraced and kissed Kika, thinking he was Isya."

"... Yaglom was a close friend with many of those who fought for human rights in the USSR, particularly with Andrei Sakharov, who gave the eulogy at the memorial service."

Additional Book References for this Edition

References are arranged roughly by the main topics covered in Yaglom's book. Naturally I could not include all the vast relevant literature. I have indicated with ** the advanced books and with * quite elementary books. We are blessed to have so many good books!

Histories of Mathematics, Biographies of Mathematicians

Mathematics Through the Ages: A Gentle History for Teachers and Others, Expanded Second Edition,* by William P. Berlinghoff and Fernando Q. Gouvêa. Mathematical Association of America: Oxton House Publishing (2003)

Mathematics and its History (2nd Edition),* by John Stillwell. Springer (2004)

Yearning for the Impossible: The Surprising Truths of Mathematics,* by John Stillwell. AK Peters (2006)

Mathematical Evolutions, ed. by Abe Shenitzer and John Stillwell. MAA (2002)

Plato's Ghost: The Modernist Transformation of Mathematics, by Jeremy Gray. Princeton University Press (2008)

Tales of Mathematicians and Physicists (2nd Edition), by Simon Gindikin, translated by A. Shuchat. Springer (2009)

Remarkable Mathematicians: From Euler to von Neumann, by Ioan James. Cambridge University Press (2003)

The Rainbow of Mathematics: A History of the Mathematical Sciences, by Ivor Grattan-Guinness. W. W. Norton (2000)

The Princeton Companion to Mathematics, edited by Timothy Gowers, June Barrow-Green, and Imre Leader. Princeton University Press (2008)

Mathematics of the 19th Century, Volume I: Mathematical Logic–Algebra–Number Theory–Probability Theory. Birkhäuser Basel 2nd Edition (2001)

Mathematics of the 19th Century: Volume II: Geometry, Analytic Function Theory. Birkhäuser Basel (1996). Both volumes edited by A. N. Kolmogorov and A. P. Yushkevich, translated by A. Shenitzer, H. Grant and O. B. Sheinin

The Map of My Life, by Goro Shimura. Springer (2008)

General Abstract Algebra

A History of Abstract Algebra, by Israel Kleiner. Birkhäuser Boston (2007)

Episodes in the History of Modern Algebra (1800-1950), Jeremy J. Gray and Karen Hunger Parshall, authors and editors. AMS (2007)

Glimpses of Algebra and Geometry (2nd Edition), by Gabor Toth. Springer (2002)

Modern Algebra and the Rise of Mathematical Structures (2nd Edition), by Leo Corry. Birkhäuser Basel (2004)

A First Course in Abstract Algebra (7th Edition), by John B. Fraleigh . Addison Wesley (2002)

Undergraduate Commutative Algebra, by Miles Reid. Cambridge University Press (1996)

Noncommutative Algebra (Graduate Texts in Mathematics),** by Benson Farb and R. Keith Dennis. Springer (1993)

Algebra (3rd Edition),** by Serge Lang. Springer (2005)

Lectures on Algebra, Volume 1,** by S. S. Abhyankar. World Scientific Publishing Company (2006)

Algebra I: Chapters 1-3.** by Nicholas Bourbaki . Springer (2008)

Algebra II: Chapters 4-7.** by Nicholas Bourbaki. Springer (2003)

Commutative Algebra: Chapters 1-7.** by Nicholas Bourbaki, translated by P. M. Cohn and J. Howie. Springer (2007)

Algebra (Graduate Studies in Mathematics),** by I. Martin Isaacs. AMS (2009)

Galois

Evariste Galois (1811-1832),* by Laura Toti Rigatelli. Birkhäuser Basel (1996)

Galois Theory for Beginners: A Historical Perspective,* by Jorg Bewersdorff, translated by David Kramer. AMS (2006)

Exploratory Galois Theory, by John Swallow. Cambridge University Press (2004)

Galois Theory (Third Edition), by Ian Stewart. Chapman & Hall/CRC (2003)

Galois' Theory of Algebraic Equations, by Jean-Pierre Tignol. World Scientific Publishing (2001)

Galois Theory,** by David A. Cox. Wiley-Interscience (2004)

Galois Theory,** by Jean-Pierre Escofier, translated by L. Schneps. Springer (2000)

Galois Theories (Cambridge Studies in Advanced Mathematics),** by Francis Borceux. Cambridge University Press (2008)

Groups as Galois Groups: An Introduction,** by Helmut Volklein. Cambridge University Press (2008)

Topics in Galois Theory (2nd Edition),** by Jean-Pierre Serre. A K Peters (2007)

Inverse Galois Theory,** by Gunter Malle and B. H. Matzat. Springer (2002)

Geometric Galois Actions: Volume 1, Around Grothendieck's Esquisse,** edited by Leila Schneps and Pierre Lochak. Cambridge University Press (1997)

Galois Groups and Fundamental Groups,** edited by Leila Schneps. Cambridge University Press (2003)

Galois Theory of Linear Differential Equations,** by Marius van der Put and Michael F. Singer. Springer (2003)

Differential Galois Theory and Lie-Vessiot Systems,** by David Blázquez-Sanz. VDM Verlag Dr. Mueller e.K. (2008)

Abel

Niels Henrik Abel and his Times,* by Arild Stubhaug, translated by Richard R. Daly. Springer (2000)

The Equation That Couldn't Be Solved: How a Mathematical Genius Discovered the Language of Symmetry,* by Mario Livio. Simon & Schuster (2006)

Abel's Proof: An Essay on the Sources and Meaning of Mathematical Unsolvability, by Peter Pesic. MIT Press (2004)

The Legacy of Niels Henrik Abel: The Abel Bicentennial, Oslo 2002,** edited by Olav A. Laudal and Ragni Piene. Springer (2004)

Abel's Theorem in Problems and Solutions: Based on the lectures of Professor V.I. Arnold,* by V. B.Alekseev, translated by Francesca Aicardi. Springer (2004)

Introduction to the Classical Theory of Abelian Functions,** by A.I. Markushevich. American Mathematical Society (2006)

Abelian Varieties with Complex Multiplication and Modular Functions,** by Goro Shimura. Princeton University Press (1997)

Complex Abelian Varieties (2nd augmented edition),** by Christina Birkenhake and Herbert Lange. Springer (2004)

Symmetry

Why Beauty Is Truth: A History of Symmetry,* by Ian Stewart. Basic Books (2007)

The Symmetries of Things, by John Horton Conway, Heidi Burgiel and Chaim Goodman-Strauss. A K Peters (2008)

Fearless Symmetry: Exposing the Hidden Patterns of Numbers, by Avner Ash and Robert Gross, with a Preface by Barry Mazur. Princeton University Press (2008)

Symmetry and the Monster: The Story of One of the Greatest Quests of Mathematics, by Mark Ronan. Oxford University Press, USA (2006)

Symmetries in Physics: Philosophical Reflections, edited by Katherine Brading and Elena Castellani. Cambridge University Press (2003)

Mirror Symmetry and Algebraic Geometry,** by David A. Cox and Sheldon Katz. American Mathematical Society (1999)

Group Theory

The Genesis of the Abstract Group Concept: A Contribution to the History of the Origin of Abstract Group Theory, by Hans Wussing. Dover Publications (2007)

Groups and Symmetry (Undergraduate Texts in Mathematics), by M. A. Armstrong. Springer (1997)

Introduction to Group Theory, by Oleg Bogopolski. European Mathematical Society (2008)

Permutation Groups, by Peter J. Cameron. Cambridge University Press (1999)

A Course in the Theory of Groups (Graduate Texts in Mathematics, 2nd edition),** by Derek J. S. Robinson. Springer (1995)

Finite Group Theory (2nd edition),** by M. Aschbacher. Cambridge University Press (2000)

Finite Group Theory (Graduate Studies in Mathematics),** by I. Martin Isaacs. American Mathematical Society (2008)

Theory of Finite Simple Groups,** by Gerhard O. Michler. Cambridge University Press (2006)

Linear Representations of Finite Groups (Graduate Texts in Mathematics),** by Jean-Pierre Serre. Springer (1996)

Representations and Characters of Groups (2nd edition),** by Gordon James and Martin Liebeck. Cambridge University Press (2001)

Buildings and Classical Groups,** by Paul B. Garrett. Chapman & Hall/CRC (1997)

Group Theory and Physics,** by S. Sternberg. Cambridge University Press (1995)

Finite Fields

Equations over Finite Fields: An Elementary Approach,* by W. M. Schmidt. Springer (2009)

Finite Fields (Encyclopedia of Mathematics and its Applications, 2nd edition), by Rudolf Lidl and Harald Niederreiter. Cambridge University Press (2008)

Algebraic Curves over a Finite Field, by J. W. P. Hirschfeld, G. Korchmaros and F. Torres. Princeton University Press (2008)

Rational Points on Curves over Finite Fields: Theory and Applications, by Harald Niederreiter and Chaoping Xing. Cambridge University Press (2001)

Error-Correcting Codes and Finite Fields. Student Edition, by Oliver Pretzel. Oxford University Press (1996)

General Geometries

The Four Pillars of Geometry,* by John Stillwell. Springer (2005)

Worlds Out of Nothing: A Course in the History of Geometry in the 19th Century, by Jeremy Gray. Springer (2007)

Geometry, by Michele Audin. Springer (2002)

Geometric Transformations for 3D Modeling (2nd edition), by Michael E. Mortenson. Industrial Press (2007)

The Shape of Space (2nd edition), by Jeffrey Weeks. Marcel Dekker (2002)

Projective Geometry

Projective Geometry: An Introduction, by Rey Casse. Oxford University Press, USA; illustrated edition (2006)

Projective Geometry (2nd edition), by H.S.M. Coxeter. Springer (2003)

Affine and Projective Geometry, by M. K. Bennett. Wiley-Interscience (1995)

Projective Geometries over Finite Fields (2nd edition), by James Hirschfeld. Oxford University Press USA (1998)

Non-Euclidean Geometries

Euclidean and Non-Euclidean Geometries: Development and History (4th edition), by Marvin Jay Greenberg. W. H. Freeman (2007)

Geometry: Euclid and Beyond, by Robin Hartshorne. Springer (2005)

Hyperbolic Geometry (London Mathematical Society Student Texts), by Birger Iversen. Cambridge University Press (1993)

Non-Euclidean Geometry (6th Edition), by H. S. M. Coxeter. Mathematical Association of America (1998)

Elementary Geometry in Hyperbolic Space, by Werner Fenchel. de Gruyter (1989)

Discontinuous Groups of Isometries in the Hyperbolic Plane,** by Werner Fenchel, Jakob Nielsen, Asmus L. Schmidt. de Gruyter (2002)

Lectures on Hyperbolic Geometry,** by R. Benedetti, Carlo Petronio. Springer (1992)

Foundations of Hyperbolic Manifolds (Graduate Texts in Mathematics),** by John G. Ratcliffe. Springer (1994)

A Gyrovector Space Approach to Hyperbolic Geometry,** by Abraham A. Ungar. Morgan and Claypool (2009)

Analytic Hyperbolic Geometry and Albert Einstein's Special Theory of Relativity,** by Abraham A. Ungar. World Scientific (2008)

Riemann

Bernhard Riemann 1826-1866: Turning Points in the Conception of Mathematics, by Detlef Laugwitz (Abe Shenitzer translator). Birkhauser-Boston 2nd edition (2008)

The Riemann Legacy: Riemannian Ideas in Mathematics and Physics, by Krzysztof Maurin. Springer (1997)

An Introduction to Riemann Surfaces, Algebraic Curves and Moduli Spaces,** by Martin Schlichenmaier. Springer, 2nd edition (2007)

Algebraic Curves and Riemann Surfaces,** by Rick Miranda. AMS (1995)

A Panoramic View of Riemannian Geometry,** by Marcel Berger. Springer (2007)

Riemannian Geometry: A Modern Introduction (Cambridge Studies in Advanced Mathematics),** by Isaac Chavel. Cambridge University Press; 2nd edition (2006)

The Riemann Hypothesis: A Resource for the Afficionado and Virtuoso Alike, edited by Peter Borwein, Stephen Choi, Brendan Rooney, Andrea Weirathmueller. Springer (2007)

Poincaré

The Mathematical Heritage of Henri Poincaré (Proceedings of Symposia in Pure Mathematics #39),** edited by Felix Browder. American Mathematical Society (1983)

Non-Euclidean Geometry in the Theory of Automorphic Functions,** by Jacques Hadamard, with historical introductions by Jeremy Gray. Translated by Abe Shenitzer. AMS (1999)

Algebraic Topology,** by Allen Hatcher. Cambridge University Press (2001). Corrected online version at
http://www.math.cornell.edu/~hatcher/AT/ATpage.html

Differential Forms in Algebraic Topology,** by R. Bott and L. W. Tu. Springer (1995)

Quaternions, Octonions, Algebras

Visualizing Quaternions (The Morgan Kaufmann Series in Interactive 3D Technology), by Andrew J. Hanson. Morgan Kaufmann (2006)

On Quaternions and Octonions: Their Geometry, Arithmetic and Symmetry, by John H. Conway and Derek A. Smith. A. K. Peters (2003)

Quaternions and Rotation Sequences: A Primer with Applications to Orbits, Aerospace and Virtual Reality, by J. B. Kuipers. Princeton University Press (2002)

Introduction to Lie Algebras (Springer Undergraduate Mathematics Series), by Karen Erdmann and Mark J. Wildon. Springer (2007)

Graduate Algebra: Non-commutative View (Graduate Studies in Mathematics),** by Louis Halle Rowen. American Mathematical Society (2008)

Clifford Algebras and the Classical Groups,** by Ian R. Porteous. Cambridge University Press (2009)

Geometric Algebra for Physicists,** by Chris Doran and Anthony Lasenby. Cambridge University Press (2007)

Lie Groups

Naive Lie Theory (Undergraduate Texts in Mathematics),* by John Stillwell. Springer (2008)

Emergence of the Theory of Lie Groups: An Essay in the History of Mathematics 1869 – 1926, by Thomas Hawkins. Springer (2000)

Lectures on Lie Groups and Lie Algebras (London Mathematical Society Student Texts), by Roger William Carter, Graeme Segal, Ian Grant Macdonald. Cambridge University Press (1995)

Lie Groups, Lie Algebras, and Representations: An Elementary Introduction, by Brian C. Hall. Springer (2004)

An Introduction to Lie Groups and the Geometry of Homogeneous Spaces, by Andreas Arvanitogeôrgos. AMS (2003)

Essays in the History of Lie Groups and Algebraic Groups, by Armand Borel. American Mathematical Society (2001)

Lie Algebras and Algebraic Groups, by Patrice Tauvel, Rupert W. T. Yu. Springer (2005)

Elementary Lie Group Analysis and Ordinary Differential Equations, by Nail H. Ibragimov. Wiley (1999)

Lie Groups: An Approach Through Invariants and Representations,** by Claudio Procesi. Springer (2009)

Lie Groups: Beyond an Introduction (2nd Edition),** by Anthony W. Knapp. Birkhauser Boston (2002)

Lie Groups (Graduate Texts in Mathematics),** by Daniel Bump. Springer (2004)

Differential Geometry, Lie Groups, and Symmetric Spaces (Graduate Studies in Mathematics),** by Sigurdur Helgason. American Mathematical Society (2001)

Lectures on Exceptional Lie Groups,** by J. Frank Adams (edited by Zafer Mahmud and Mamoru Mimora). University of Chicago Press (1996)

Simple Groups of Lie Type (corrected edition),** by Roger W. Carter. Wiley (1989)

Infinite-dimensional Lie Algebras,** by Minoru Wakimoto. American Mathematical Society (2000)

Klein's Erlangen Program, Kleinian Groups

Indra's Pearls: The Vision of Felix Klein, by David Mumford, Caroline Series, and David Wright. Cambridge University Press; illustrated edition (2002)

Psychogenesis and the History of Science, by Jean Piaget, Rolando García. Columbia University Press (1988)

The Eightfold Way: The Beauty of Klein's Quartic Curve, by Silvio Levy. Cambridge University Press (2001)

Geometry, by V. V. Prasolov and V. M. Tikhomirov. American Mathematical Society (2001)

The Symbolic Universe: Geometry and Physics 1890–1930, by Jeremy Gray, author and editor. Oxford University Press (1999)

Differential Geometry: Cartan's Generalization of Klein's Erlangen Program (Graduate Texts in Mathematics),** by R. W. Sharpe. Springer; Corrected edition (2000)

Kleinian Groups and Uniformization in Examples and Problems (Translations of Mathematical Monographs), by Samuil Leibovich

Krushkal', Boris Nikolaevich Apanasov, Nikolai Aleksandrovich Gusevskii. AMS (1986)

Kleinian Groups,** by Bernard Maskit. Springer (2004)

Hyperbolic Manifolds and Kleinian Groups,** by Katsuhiko Matsuzaki, Masahiko Taniguchi. Oxford University Press, USA (1998)

Note: In number theory, the "Langlands program" involves Lie groups. For every Lie group, you have an associated number theory and the Langlands program. It has influenced much of the work in algebraic number theory in recent decades. The study of modular forms, in which Klein and Poincaré did such important work, is critical for the Langlands program, as it was for Andrew Wiles' proof of the main part of the Shimura-Taniyama conjecture relating modular forms to elliptic curves, implying Fermat's Last Theorem.

Modular Forms and Fermat's Last Theorem, edited by G. Cornell, J. H. Silverman, G. Stevens. Springer paperback edition (2009)

A number of typographical errors were found while reading the original edition. I have added an errata for the ones I found.

My thanks to Marvin Jay Greenberg, Emeritus Professor of Mathematics at the University of California at Santa Cruz, for his assistance with this foreword.

Richard Bozulich
Shonan Institute of Technology
Fujisawa, Japan

Errata

Page 11, line 2: Replace "maps of for the trapezoid . . ." with "maps for the trapezoid . . ."

Page 13, line 17: Replace the equation $2x^5 - 15x^4 + 29x^3 + 6x^2 + 40x = 0$ with $2x^5 - 15x^4 + 29x^3 + 6x^2 - 40x = 0$

Page 22, line 5: Replace "Their position was that of . . ." with "Their positions were that of . . ."

Page 48, second line from the bottom: Replace "PQR is more then π . . ." with "PQR is more than π . . ."

Page 77, line 7: Replace "in the city gymnasium He moved . . ." with "in the city gymnasium. He moved . . ."

Page 91, second paragraph line 3: Replace "to the Americans B. Peirce and C. Peirce, who . . ." with "to the Americans B. Pierce and C. Pierce, who . . ."

Page 91, second paragraph line 17: Replace "American Charles Peirce . . ." with "American Charles Pierce . . ."

Page 91, second paragraph line 19: "algebras, as Peirce preferred . . ." Should read "algebras, as Pierce preferred . . ."

Page 97, fourth paragraph lines 8, 9, 12, 15 and 18: Replace all occurrences of "Peirce" with "Pierce"

Page 97, fifth paragraph line 1: Replace "Peirce" with "Pierce"

Page 141, note 15, line 3: Replace "$(u + v)^3$" with "$(u^3 + v^3)$"

Page 141, note 15, line 5, Replace "-9" with "$-q$"

Page 156, first paragraph line 2: Replace "Decartes" with "Desargues"

Page 170, line 8 from bottom: Replace "revolutionatory" with "revolutionary"

Page 180, note 165 line 13: Replace J.-L. Lagrange with "J.L. Lagrange"

Page 203, note 217, line 8: Replace "T. A. Wolf' with "J. A. Wolf'

Page 216, line 6 from bottom: Replace "centro-Euclidean" with "Euclidean"

Foreword

This book on the history of mathematics may, to a certain extent, be regarded as a biography. However, it does not tell the story of the life of a specific person (or the stories of two people, as the title may lead one to think), but the story of the evolution of a rather general concept, that of *symmetry*. It is my firm belief that of all the general scientific ideas which arose in the nineteenth century and were inherited by our century, none contributed so much to the intellectual atmosphere of our time as the idea of symmetry. This is confirmed by the abundance of books in different categories and genres devoted to symmetry in physics, to symmetry in chemistry, to symmetry in biology and so on, or to the general (occasionally philosophical) concept of symmetry. The major role played by symmetry notions in modern education is attested even by the cover designs of many textbooks, for instance H.R. Jacobs's *Geometry* (Freeman, 1974). Symmetry also plays a role in art, as evidenced by, say, the interest shown for the works of the well-known Dutch "mathematical" designer Maurits Cornelis Escher. Thus it seems fitting to describe the rise and evolution of ideas of mathematical symmetry. This is done in this book, which is intended for a wide range of readers, including nonexperts with an interest in the history of mathematics and in general scientific problems.

It is natural to link an account of the genesis and development of the idea of symmetry with the names of two mathematicians: the German Felix Klein and the Norwegian Sophus Lie, who played leading roles in discerning the importance of the relevant notions and created the mathematical apparatus capable of reflecting them. The history of the theory of symmetry, at least its mathematical part, is not very long. However, the very origins of mathematics involved the idea of symmetry, as indicated by the first manifestations of mathematical thought dating back to the Paleolithic period and due to the predecessors of modern man known as the Cro-Magnons. Here we find objects with geometric patterns displaying an exceptional sense of form and note the first attempts, closely related to symmetry, to make these patterns more regular: bracelets with numbers depicted as patterns of dashes, cuts or holes based on certain symmetry relationships, with an obvious preference for the numbers 7 and 14, underlying many rituals. Passing to the origins of mathematical science (the Greek mathematics of proofs), we again see the role of symmetry in the Ionic school of Thales of Miletus and the south Italian school of the Pythagoreans. A list of theorems attributed to Thales (congruence of the base angles in an isoceles triangle; congruence of vertical angles; the fact that the diameter divides a circle into congruent parts and that the angle subtended by a diameter is a right angle; congruence of two triangles with equal side and equal adjacent angles, etc.) clearly shows that their proofs, as given in the Ionic school, were based on symmetry and on motion ("isometry"), because such proofs are undoubtedly most natural for these theorems. It is also known that the Pythagoreans sought to reveal universal harmony, one of whose manifestations was the symmetry of numbers and of numerical relations; hence their great interest in numbers with some sort of internal symmetry. However, the more evolved mathematics of the Athenian and Alexandrian periods rejected both the emphasis on geometric symmetry typical of the Ionic school and the numerical mysticism of the first Pythagoreans. Moreover, the scholars of the post-Aristotelian period regarded the wide use of graphic illustrations and symmetry notions as Thales' weakness, which had to be overcome. Euclid's *Elements*, the canonical collection of the mathematical knowledge of the high classical age, contains virtually no mention of symmetry. One gets the clear impression that Euclid tried to avoid any mention of motion—though in this he was not consistent. The symmetry concept is closely associated with the last books of Euclid's *Elements* devoted to regular polyhedra. But even these books (which were added to the main body of the *Elements* after Euclid and demonstrate the tenacity of Pythagorean traditions in ancient science) contain no direct references to symmetry and motion (isometry).

The subsequent development of mathematics, under Euclid's influence, hardly changed the sceptical attitude toward the notion of symmetry and its use. Of the European scholars of the modern age who revived the study of mathematics and of the natural sciences—after a break of more than a thousand years—Johann Kepler was closest to the Pythagoreans, and his keen interest in, say, regular polyhedra was not accidental. However, Kepler was primarily an astronomer and not a mathematician. Also, his influence on later generations of scholars was not very great. Newton's antipathy toward Kepler, reflected, for instance, in the almost total lack of reference to Kepler even in works based on Kepler's achievements, was due to the sharp difference in the intellectual qualities of these two founders of European science. It is striking that the extreme mystic Newton referred more readily to the rationalist Galileo than to Kepler, a mystic like himself! In its subsequent development, European science firmly acknowledged Newton rather than Kepler as its teacher and ideologist. Thus any account of the rise of ideas of symmetry in modern mathematics may well begin at the turn of the nineteenth century, as does the present book.

While sufficient attention is paid in the book to the origins of the concept of symmetry in the "new" European mathematics, the further development of the relevant ideas is not exhaustively treated here. The ideas derived from Klein and Lie had a profound impact on all of twentieth-century mathematics; in the present book the narrative is largely restricted to the nineteenth century. We deal neither with the rather extensive nineteenth-century research in the natural sciences related to the rise of scientific crystallography nor with the massive twentieth-century applications of symmetry in theoretical physics in general and in the theory of elementary particles in particular. But then this simple introductory book does not claim to be in any sense exhaustive. Our purpose is to kindle the reader's interest and to prompt him or her to explore the relevant literature.

In order to read the book it is enough to know high school mathematics. The book is meant for different groups of readers, such as college undergraduates majoring in mathematics, high school students, high school and college teachers, and persons without mathematical training but with an interest in the history of science and in general scientific problems. This explains the unusual structure of the book: the main text is followed by numerous and often extensive notes which contain, among other things, most of the numerous references to the literature. When reading the book for the first time, the reader is advised to ignore the notes or to make selective use of them. The bibliographical references are intended only for those readers who expect to continue to study the questions considered in the book.

The possible variety of interests and backgrounds of the book's prospective readers has resulted in the diversity of its contents. Some parts may interest the mathematician, while other parts are addressed to people who are more interested in history in general and in the history of science in particular. This is why the levels of difficulty of the references vary greatly.

The formulas displayed in the book are denoted by two numbers in parentheses, separated by a dot, to the right of the formula; thus, the number (X.Y) stands for formula Y in Chapter X.

The book is based on the author's lectures to graduate-level students majoring in "pure" mathematics at Yaroslavl University, most of whom subsequently go on to teach in secondary schools; this probably influenced the choice of material and the nature of the narrative. The book was significantly influenced by discussions on the subject with the author's friends and colleagues, in particular S.C. Gindikin, B.A. Rosenfeld, and A.M. Yaglom. However, the author assumes full responsibility for the book's content, in particular for the thoughts which may occasionally seem controversial. The author would like to express his gratitude to the translator, Sergei Sossinsky, for his precise rendering of the Russian original into English, to my colleague Alexei Sossinsky for his assistance, and to Hardy Grant and Abe Shenitzer for editing the English text.

I.M. YAGLOM

Contents

CHAPTER 1

The Precursors: Evariste Galois and Camille Jordan

On the morning of May 30, 1832, a French peasant taking vegetables to the Paris market picked up a wounded youth and transported him to a hospital. One day later the young man died, in the presence of his younger brother—and it was the dying man who had to offer consolation to his brother.

The deceased, not yet twenty-one, was Evariste Galois (1811–1832), a mathematician and a well-known revolutionary, recently released from prison. It is believed that the duel in which he had been fatally wounded had been instigated by the police.[1]

Galois's short life was not marked by success. Twice he tried to enter the best college in France, the famed École Polytechnique in Paris, but he failed both times, although the examiners were no doubt mathematicians of much lower caliber than the examinee;[2] he was expelled for political reasons from the École Normale, a second-best school at the time, after his first year of study;[3] his scientific achievements were unrecognized.[4] Galois wrote about his results to the leading contemporary mathematicians, the academicians Augustin Louis Cauchy (1789–1857) and Siméon Denis Poisson (1781–1840), but Cauchy did not answer at all, while Poisson found the paper incomprehensible and returned it to the author. Galois was certain that Cauchy, a conservative and a royalist, had deliberately suppressed the results obtained a by confirmed republican. However, he was being unfair: Cauchy could not have understood Galois's results. Indeed, no one was in a position to appreciate them at the time. If Cauchy had read Galois's letter, his response would most likely have been similar to Poisson's. But at that time Cauchy was concerned with other matters. He had left France, refusing to pledge allegiance to Louis Philippe of Orléans, who, in 1830, had replaced the Bourbons, to whom Cauchy had always been loyal.[5] (The Bourbons, in turn, had honored Cauchy: Charles X had even bestowed the title of baron on him.) Cauchy did not return to Paris until many years later, having obtained special permission not to pledge allegiance to the new government. Undoubtedly, Cauchy had not read Galois's letter. As we shall see, this proved to be very fortunate indeed.

Cauchy died in 1857. In the 1860s it was decided to publish his collected works,[6] and a leading mathematician of the time, Camille Jordan (1832–1922),

Evariste Galois

was appointed to examine his papers in order to find unpublished works which could be included in the new edition.

Jordan failed to find any unpublished works by Cauchy, but among the latter's papers he discovered Galois's letter, which had lain idle for more than thirty years and had apparently never been read; he was amazed by it. In the interim, major successes had been achieved in mathematics; the groundwork had largely been laid[7] for Galois's work to be recognized, and Jordan was just the man to give credit where it was due. Ideas close to those contained in Galois's remarkable letter had probably interested Jordan at an earlier date, and now they seriously engaged his attention. Jordan attempted to find all of Galois's works published during his lifetime or, as most had been, posthumously, and a number of Jordan's papers in the 1860s were devoted to explaining and elaborating the same ideas. Eventually, Jordan decided to write a large monograph about that branch of mathematics. The book came out in 1870; it was entitled *Traité des substitutions et des équations algèbriques.* It is difficult to overestimate its importance in popularizing and elaborating Galois's ideas.

What was Galois's contribution to mathematics? His main result is related to the important question of the solvability of algebraic equations in radicals. But however important the theorems proved by Galois, the methods by which these results were obtained were more important still. It is not only (and not so much) *what* Galois proved but, mainly, *how* he proved it. In order to explain the meaning of his results we must turn to the history of algebra and, in particular, to the progress in the theory of (algebraic) equations before Galois's time.

It is well known that the roots of any quadratic equation $x^2 + px + q = 0$ may be found from the simple formula $x = -p/2 \pm \sqrt{(p/2)^2 - q}$. This formula was known to the ancient Babylonians: the cuneiform inscriptions serving to instruct future priests many thousands of years ago contain large groups of problems on quadratic equations solved with the aid of tables of square roots. These problems were often of a geometric nature but could be reduced to quadratic equations by means of the Pythagorean theorem on right triangles, which was known to the Babylonians (not by that name, of course). In ancient Greece, where geometry prevailed over algebra, the method for solving quadratic equations was given in geometric form: the formula was replaced by rules (different for different signs of the equation's coefficients) for *constructing* the segment x from known segments p and q such that $x^2 \pm px \pm q^2 = 0$. The knowledge of Greek mathematicians, extensive in geometry and even in the elements of the differential and integral calculus, was rather limited in algebra. They went no further than the solution of quadratic equations, despite the fact that the geometric problems which they considered included quite a few involving the solution of cubic equations.[8]

A lively interest in cubic equations was shown by medieval Arab mathematicians,[9] who generally paid much more attention to algebra than the Greeks. The word "algebra" itself is of Arabic origin, deriving from a term for a specific method of solving equations used by Arab mathematicians (transferring terms of an equation from one side to the other with a change of sign). Some cubic equations were also considered by medieval European scholars, in particular by the most outstanding medieval mathematician, the Italian merchant Leonardo Pisano (1180–1240), better known as Fibonacci (which means son of the good-natured man—Bonacco was the nickname of Leonardo's father). However, no decisive breakthroughs were made, perhaps because the creative potential of these mathematicians was checked by excessive respect for the ancient Greeks, an a priori belief that the Greeks could not be surpassed. The solution of cubic and quartic equations, achieved by mathematicians of the Renaissance, was important mainly because it finally put an end to that extremely harmful delusion.

The rule of Frederick II Hohenstaufen (1194–1250) in the Kingdom of the Two Sicilies was a rehearsal of sorts for the Renaissance. Frederick II became king in 1197 at the age of three, and emperor of the Holy Roman Empire in 1215. But he disliked bleak Germany and was in love with southern Italy. He also disliked the tournaments relished by medieval knights (in particular by his grandfather Frederick I Barbarossa), in which armed men mutilated each other; in Italy he patronized less bloody competitions in which the opponents did not exchange blows with sword or lance but foiled each other with mathematical problems. It was at these bouts that Leonardo Pisano's talent first flourished; there he showed his ability to solve cubic equations (probably chosen by Leonardo himself, because his constant rival Johannes of Palermo, a Sicilian merchant and later a university professor, was Leonardo's good friend). The tradition of mathematical tournaments was continued in Renaissance Italy, and it played a considerable role in the first serious successes scored by European science.

Apparently, the cubic equation

$$x^3 + px + q = 0 \tag{1.1}$$

(any cubic equation can be reduced to the form (1.1)[10]) was solved for the first time by Scipione del Ferro (1456–1526), a professor at Bologna University, which was one of the leading and best known universities of Northern Italy.[11] Del Ferro communicated the solution to a relative, Anton Maria Fior. Having gained possession of the formula for solving cubic equations, Fior challenged the first mathematician of Italy, Niccolo Tartaglia (1500–1557),[12] to a mathematical tournament. At first, Tartaglia was not worried in the least, knowing Fior to be a mediocre mathematician. However, not long before the competition was to begin, he was told that Fior possessed a formula for solving any cubic equation, an invaluable asset for competitions of that kind, which he had obtained from his relative del Ferro. Urged on by vanity and the fear of being defeated, Tartaglia soon found the same formula on his own. As a result, he vanquished Fior. First Tartaglia very quickly solved all the problems offered by Fior (they all involved the solution of cubic equations), and after that the upset Fior could not solve a single one of Tartaglia's problems.

On learning of Tartaglia's discovery, another outstanding mathematician of the time, Girolamo Cardano (1501–1576),[13] was anxious to include it in the algebra textbook he was writing (*Ars magna*[14]). He succeeded in luring Tartaglia to a small inn in a provincial town where, intimidated by Cardano, Tartaglia described in Latin verse[15] the key to the formula for the solution of equation (1.1). Tartaglia claimed that Cardano promised not to publish the corresponding result, which Tartaglia was saving for the book that he himself was preparing. We can imagine his indignation when he saw the formula for solving the cubic equation (1.1) in Cardano's *Ars magna*. Here is how it appears in modern form:[16]

$$x = \sqrt[3]{-q/2 + \sqrt{(q/2)^2 + (p/3)^3}} + \sqrt[3]{-q/2 - \sqrt{(q/2)^2 + (p/3)^3}}\,. \tag{1.2}$$

This formula is still called the *Cardano formula*, despite the fact that Cardano made no claims to its discovery and actually wrote that he had learned it from Tartaglia.[17] (Some of the difficulties connected with formula (1.2), due to the fact that real roots of equation (1.1) are often given by the formula in the form of a combination of expressions involving complex numbers,[18] were explained by the last of Bologna University's outstanding mathematicians—Raffaele Bombelli (c. 1526–1573) in his book *L'Algebra* written about 1560 and published in 1572.)

Cardano's *Ars magna* contains another outstanding result: rules for solving any quartic equation[19]

$$x^4 + px^2 + qx + r = 0, \tag{1.3}$$

rules obtained by Cardano's pupil Ludovico Ferrari (1522–1565).[20] Ferrari rewrote (1.3) as

$$(x^2 + p/2)^2 = x^4 + px^2 + (p/2)^2 = -qx - r + (p/2)^2,$$

added the expression $2(x^2 + p/2)y + y^2$, where y is an unspecified number, to both sides of the equation and obtained the equality

$$(x^2 + p/2 + y)^2 = 2yx^2 - qx + (y^2 + py - r + (p/2)^2).$$

Here we have a complete square on the left-hand side; on the right-hand side we have the quadratic trinomial $Ax^2 + Bx + C$ in the unknown x, which also becomes a complete square if $B^2 = 4AC$, i.e., if

$$q^2 = 2y(4y^2 + 4py - 4r + p^2). \tag{1.4}$$

Equation (1.4) is a cubic equation in the unknown y, now called *Ferrari's resolvent* for equation (1.3). If y_0 is a root of equation (1.4) (which can be found by using formula (1.2)), then (1.3) turns into a combination of two quadratic equations

$$x^2 + p/2 + y_0 = \pm\sqrt{2y_0}(x - q/(4y_0)),$$

whose roots coincide with the roots of the initial equation (1.3).

Thus the problem of solving all cubic and quartic equations proved not to be very difficult; the discovery of the relevant formulas was the first triumph achieved by European mathematical thought, awakening from the thousand-year slumber of the Middle Ages.[21] It was only natural that the question of solving the (general) quintic equation arose at this point. However, several centuries of trials yielded only numerous incorrect solutions: no one could find the correct one!

The hypothesis that the general fifth-degree equation cannot be solved at all, i.e., there is no formula similar to (1.2) for finding the roots of the equation from its coefficients by means of a finite number of algebraic operations (addition and subtraction, multiplication and division, raising to a power and extracting roots), was first put forward by Joseph Louis Lagrange (1736–1814), who came from a French family which had settled in Italy and become Italianized to some extent. Lagrange's brilliant mathematical abilities became apparent when he was very young. At 19 he was appointed professor of the Artillery Academy in his native Turin,[22] and a year later he took an active part in founding the Turin Academy of Sciences (actually, the scientific society organized by Lagrange and his friends acquired this name somewhat

Joseph Louis Lagrange

later). The Academy's publications contained many of Lagrange's papers on mathematics, mechanics, and physics. In 1759, on the recommendation of the famous Leonhard Euler (1707–1783),[23] the twenty-three-year-old Lagrange was elected foreign member of one of the most influential scientific societies in Europe, the Berlin (Prussian) Academy of Sciences. Following Euler's departure from Berlin to St. Petersburg (to the St. Petersburg Academy of Sciences, with which Euler was connected practically all his life), King Frederick II of Prussia, advised by Euler and encouraged by the Parisian Jean le Rond d'Alembert (see Note 176 below) whom he deeply respected, appointed the thirty-year-old Lagrange head of the Berlin Academy's mathematical class (a post held previously by Euler).[24] It was during Lagrange's Berlin period, which lasted until 1787, that he carried out his fundamental studies in algebra.

Lagrange devoted a long memoir (over 200 pages) to the theory of equations. *Réflexions sur la résolution algèbrique des équations* (1770–73) served as the starting point for Galois as well as for Ruffini, Abel, and Cauchy (see Note 7). Lagrange began his memoir by critically reviewing all previous attempts to solve quintic equations, in order to determine the general reasons for their failure. He pointed out that the reduction of cubic equations to quadratics (see Note 15) and of quartic equations to cubics is essentially based on a common idea: it consists in writing out *Lagrange's resolvent* and then finding Tartaglia's resolvent in the case of cubic equations (see Note 15) and Ferrari's resolvent in the case of quartic equations (as explained in the text above). However, an application of this method to a quintic equation transforms it into an equation of the *sixth* degree and, generally, for all $n \geqslant 5$, the degree of Lagrange's resolvent for an nth degree equation proves to be higher than the degree of the initial equation. This led Lagrange to doubt the existence of a formula for solving nth degree equations for $n \geqslant 5$. It is particularly important that the main role in Lagrange's studies was played by certain permutations of the equation's roots. Lagrange even made a truly prophetic statement to the effect that the theory of permutations was the real crux of the question (of the solution of algebraic equations in radicals). Subsequently, that rather vague assumption found its brilliant confirmation in the works of Galois. It should be noted that although Lagrange did not know the term "group" (which will be dealt with below) and did not introduce the notion anywhere, he was led to it by the study of permutations of roots. This is why one of the first theorems of group theory is named after him.[25]

The first proof that it is impossible to solve the general equations of the fifth and higher degrees in radicals was given by the Italian doctor and outstanding amateur mathematician Paolo Ruffini (1765–1822); it was set forth in the algebra textbook he published at his own expense in 1799 under the rather long title *Teoria generale delle equazioni, in cia si dimonstra impossibili la soluzione algebraica delle equazioni generali di grado superiore al quatro*. This remarkable book was hardly noticed outside Italy;[26] while in Italy it encountered intense opposition on the part of mathematicians headed by the authoritative Gianfrancesco Malfatti (1731–1807), professor at the university of Ferrara, known for his many unsuccessful attempts to find a formula for the solution of quintic equations.[27] Evidently, mathematicians were displeased by a doctor's intrusion into a field they regarded as their own; nevertheless, at the end of his life Ruffini became a professor of mathematics at the university of Modena. Ruffini's proof that the general quintic equation can not be solved was not irreproachable, but the author himself probably knew that best of all. He later made several attempts to improve the proof in a long series of papers (1801–1813), but he was only partly successful. This notwithstanding, it is often said that modern algebra began with the appearance of Ruffini's works, which were not appreciated by anyone in their time.

A "flawless" proof of a formula, involving only the operations of addition, subtraction, multiplication, division, raising to a power, and extracting roots,

Niels Henrik Abel

for finding the solutions of the general quintic equation $ax^5 + bx^4 + cx^3 + dx^2 + ex + f = 0$ in terms of the coefficients a, b, c, d, e, and f was presented in 1824–1826 by one of the greatest mathematicians of the nineteenth century, a young Norwegian (still a student at the time), Niels Henrik Abel (1802–1829), whose life, like Galois's, was profoundly tragic.[28] In Abel's lifetime (as well as later, see Chapters 2 and 8) Norway was extremely provincial; there were no qualified persons who could guide his studies. It was very fortunate that there happened to be a good teacher at his school who recognized the pupil's talent and drew his attention to the works of Newton, Euler, and Lagrange. There were no mathematicians at Christiania (Oslo) University who could read his papers; in particular, no one was able to find the mistake in the formula for solving the general quintic equation given by Abel in 1823.[29] However, Abel soon realized that his solution had been erroneous and, in 1824, published as a separate booklet an extremely concise proof that the general quintic could not be solved in radicals.[30] Abel's patrons, who admired the poor young man's diligence and indubitable talent but lacked the competence to check or guide his research, succeeded in obtaining for him a scholarship from the Norwegian government. This enabled Abel, who had suffered from poverty all his life, to visit Germany and France, consult mathematicians in those countries, and refine his knowledge at famous universities. That trip proved very profitable for Abel despite the fact that, due to his modesty and shyness, he was unable to establish personal relations with prominent French and German scientists. It is striking that virtually the only person to acknowledge Abel's talent was not a professional scientist but the prominent German engineer, entrepreneur, and amateur mathematician August Leopold Crelle (1780–1855), an excellent judge of people, who had no substantial scientific achievements to his credit and therefore was not highly respected in academic circles. Crelle was a member of the Berlin Academy of Sciences, although he was elected more for his accomplishments as an engineer and for his organizational activities than for his purely scientific results. He was also a very rich man: at the time, most German railways were built according to his proposals. Crelle's profound faith in Abel and in J. Steiner, a Swiss amateur unknown to established scientists (more about him below), prompted him to found the first specialized mathematical journal in Germany. It was called *Journal für reine und angewandte Mathematik*; contrary to Crelle's intentions, the magazine soon came to be ironically called in scientific circles *Journal für reine unangewandte Mathematik*. It went on play a major role in German science. The first volumes were filled with Abel's (and Steiner's) papers. In particular, the first issue of the *Crelle Journal*, as almost immediately it began to be called for short, contained, among other works, a long French memoir by Abel called, *Démonstration de l'impossibilité de la résolution algébrique des équations générales qui passent le quatrième degré* (1826), which made his results accessible to all mathematicians.

Abel's paper published in Crelle's journal attracted the attention of the famous C. Jacobi (see Note 240); it was Jacobi who introduced such terms as "Abelian integrals," "Abelian functions," and the like, and of other German scientists. Due to their efforts Abel was elected professor at Berlin University in 1828, but the official notice reached Christiania (Oslo) several days after his death from tuberculosis at the age of twenty-seven (Abel did get a private communication of his election—a last consolation before his death).

Ruffini's works apparently remained unknown to Galois; however, he knew Abel's papers and valued them highly. But the Ruffini-Abel theorem only asserted the absence of a general formula for the solution of every quintic equation, but failed to prove the existence of specific equations whose roots could not be expressed by means of radicals (they could, conceivably, be so expressed by means of a formula appropriate only for the given equation but not for all the others). Also, it did not determine whether a given equation could or could not be solved in radicals,[31] nor indicate how to find the solution if one existed. It was Galois who first answered all these questions, and the methods and notions he applied were destined to play an outstanding part in all of nineteenth and twentieth-century mathematics.

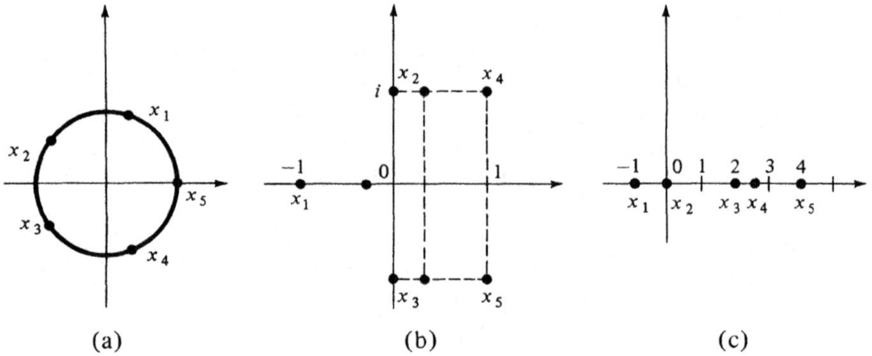

FIGURE 1

Galois based his works on certain appraisals of the "degree of symmetry" of an algebraic equation. It is obvious that the quintic equation $x^5 - 1 = 0$ whose (complex) roots are depicted in Fig. 1(a) is "more symmetric" than the equation $x^5 - x^4 + x^3 + x^2 + 2 = 0$ (see Fig. 1(b), in which the roots of this equation are shown), and the latter is "more symmetric" than the equation $2x^5 - 15x^4 + 29x^3 + 6x^2 - 40x = 0$ with the roots $x_1 = -1, x_2 = 0, x_3 = 2, x_4 = 2\frac{1}{2}, x_5 = 4$ (Fig. 1(c)). Similarly, a square, say, is more symmetric than an equilateral trapezoid (Fig. 2(b)), and the latter is more symmetric than the scalene quadrangle depicted in Fig. 2(c). The mathematical degree of symmetry of a polygon is judged by the set of distance-preserving maps of the polygon onto itself. Thus, for the square $A_1A_2A_3A_4$ (see Fig. 2(a)), that set includes the rotation of the square by 90° about its centre O which sends the vertices A_1, A_2, A_3, and A_4 into A_2, A_3, A_4, and A_1 respectively (we will write this rotation as $A_1A_2A_3A_4 \to A_2A_3A_4A_1$ or denote it by the permutation $\binom{1\,2\,3\,4}{2\,3\,4\,1}$, which indicates that vertex number 1 is sent to vertex number 2, vertex number 2 to vertex number 3, and so on); the rotation $A_1A_2A_3A_4 \to A_3A_4A_1A_2$, or $\binom{1\,2\,3\,4}{3\,4\,1\,2}$, about the point O by 180° (or reflection in O); the rotation $A_1A_2A_3A_4 \to A_4A_1A_2A_3$, or $\binom{1\,2\,3\,4}{4\,1\,2\,3}$, about O by 270°; the reflections $A_1A_2A_3A_4 \to A_1A_4A_3A_2$, or $\binom{1\,2\,3\,4}{1\,4\,3\,2}$, and $A_1A_2A_3A_4 \to A_3A_2A_1A_4$, or $\binom{1\,2\,3\,4}{3\,2\,1\,4}$, in the diagonals A_1A_3 and A_2A_4; the reflections $A_1A_2A_3A_4 \to A_4A_3A_2A_1$, or $\binom{1\,2\,3\,4}{4\,3\,2\,1}$, and $A_1A_2A_3A_4 \to A_2A_1A_4A_3$, or $\binom{1\,2\,3\,4}{2\,1\,4\,3}$, in the midlines KL and MN and, of course, the identity transformation $A_1A_2A_3A_4 \to A_1A_2A_3A_4$, or $\binom{1\,2\,3\,4}{1\,2\,3\,4}$, which does not move any of the quadrangle's vertices. In short, the set of transformations of a square onto itself is given by the 8 permutations

$$\pi_1 = \begin{pmatrix}1234\\2341\end{pmatrix}, \quad \pi_2 = \begin{pmatrix}1234\\3412\end{pmatrix}, \quad \pi_3 = \begin{pmatrix}1234\\4123\end{pmatrix}, \quad \rho_1 = \begin{pmatrix}1234\\1432\end{pmatrix},$$

$$\rho_2 = \begin{pmatrix}1234\\3214\end{pmatrix}, \quad \sigma_1 = \begin{pmatrix}1234\\4321\end{pmatrix}, \quad \sigma_2 = \begin{pmatrix}1234\\2143\end{pmatrix}, \quad \text{and} \quad \varepsilon = \begin{pmatrix}1234\\1234\end{pmatrix}, \tag{1.5}$$

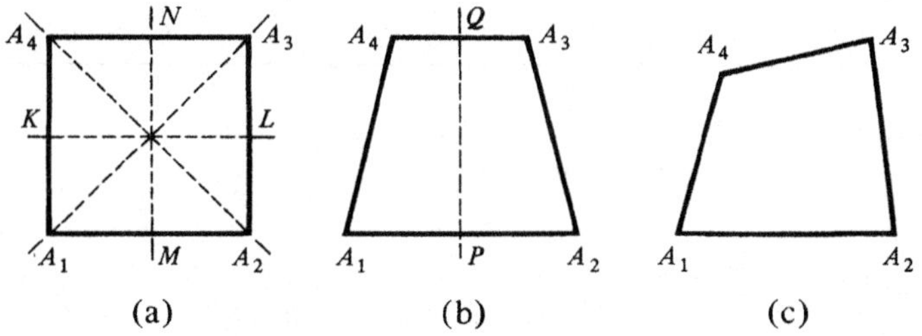

FIGURE 2

and is thus relatively varied. On the other hand, the set of distance-preserving maps of for the trapezoid $A_1A_2A_3A_4$ in Fig. 2(b) onto itself is much poorer, for it consists of just the reflection $A_1A_2A_3A_4 \to A_2A_1A_4A_3$, or $\binom{1\,2\,3\,4}{2\,1\,4\,3}$, and the identity map; and the set of distance-preserving maps of the scalene quadrangle in Fig. 2(c) onto itself consists of the identity map alone.

Similarly, according to Galois, the degree of symmetry of an nth-degree equation $f(x) = 0$ with rational coefficients is described by the set of permutations of its roots x_1, x_2, ..., x_n that preserve all the algebraic relationships between them (expressed by equations of the form $P(x_1, x_2, \ldots, x_n) = 0$ where $P(x_1, x_2, \ldots, x_n)$ is a polynomial in n variables x_1, x_2, ..., x_n with integer coefficients which may, of course, depend on some, rather than all n variables). Thus, in the case of the so-called cyclic polynomial $x^5 - 1 = 0$, all the relations between the roots reduce to the equality $x_5 = 1$ and the relations $x_1^2 = x_2$, $x_1^3 = x_3$, $x_2^2 = x_4$, and so on. The set of these relations is invariant under the permutations $(x_1, x_2, x_3, x_4, x_5) \to (x_2, x_4, x_1, x_3, x_5)$, or $\binom{1\,2\,3\,4\,5}{2\,4\,1\,3\,5}$, (for instance, under this permutation the relation $x_1^2 = x_2$ becomes $x_4^2 = x_3$); $(x_1, x_2, x_3, x_4, x_5) \to (x_3, x_1, x_4, x_2, x_5)$, or $\binom{1\,2\,3\,4\,5}{3\,1\,4\,2\,5}$; $(x_1, x_2, x_3, x_4, x_5) \to (x_4, x_3, x_2, x_1, x_5)$, or $\binom{1\,2\,3\,4\,5}{4\,3\,2\,1\,5}$; and, of course, the identity substitution $(x_1, x_2, x_3, x_4, x_5) \to (x_1, x_2, x_3, x_4, x_5)$, or $\binom{1\,2\,3\,4\,5}{1\,2\,3\,4\,5}$. Thus, the set in question consists of only 4 permutations:

$$\tau_1 = \begin{pmatrix} 1\,2\,3\,4\,5 \\ 2\,4\,1\,3\,5 \end{pmatrix}, \quad \tau_2 = \begin{pmatrix} 1\,2\,3\,4\,5 \\ 3\,1\,4\,2\,5 \end{pmatrix}, \quad \tau_3 = \begin{pmatrix} 1\,2\,3\,4\,5 \\ 4\,3\,2\,1\,5 \end{pmatrix}, \quad \text{and}$$

$$\varepsilon = \begin{pmatrix} 1\,2\,3\,4\,5 \\ 1\,2\,3\,4\,5 \end{pmatrix}. \tag{1.6}$$

All the algebraic relations between the roots of the equation $x^5 - x^4 + x^3 + x^2 + 2 = 0$ reduce to the equalities $x_1 = -1$; $x_2^2 = -1$; $x_2 + x_3 = 0$; $x_4^2 - 2x_4 + 2 = 0$; $x_4 + x_5 = 2$. The set of these relations is invariant under the identity permutation as well as under the permutations t_1: $(x_1, x_2, x_3, x_4, x_5) \to (x_1, x_3, x_2, x_4, x_5)$, or $\binom{1\,2\,3\,4\,5}{1\,3\,2\,4\,5}$; t_2: $(x_1, x_2, x_3, x_4, x_5) \to (x_1, x_2, x_3, x_5, x_4)$, or $\binom{1\,2\,3\,4\,5}{1\,2\,3\,5\,4}$ and t_3: $(x_1, x_2, x_3, x_4, x_5) \to (x_1, x_3, x_2, x_5, x_4)$, or $\binom{1\,2\,3\,4\,5}{1\,3\,2\,5\,4}$. Thus, the family of permutations preserving the set of all algebraic relations between the roots of the equation under consideration consists of

$$t_1 = \begin{pmatrix} 1\,2\,3\,4\,5 \\ 1\,3\,2\,4\,5 \end{pmatrix}, \quad t_2 = \begin{pmatrix} 1\,2\,3\,4\,5 \\ 1\,2\,3\,5\,4 \end{pmatrix}, \quad t_3 = \begin{pmatrix} 1\,2\,3\,4\,5 \\ 1\,3\,2\,5\,4 \end{pmatrix},$$

$$\text{and, of course,} \quad \varepsilon = \begin{pmatrix} 1\,2\,3\,4\,5 \\ 1\,2\,3\,4\,5 \end{pmatrix}. \tag{1.7}$$

Finally, since the equalities $x_1 = -1$, $x_2 = 0$, $x_3 = 2$, $2x_4 = 5$, $x_5 = 4$ are among the algebraic relations linking the roots of the equation $2x^5 - 15x_4 + 29x_3 + 6x^2 - 40x = 0$, the only permutation preserving all these relations is the identity $\varepsilon = \binom{1\,2\,3\,4\,5}{1\,2\,3\,4\,5}$.

It is clear that the set of distance-preserving maps I_F of the quadrangle $F \equiv A_1A_2A_3A_4$ (or the n-sided polygon $F \equiv A_1A_2A_3 \cdots A_n$, or of an arbitrary figure F) onto itself must contain the identity map (the identity permutation $\varepsilon = \binom{1\,2\,3\,\cdots\,n}{1\,2\,3\,\cdots\,n}$ of the vertices of the polygon). Further, if $\sigma = \binom{1\;\;2\;\;\cdots\;\;n}{a_1\,a_2\,\cdots\,a_n}$ and $\tau = \binom{1\;\;2\;\;\cdots\;\;n}{b_1\,b_2\,\cdots\,b_n}$ are two vertex permutations corresponding to distance-preserving maps of the polygon $F = A_1A_2 \cdots A_n$ onto itself, then their product

$$\tau\sigma = \begin{pmatrix} 1 & 2 & \cdots & n \\ b_1 & b_2 & \cdots & b_n \end{pmatrix} \cdot \begin{pmatrix} 1 & 2 & \cdots & n \\ a_1 & a_2 & \cdots & a_n \end{pmatrix} = \begin{pmatrix} 1 & 2 & \cdots & n \\ b_{a_1} & b_{a_2} & \cdots & b_{a_n} \end{pmatrix}$$

is contained in the set I_F (if both σ and τ map F onto itself, then so does their product $\tau\sigma$ (first σ, then τ!)). Finally, if $\pi = \binom{1\;\;2\;\;\cdots\;\;n}{p_1\,p_2\,\cdots\,p_n}$ is the permutation corresponding to a distance-preserving map of the polygon $F \equiv A_1A_2 \cdots A_n$ onto itself, then the inverse map corresponds to the inverse permutation

$$\pi^{-1} = \begin{pmatrix} p_1 & p_2 & \cdots & p_n \\ 1 & 2 & \cdots & n \end{pmatrix} \left(= \begin{pmatrix} 1 & 2 & \cdots & n \\ q_1 & q_2 & \cdots & q_n \end{pmatrix} \right).$$

Here q_1, say, is a number such that $p_{q_1} = 1$; it should be pointed out that in the notation for the permutation $\pi = \binom{1\;\;2\;\;\cdots\;\;n}{p_1\,p_2\,\cdots\,p_n}$ what matters are only the columns $\binom{1}{p_1}, \binom{2}{p_2}, \ldots, \binom{n}{p_n}$ and not their order, so that the permutation π can also be written as $\pi = \binom{i_1\;\;i_2\;\;\cdots\;\;i_n}{p_{i_1}\,p_{i_2}\,\cdots\,p_{i_n}}$, where $(i_1, i_2, \ldots, i_n)$ is an arbitrary permutation of the numbers $(1, 2, \ldots, n)$. Indeed, the map π sends the polygon $F \equiv A_1A_2 \cdots A_n$ into the same polygon $F' = A_{p_1}A_{p_2} \cdots A_{p_n}$, while the inverse map π^{-1} sends F' into F, i.e., it is also a distance-preserving map of F onto itself.

It is clear that the set $\mathscr{G}$ of permutations of the roots of the algebraic equation $f(x) = 0$ which transform all algebraic relations between the roots into (other) relations between roots of the same equation has the following three properties:

(1) $\mathscr{G}$ contains the (unique) identity permutation $\varepsilon = \binom{1\,2\,\cdots\,n}{1\,2\,\cdots\,n}$;

(2) together with every two permutations $\sigma = \binom{1\;\;2\;\;\cdots\;\;n}{a_1\,a_2\,\cdots\,a_n}$ and $\tau = \binom{1\;\;2\;\;\cdots\;\;n}{b_1\,b_2\,\cdots\,b_n}$ (where the equality $\tau = \sigma$ is not excluded), $\mathscr{G}$ contains their product $\tau\sigma = \binom{1\;\;2\;\;\cdots\;\;n}{b_1\,b_2\,\cdots\,b_n} \cdot \binom{1\;\;2\;\;\cdots\;\;n}{a_1\,a_2\,\cdots\,a_n} = \binom{1\;\;2\;\;\cdots\;\;n}{b_{a_1}\,b_{a_2}\,\cdots\,b_{a_n}}$;

(3) together with every permutation $\sigma = \binom{1\;\;2\;\;\cdots\;\;n}{a_1\,a_2\,\cdots\,a_n}$ the set $\mathscr{G}$ also contains the

inverse permutation $\sigma^{-1} = \binom{a_1\ a_2\ \cdots\ a_n}{1\ \ 2\ \ \cdots\ \ n} = \binom{1\ \ 2\ \ \cdots\ \ n}{\alpha_1\ \alpha_2\ \cdots\ \alpha_n}$, where $\alpha_{a_i} = i$, $i = 1, 2, \ldots, n$.

Galois called any set of permutations satisfying conditions (1), (2), and (3) a **group** of permutations. Essentially, the notion of a group (now regarded as one of the most important notions in all of mathematics[32]) had appeared before Galois.[33] Its presence in Lagrange's works has already been pointed out. Ruffini's and Abel's research had also involved profound group-theoretic ideas, although these were not set down in a clear fashion.[34] On the other hand, Galois's understanding of the role of group theory in the study of (algebraic) equations was much clearer than that of his precursors; the very fact that the new notion was given a specific name (and that the terminology of group theory had been worked out) was undoubtedly of major importance. The main idea, clearly outlined by Galois, consisted in characterizing each equation by the "degree of symmetry" determined by the group of permutations of the roots which leave the algebraic relations between the roots of the equation unchanged. Galois called this group the group of the equation; it is now known as the *Galois group* (of the equation). The simplest (smallest) such group is, of course, the group consisting of the identity permutation.The example of the equation $2x^5 - 15x^4 + 29x^3 + 6x^2 + 40x = 0$ shows that to the simplest Galois group there correspond the simplest equations—those whose solutions are rational, i.e., can be written without using radicals.

The notion of a group in the theory of equations was insufficient for Galois; equally important for him was the more complex notion of a *field*, also essentially originating in Lagrange's work. While a number of mathematicians, including Abel, had studied fields, it was Galois who named the concept and defined it rigorously. For Galois, a number **field** was a set of numbers closed with respect to the operations of addition and multiplication (i.e., a set such that the sum and product of any two numbers in that set also belongs to the set); this set of numbers must contain the numbers 0 and 1, as well as the difference and quotient (with nonzero denominator) of any two numbers in it. The best known examples of fields are the field Q of rational numbers, the field R of real numbers, and the field C of complex numbers; an intermediate field between Q and R (or C) can be obtained by extending Q by "adjoining" the root of some equation which cannot be solved in Q (if the role of such an equation is played by the quadratic equations $x^2 - 2 = 0$ or $x^2 + 1 = 0$, then we obtain the respective fields of numbers of the form $a + b\sqrt{2}$ and $a + bi$, where $i^2 = -1$ [35])—such an "algebraic extension of fields," which Galois considered following Lagrange, played a major role in his constructions. The notion of a field arose in connection with the primary concept of the (Galois) group of an equation: in the definition of a Galois group $\mathcal{G}$ one must indicate over what field F the group is to be considered. Specifically, the Galois group preserves the set of algebraic relations between the roots $x_1, x_2, \ldots, x_n$ of the initial equation $f(x) = 0$ which are expressed by the condition that certain polynomials $P(x_1, x_2, \ldots, x_n)$ in n variables with coefficients from the given field F are equal to zero. The fundamental theorem of algebra, which asserts that in the field C of complex numbers every nth-degree equation $f(x) = 0$ has exactly n roots $x_1 = c_1, x_2 = c_2, \ldots, x_n = c_n$, implies that the Galois group of any equation is trivial over that field, i.e., consists of the identity substitution ε only. The theory constructed by Galois comes down to the parallel consideration of two processes: the extension of the main field F (containing the coefficients in the relations $P(x_1, x_2, \ldots, x_n)$ between the roots of the equation) and the simultaneous reduction of the Galois group $\mathcal{G}$.

In order to understand the rather complex constructions arising in the branch of algebra now known as *Galois theory* (many substantial books[36] are devoted to it; courses on Galois theory are studied in the mathematics departments of all the universities in the world[37]), Galois had to gain a deep insight into the theory of groups and fields. He introduced the basic terminology of group theory including such terms as *group*, *subgroup* (a subset of the group's elements, itself forming a group with respect to the operation of "multiplication of elements" in the given group), and *order* of a group (the number of elements in a group; see Note 25). He also introduced such important notions as *normal* subgroup (see below) and singled out such important classes of groups as *simple* groups and *solvable* groups[38] (however, it should be kept in mind that some of Galois's definitions were rather sketchy and most of his theorems were not proved[39]). Galois's main result consisted in describing the Galois groups of equations that are solvable in radicals; he found the necessary and sufficient conditions for such solvability, and it was precisely the groups of solvable equations which he called *solvable*.

It is clear that the set of distance-preserving maps of a polygon F (or an arbitrary figure F) onto itself is also a group; nowadays that group is called the *symmetry group* of that figure. Of course, the maps included in the symmetry group of F do not necessarily have to be thought of as permutations; thus, for example, the symmetry group (1.5) of the square $F \equiv A_1A_2A_3A_4$ consists of four rotations π_1, π_2, π_3, and ε (where ε is a rotation by 360° or the identity map) and four reflections ρ_1, ρ_2, σ_1 and σ_2 with the following "multiplication table" of the group's elements:

(Second factor)

(First factor)	ε	π_1	π_2	π_3	ρ_1	ρ_2	σ_1	σ_2
ε	ε	π_1	π_2	π_3	ρ_1	ρ_2	σ_1	σ_2
π_1	π_1	π_2	π_3	ε	σ_2	σ_1	ρ_1	ρ_2
π_2	π_2	π_3	ε	π_1	ρ_2	ρ_1	σ_2	σ_1
π_3	π_3	ε	π_1	π_2	σ_1	σ_2	ρ_1	ρ_2
ρ_1	ρ_1	σ_1	ρ_2	σ_2	ε	π_2	π_1	π_3
ρ_2	ρ_2	σ_2	ρ_1	σ_1	π_2	ε	π_3	π_1
σ_1	σ_1	ρ_2	σ_2	ρ_1	π_3	π_1	ε	π_2
σ_2	σ_2	ρ_1	σ_1	ρ_2	π_1	π_3	π_2	ε

(1.5′)

More generally, a group (an arbitrary group not necessarily consisting of permutations) is any (finite or infinite) family $\mathcal{G} = \{\alpha, \beta, \gamma, \ldots; \varepsilon\}$ of elements (the element ε of the group plays a special role) for which there is defined a "multiplication" that assigns to every two elements α and β of the group a third element, their "product" $\delta = \alpha\beta$, in such a way that the following requirements hold:

(1) $(\alpha\beta)\gamma = \alpha(\beta\gamma)$ for all $\alpha, \beta, \gamma \in \mathcal{G}$ (associativity);
(2) $\alpha\varepsilon = \varepsilon\alpha = \alpha$ for all $\alpha \in \mathcal{G}$ (the element ε is known as the *identity* element of the group);
(3) for every $\alpha \in \mathcal{G}$ there exists an element $\alpha^{-1} \in \mathcal{G}$, such that $\alpha\alpha^{-1} = \alpha^{-1}\alpha = \varepsilon$ (the element α^{-1} is known as the *inverse* of α).

If, moreover,

(4) $\alpha\beta = \beta\alpha$ for all $\alpha, \beta \in \mathcal{G}$ (commutativity), then the group $\mathcal{G}$ is said to be *commutative.*

This "abstract" approach to the notion of a group (in which neither the nature of the group elements nor the meaning of the "group operation" ("multiplication") are specified) originated in the work of Cauchy mentioned in Note 7. The idea of defining a group by the "multiplication table" of its elements, similar to table (5′) (of course such a table can be written out only for a finite group, that is, a group with a finite number of elements) is due to the English mathematician A. Cayley, whose name will appear many times in this book. Such tables are called *Cayley tables.* It is clear that a multiplication table of the type (1.5′) defines a group if and only if it satisfies certain conditions, corresponding to the properties (1)–(3): thus, for example, it must begin with the "identity row" and "identity column" corresponding to the element ε which repeat the row and column of factors (this corresponds to the conditions $\varepsilon\alpha = \alpha\varepsilon = \alpha$ for all α);[40] the element ε must appear once and only once in each row and column, etc.

Commutative groups—under another name—played an important role in the investigations of Ruffini and, especially, of Abel. Such groups are now called *abelian,* after Abel.[41]

It is now clear that the "greater symmetry" of a square as against an isosceles trapezoid, or, even more, a scalene quadrangle, is expressed by the larger size of the square's symmetry group: the group (1.5) of symmetries of a square contains eight elements while the symmetry group of an isosceles trapezoid consists of only two isometries, and the symmetry group of a scalene quadrangle contains only the identity map ε (which is an element of the symmetry group of every figure). But now our conclusion about the greater symmetry of the "cyclic" equation $x^5 - 1 = 0$ compared to that of the equation $x^5 - x^4 + x^3 + x^2 + 2 = 0$, which seemed obvious at the start, must be re-examined: the Galois groups (over the field of rational numbers) of these equations contain the same number (four) of elements; these groups—(1.6) and (1.7)—differ only in their "multiplication tables" (their Cayley tables)[42]:

	ε	τ_1	τ_2	τ_3
ε	ε	τ_1	τ_2	τ_3
τ_1	τ_1	τ_3	ε	τ_2
τ_2	τ_2	ε	τ_3	τ_1
τ_3	τ_3	τ_2	τ_1	ε

(1.6′)

and

	ε	t_1	t_2	t_3
ε	ε	t_1	t_2	t_3
t_1	t_1	ε	t_3	t_2
t_2	t_2	t_3	ε	t_1
t_3	t_3	t_2	t_1	ε

(1.7′)

Now let us again consider an arbitrary group $\mathscr{G}$. In group theory (and in group-theoretic constructions of Galois groups) an especially important role is played by the so-called normal subgroups. Suppose

$$a = \begin{pmatrix} 1 & 2 & 3 & \cdots & n \\ a_1 & a_2 & a_3 & \cdots & a_n \end{pmatrix} \quad \text{and} \quad \alpha = \begin{pmatrix} 1 & 2 & \cdots & n \\ \alpha_1 & \alpha_2 & \cdots & \alpha_n \end{pmatrix}$$

are any two permutations. The image of the permutation a (of the permutation $i \to a_i$) under the permutation α (the permutation $i \to \alpha_i$) is the permutation $a'\colon \alpha_i \to \alpha_{a_i}$; thus, for example, the permutation $\alpha = \left(\begin{smallmatrix} 1 & 2 & 3 & 4 \\ 4 & 2 & 1 & 3 \end{smallmatrix}\right)$ sends the permutation $a = \left(\begin{smallmatrix} 1 & 2 & 3 & 4 \\ 4 & 3 & 2 & 1 \end{smallmatrix}\right)$ into the permutation $a' = \left(\begin{smallmatrix} 4 & 2 & 1 & 3 \\ 3 & 1 & 2 & 4 \end{smallmatrix}\right)$ or $a' = \left(\begin{smallmatrix} 1 & 2 & 3 & 4 \\ 2 & 1 & 4 & 3 \end{smallmatrix}\right)$. (Recall that the order of the columns of the permutation is immaterial.) More generally, the image of the map $a\colon x \to f(x)$, under the map $\alpha\colon x \to \varphi(x)$ is the map $a'\colon \varphi(x) \to \varphi(f(x))$; it is not too hard to see that $a' = \alpha a \alpha^{-1}$. Similarly, one says that the permutation (transformation) α sends the set of permutations (transformations) $\{a, b, c, \ldots\}$ into the set $\{a', b', c', \ldots\} = \{\alpha a \alpha^{-1}, \alpha b \alpha^{-1}, \alpha c \alpha^{-1}, \ldots\}$. The subgroup $\mathscr{H}$ of a group $\mathscr{G}$ of permutations (transformations) is said to be *normal* if all the permutations (transformations) in $\mathscr{G}$ send $\mathscr{H}$ into itself.

This definition can be restated as follows. Suppose $\mathscr{G}$ is a group and $\mathscr{H}$ is its subgroup consisting of the elements ε (the identity permutation, or identity element of the group) κ, λ, ... For each element α of the ("large") group $\mathscr{G}$ define the set

$$\alpha\mathscr{H} = \{\alpha\varepsilon = \alpha, \alpha\kappa, \alpha\lambda, \ldots\}$$

of elements in $\mathscr{G}$. If $\alpha \in \mathscr{H}$, then, obviously, all the elements from $\alpha\mathscr{H}$ belong to $\mathscr{H}$; it is also easy to check that, in this case, $\alpha\mathscr{H}$ coincides with $\mathscr{H}$. If α does not belong to $\mathscr{H}$, then none of the elements of the set $\alpha\mathscr{H}$ belongs to $\mathscr{H}$. Similarly, if α and β are two elements in $\mathscr{G}$, then either the sets

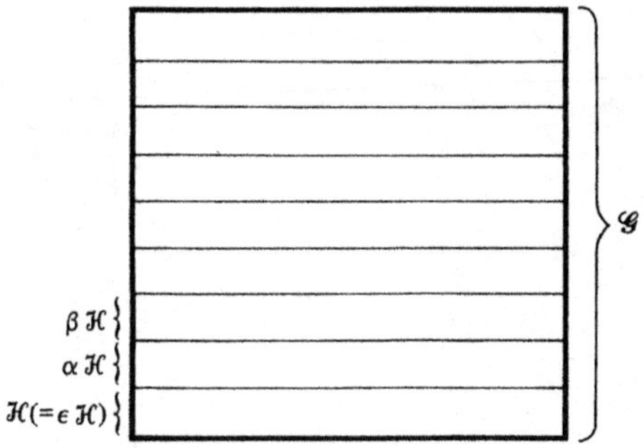

FIGURE 3

$$\alpha\mathscr{H} = \{\alpha, \alpha\kappa, \alpha\lambda, \dots\}$$

and

$$\beta\mathscr{H} = \{\beta, \beta\kappa, \beta\lambda, \dots\}$$

of elements in $\mathscr{G}$ coincide (this will be the case if $\alpha^{-1}\beta \in \mathscr{H}$ and therefore $\beta^{-1}\alpha \in \mathscr{H}$ since, as can easily be checked, $\beta^{-1}\alpha = (\alpha^{-1}\beta)^{-1}$) or $\alpha\mathscr{H}$ and $\beta\mathscr{H}$ have no common elements.

Now consider all possible sets of elements of the form $\mathscr{H}(=\varepsilon\mathscr{H})$, $\alpha\mathscr{H}$, $\beta\mathscr{H}$, $\gamma\mathscr{H}, \dots,$ where $\varepsilon, \alpha, \beta, \gamma, \dots$ are all the elements of $\mathscr{G}$. Some of these sets coincide, and some have no common elements. Thus we obtain a partition of the entire group into "classes" $\alpha\mathscr{H}, \beta\mathscr{H}, \dots$ etc. (including the subgroup $\mathscr{H}$ itself which may be written in the form $\varepsilon\mathscr{H}$ or in the form $\gamma\mathscr{H}$ for any element γ in $\mathscr{H}$). Such a partition of $\mathscr{G}$ into nonoverlapping subsets (see the schematic Fig. 3) is known as a partition of $\mathscr{G}$ into *left cosets* of the group $\mathscr{G}$ with respect to the subgroup $\mathscr{H}$. Similarly, one defines *right cosets of* $\mathscr{G}$ by $\mathscr{H}$:

$$\mathscr{H}(=\mathscr{H}\varepsilon),\ \mathscr{H}\alpha,\ \mathscr{H}\beta, \dots.$$

A normal subgroup $\mathscr{H}$ of the group $\mathscr{G}$ can be defined by the property that the set of left cosets of $\mathscr{G}$ by $\mathscr{H}$ coincides with the set of right cosets (so that in this case we can speak of *cosets* of $\mathscr{G}$ by $\mathscr{H}$ without using the adjectives "left" and "right"). Here it is possible to introduce an "arithmetic" of sorts into the set of cosets itself, since if $\mathscr{K}$ and $\mathscr{L}$ are two cosets of $\mathscr{G}$ by $\mathscr{H}$ (where $\mathscr{H}$ is a normal subgroup of the group $\mathscr{G}$) then the set of all possible products $\kappa\lambda$ where $\kappa \in \mathscr{K}$ and $\lambda \in \mathscr{L}$, is itself a coset (which can be viewed as the product of the classes $\mathscr{K}$ and $\mathscr{L}$ and denoted by $\mathscr{K}\mathscr{L}$). It is easy to verify that the set of cosets of the group $\mathscr{G}$ by its normal subgroup $\mathscr{H}$ is itself a group whose identity element is the subgroup $\mathscr{H}$ while the element inverse to the class $\alpha\mathscr{H}$ is the class $\alpha^{-1}\mathscr{H}$; the "group of cosets" is called the *quotient group* of $\mathscr{G}$ by $\mathscr{H}$ and is denoted by $\mathscr{G}/\mathscr{H}$. It seems that this construction was already known to Galois. This is a guess, however, since there is no mention of it in Galois's notes. The notion of a quotient group (together with the term "quotient group" and the symbol $\mathscr{G}/\mathscr{H}$) was introduced by C. Jordan.

Camille Jordan

It is clear that for a commutative (abelian) group $\mathscr{G}$ there is no difference between left and right cosets (since for such a group $\alpha\nu = \nu\alpha$ for all $\alpha, \nu \in \mathscr{G}$ and therefore $\alpha\mathscr{H} = \mathscr{H}\alpha$ for any $\alpha \in \mathscr{G}$ and any subgroup $\mathscr{H}$ of the group $\mathscr{G}$); therefore each subgroup of a commutative group is normal. It is also clear that any subgroup $\mathscr{H}$ of index 2 in the group $\mathscr{G}$, i.e., such that the family of cosets of $\mathscr{G}$ with respect to $\mathscr{H}$ consists of just two elements, namely the subgroup $\mathscr{H}$ itself and the elements of $\mathscr{G}$ not in $\mathscr{H}$, is normal; indeed, here we have only one possibility of partitioning $\mathscr{G}$ into a family (a pair) of non-intersecting cosets regardless of whether we have in mind "left" cosets of "right" ones. In particular, the possibility of partitioning all permutations $\binom{1\ 2\ \cdots\ n}{i_1\ i_2\ \cdots\ i_n}$ into "even" and "odd" permutations[43] goes back to Lagrange. The product of any two even permutations is again even, and the inverse of an even permutation is even. Accordingly, the family of all even permutations constitutes a subgroup of the group of all permutations (the group S_n of all permutations of n elements is now called the *symmetric group of degree n* and the corresponding group A_n of even permutations is called the *alternating group of degree n*); since this is a subgroup of index 2, it is always normal. The quotient group of a group $\mathscr{G}$ by its subgroup $\mathscr{H}$ of index 2 (say, the quotient group S_n/A_n) consists of only two elements; its "structure" is obviously that of the simplest group of the two numbers 1 and -1 with the multiplication table:

$$\begin{array}{c|cc} & 1 & -1 \\ \hline 1 & 1 & -1 \\ -1 & -1 & 1 \end{array} \tag{1.8}$$

Another example of a normal subgroup is the subgroup of all translations in the group of isometries of the plane or the subgroup of all isometries of the plane in the group of similitudes of the plane (why?). The quotient groups $\mathscr{I}/\mathscr{T}$ and $\mathscr{S}/\mathscr{I}$, where $\mathscr{T}$ is the group of translations, $\mathscr{I}$ the group of isometries and $\mathscr{S}$ the group of similitudes, have the same "structure" as, respectively, the group $\mathscr{I}_0$ of "centered" isometries, i.e., isometries which leave a point O of the plane fixed, and the group R_+ of all positive real numbers (the ratios that characterize the individual transformations $\sigma \in \mathscr{S}$) under multiplication (the proof is left to the reader).[44]

It is clear that each group $\mathscr{G}$ is a normal subgroup of itself (here the "partition into cosets" consists of only one element $\mathscr{G}$, and surely, in this case, there can be no difference between left and right cosets). Another normal subgroup of any group $\mathscr{G}$ is its so-called "trivial" (identity) subgroup $\mathscr{N}$ consisting of the identity element (it is clear that for the family $\mathscr{N} = \{\varepsilon\}$ with the "group operation" $\varepsilon \cdot \varepsilon = \varepsilon$ all the conditions characterizing a group hold); here the partition of $\mathscr{G}$ into (right or left) cosets is simply the partition of the group into its distinct elements. Galois called a group $\mathscr{G}$ with no "nontrivial" normal subgroups, that is, with no normal subgroups different from $\mathscr{G}$ and $\mathscr{N}$, *simple*—a concept that reminds one of the definition of a prime (natural) number n as a number with no "nontrivial" divisors, that is, with no divisors other than 1 and n. For $n \geqslant 3$, the symmetric group S_n (the group of all permutations of n elements) is not simple (it is "composite"), since it contains the (nontrivial) normal subgroup A_n. But is A_n itself simple? Galois showed that the group A_4 of order 12 (i.e., the group contains 12 elements) is not simple and that all the other alternating groups A_n, $n \neq 4$, are simple. The difference, in this respect, between the groups A_4 (nonsimple) and A_5 (simple) reflects the difference between the general quartic equation (which is solvable in radicals) and the general quintic equation (which is not). Thus Galois touched on the topic of finite simple groups (cf. Note 260).

Little could he anticipate the explosive development of this complex topic in the second half of the twentieth century![45]

At the present time, a field (not necessarily numerical) is defined as a set $\mathscr{M} = \{\alpha, \beta, \gamma, \ldots\}$ of arbitrary elements with two operations ("addition" and "multiplication"). These operations assign to every two elements $\alpha, \beta \in \mathscr{M}$ two new elements denoted respectively by $\alpha + \beta$ (the "sum" of α and β) and $\alpha\beta$ (their "product"), and satisfy a number of conditions. One condition is that the elements of the field must form a commutative group with respect to addition. It is convenient to call the neutral element of this group, i.e., the element which does not change any element to which it is added (previously this element was referred to as the "unit" element) the "zero element" or simply the "zero" of the field and to denote it by, say, the Greek letter o:

$$\alpha + o = \alpha, \quad \text{for all } \alpha \in \mathscr{M}$$

(sometimes the zero element is denoted by the number 0). A second condition is that all the nonzero elements of the field must form a (commutative)[46] group with respect to multiplication; the fact that we excluded zero here is, of course, necessary, since it follows from the properties of the field (including the property of distributivity which will be mentioned below) that for each element α of the field we have $\alpha o = o$. A third condition is that addition and multiplication of the elements of the field are related by the distributive law:

$$(\alpha + \beta)\gamma = \alpha\gamma + \beta\gamma, \quad \text{for all } \alpha, \beta, \gamma \in \mathscr{M},$$

Galois showed his deep insight into the theory of fields by producing a list of all possible *finite* fields. The simplest of these is, of course, the field consisting of two elements, which may be denoted by the numbers 0 and 1, satisfying the following rules[47]:

addition table:

+	0	1
0	0	1
1	1	0

multiplication table: (1.9)

	0	1
0	0	0
1	0	1

It turns out that a field with n elements can exist if and only if the number n is of the form p^q, where p is a prime: in this case for every n of the form p^q there is exactly one (up to the manner of denoting the elements) field of order $n = p^q$. All such fields are called *Galois fields.* We note that Galois fields, which have hitherto been viewed as somewhat "exotic", have recently acquired great significance as a result of many applications: they can be used very efficiently in *coding theory*, which studies the most effective means of channeling information by, say, radio or telegraph. In particular, for channels by means of which one can send $n = p^q$ different signals (say, $4 = 2^2$ or $8 = 2^3$ signals) the theory of fields yields a convenient system for grouping the signals for coding different letters, so that the (coded) letters are very different from each other and cannot be confused, whereas for channels for which the number m of possible signals is not of the form p^q (say with $6 = 2 \cdot 3$ signals) such a convenient coding system does not exist.

It was, above all, a realization of the varied (and sophisticated) uses of groups that led Jordan to study the work of Galois. This study inspired his *Treatise on permutations ...*, which was the first systematic textbook on Galois theory and, at the same time, the first systematic textbook on group theory in world literature. This remarkable book[48] introduces and studies all the main group-theoretic terms and notions which Galois had no time to discover: the notion of a quotient group (see above) and the so-called *normal series* of a group $\mathscr{G}$, consisting of a nested sequence of normal subgroups

$$\mathscr{G} \supset \mathscr{H}_1 \supset \mathscr{H}_2 \supset \cdots \supset \mathscr{N} \tag{1.10}$$

(the quotient groups $\mathscr{G}/\mathscr{H}_1, \mathscr{H}_1/\mathscr{H}_2, \ldots$ etc. are called the *factors* of the normal series (10)); the main part of the so-called Jordan–Hölder theorem on normal series,[49] the notions of transitivity (and intransitivity) and primitiveness (and imprimitiveness) (due to Ruffini but not defined precisely by him), and so on. This list does not exhaust Jordan's achievements in the study of group theory. In the sequel we shall have occasion to return to yet another of his important contributions—a contribution which is very close to the main topic of the present book (see Chapter 6 and Note 229).

CHAPTER 2

Jordan's Pupils

In the period when Camille Jordan was absorbed by his book and inspired by relevant ideas, two young mathematicians were studying with him. They had finished their university studies and had come to Paris to enlarge their vision and begin independent research. They were the Norwegian Sophus Lie and the German Felix Klein. Their position was that of Jordan's postgraduate students, and they proved to be fine pupils indeed. As fate would have it, Lie's and Klein's studies with Jordan lasted for a very short time. However, they struck deep roots, and the ideas of Galois and Jordan played a crucial role in the subsequent scientific careers of both mathematicians.

Sophus Lie was born in 1842 into the family of a pastor in Norway. His childhood was passed in his parents' home on the shore of the ocean near Bergen. He travelled the length and breadth of the country on foot and all his life retained a passionate love for the beauty of Norwegian fiords and Norway's natural scenes. At school Lie mastered all subjects equally well, and after finishing school was at first unable to choose an occupation. His father wanted him to follow in his footsteps and become a pastor, and Sophus gave serious thought to studying theology. It was much later, after considerable thought and not without painful doubts, that he undertook the study of mathematics and natural sciences. At first, his studies at Christiania University failed to put an end to his doubts. The breakthrough came in 1868, when Lie read the works of V. Poncelet and J. Plücker (to which we will return below). These outstanding geometers made the strongest impression on the young Lie. Their works led to his first publications, which were followed by a continuous stream of papers, uninterrupted for several decades. To continue his education Lie moved to Berlin in 1870. There he met and immediately made friends with Klein, who was seven years his junior; the first joint work by Lie and Klein, described below, comes from the same year. The close personal and scientific relationship between Lie and Klein, which began then in Berlin, played a major role in the life of both mathematicians and continued until Lie's death.

The two made a visit to Paris, prompted by their desire to meet Jordan and also Gaston Darboux (1824–1917). Darboux was the best known specialist in

Gaston Darboux

differential geometry, which applies the differential calculus to the study of local properties (i.e., properties dealing only with small neighborhoods of a point) of curves and surfaces.[50] Darboux's voluminous and profound works (mention should be made, above all, of *Leçons sur la théorie générale des surfaces et les applications géométriques du calcul infinitésimal*, Vol. 1–4, Paris, Gauthier-Villars, 1887–1896; second edition 1914–1925) influenced both Klein and, especially, Lie. In particular, many of Lie's works were inspired by the approach of the *General Theory of Surfaces*, which organically combines differential geometry and the theory of differential equations. Here geometric questions are very efficiently reduced to analytic ones, and both approaches are used to study differential equations. All of this compels us to write about Darboux in greater detail.

Darboux was born in Nîmes in the south of France, but his whole life as a researcher and teacher was associated with Paris. He lived continuously in Paris from the age of eighteen and played an outstanding role in its intellectual life, above all as head of *L'Institut* and, as such, a member of the *French Académie* (see Notes 65 and 176). Darboux's name is linked to a considerable degree with the flourishing of the École Normale, as well as the tradition whereby all outstanding French mathematicians taught in secondary school after graduating from college. Darboux was virtually the first outstanding mathematician to study at the École Normale (a teachers' college), then a school less well known than the École Polytechnique (see Chapter 1). Subsequently he taught at the École Normale for many years. The respect enjoyed by Darboux even in government circles soon proved to be of great use to Lie (see below). In other cases, however, Darboux's influence was less favorable: for example Darboux, somewhat conservative in his mathematical tastes, opposed the defense of Henri Léon Lebesgue's (1875–1941) doctoral thesis. Only the influence of Emile Picard (1856–1941), Darboux's future successor as president of the *Institut*, sanctioned the defense of Lebesgue's thesis, which was to play an outstanding role in twentieth-century mathematics.

Klein and Lie were not destined to remain in Paris for a long time; nevertheless, the personal contacts of both mathematicians with Jordan (and Darboux) played an enormous role in their subsequent research (see Chapter 8). Actually, the two friends were planning to stay in Paris long enough to become familiar with the main achievements of the French mathematical school, and then move on to London for contacts with English mathematicians. The Franco–Prussian war broke out in 1871 and the German Klein had to leave France in a hurry. (O idyllic age when Klein was not even detained in Paris and could freely leave for Germany!) He intended to reenter France with the Prussian troops, but his military career was aborted—he contracted typhus and, in the meantime, France was rapidly routed. Left without his friend, Lie—who was an experienced hiker—decided to take advantage of the forced interruption in their studies to make a trek through all of France, the Alps, and Italy. But in the wartime atmosphere the plan proved to be rather unfortunate. Because of his poor French, conspicuous height and handsome but purely Nordic appearance,[51] Lie was immediately arrested as a German spy and imprisoned. Apparently, French patriots found Lie's manner of looking around in an abstracted way (he was then thinking through some mathematical problem) and then fevereshly scribbling in a little notebook (he was making mathematical jottings—in Norwegian) extremely suspicious. He spent about a month in the prison of Fontainebleau (just southwest of Paris—pretty far from the Alps!). As soon as he learned of Lie's arrest, Darboux used all his contacts in order to have Lie freed. But conditions in prison were not particularly bad, and Lie spent the time pondering over some aspects of Plücker's line geometry, to which his attention had been drawn by Klein and about which we shall say more below. Upon being freed from prison, Lie, the tireless hiker, went on his trek through France and Italy.

While Lie could be described as a typical nineteenth-century scholar, his friend and colleague Felix Klein was a very different individual, both in his attitude to science and in character. A born leader, a brilliant polemicist, a great teacher, and an excellent organizer, capable of implementing the most complex schemes and undertakings, Klein was a precursor of twentieth-century science. Klein combined the qualities of organizer, teacher, and researcher to a remarkable extent. (Some of his modern counterparts are, say, the Parisian Jean Alexandre Dieudonné (born in 1906), one of the leaders of the Bourbaki group, and the Moscow physicist Lev Landau (1908–1968).)

In the relationship between Lie and Klein, the latter, who was the younger of the two, played the role of the elder. Thus it was on Klein's initiative that the friends set out from Berlin to Paris and London (at that time they failed to reach London), it was Klein who (years later) suggested that Lie move from Norway to Germany (Leipzig), and so on. Klein's leadership was readily accepted by the unassuming and kindhearted Lie although, perhaps, deep down he felt slighted. Like any truly outstanding scientist, Lie was well aware of the value of his work and was proud of it (his works will be dealt with in detail in Chapter 6). Lie also knew that from the purely scientific standpoint his influence on Klein was greater than Klein's on him. In any case, it was a question of scientific primacy which provoked the only conflict between Lie and Klein in their otherwise remarkably smooth friendship.[52]

Klein was born in 1849 in Düsseldorf into the family of an official in the finance department. His father held extremely conservative, old-Prussian views; Felix adhered to some while flatly rejecting the others. In accordance with his father's wishes, Klein studied at a classical gymnasium. There, much attention was paid to ancient languages and very little to mathematics and the natural sciences. The deep antipathy Klein developed for the gymnasium[53] played an important role in his future pedagogical views. After graduating from the gymnasium, Klein entered the university in Bonn. There he was immediately noticed and singled out by Julius Plücker (1801–1868), who headed the departments of (experimental) physics and (pure) mathematics. In 1866, the seventeen-year-old Klein became Plücker's assistant in the physics department.

Plücker intended to make a physicist out of Klein. The latter showed a lively interest in physics (he was very "physics-minded"—more about that below) and had no objections to Plücker's intentions. But these plans were not destined to be carried through. In 1868 Plücker died, and it fell to Klein's lot to carry out the painstaking job of preparing for publication his mentor's unfinished works, above all the second part of the remarkable *Neue Geometrie des Raumes, gegründet auf die Betrachtung der geraden Linien als Raumelement*. Work on the book (published in 1869) inspired Klein, and his first series of independent papers grew out of it, contributing to Klein's development as a mathematician.

After Plücker's death, Klein lost his post as an assistant, left Bonn and went to Göttingen and to Berlin where he became acquainted with the young but very influential Göttingen mathematician Rudolf Friedrich Alfred Clebsch (1833–1872), the physicist Wilhelm Weber (1804–1891), a friend and colleague of the great Gauss,[54] and the head of the Berlin school of mathematics Karl Theodor Wilhelm Weierstrass (1815–1897). It should be noted that whereas Klein's relations with Clebsch and Weber were quite friendly from the beginning, his relationship with Weierstrass was marked from the outset by barely concealed antipathy on both sides. The roots of that hostility lay in the total incompatibility of Klein's and Weierstrass's scientific positions. This deserves a more detailed explanation.

It is now well known that the human brain is not symmetric, and that the left and right hemispheres of the cerebrum each have their specific functions. The great interest focussed at present on a range of a questions having to do with that asymmetry was reflected, in particular, in the award of the 1981 Nobel Prize in biology and medicine to the American psychologist Roger Sperry for research in that field. In the standard case of the right-handed person, the left hemisphere is responsible for analytic, logical thinking, while the right hemisphere provides the "pictorial," synthetic vision of the world (of course, here we describe the differences between the hemispheres roughly and incompletely). With the same degree of approximation as above, it can be said that the left hemisphere, undoubtedly linked with speech, writing,[55] as well as computation and the use of the set of natural numbers in general, controls the algebraic aspects of mathematics (since algorithmic procedures have to do with linearly arranged algebraic formulas, they undoubtedly belong to the domain of the left hemisphere), while the right hemisphere is associated with geometric vision, with diagrams and pictures. Possibly, however, it may be more correct to link the left hemisphere with logic, and the right one with physics, bearing in mind the global approach to nature and to the phenomena of the natural sciences characteristic of most physicists.

The above may go part of the way in explaining the striking fact of the existence of two types of mathematicians, opposite in some respects: algebraists, whose thinking has to do primarily with logic, formulas, and algorithmic procedures; and geometers or physicists, who proceed mostly from graphic and visual impressions rather than from formulas. The existence of these two types of poorly correlated approaches to mathematics, as well as the existence of scientists for whom one or the other dimension in mathematics is the dominant one, was pointed out in a lecture[56] by the eminent mathematician Hermann Weyl (1885–1955) (we shall come across his name again below).[57] In that lecture Weyl mentioned Klein and B. Riemann (see below) as examples of "physicists" and Weierstrass as an "algebraist." An earlier example is presented by the founder of the differential and integral calculus, the great physicist Isaac Newton (1642–1727) on the one hand, and the great logician Gottfried Wilhelm Leibniz (1646–1716) on the other. It is plausible that the mutual antipathy developed by Klein and Weierstrass was fostered by analogous differences in their respective scientific outlooks.[58]

Klein's "physical" thinking has already been mentioned; it was reflected in many of his research papers, for example, in the remarkable *Lectures on Riemann Surfaces* (a course of lectures delivered in Göttingen and circulated in mimeographed form) in which Klein took the liberty of considering the distribution of electric charges along a conductor shaped as an abstract Riemann surface of extremely complex topological structure in order to prove purely mathematical theorems. Klein's teaching (see Note 60) was also characterized by a physical and graphic approach, and, in consequence, by a certain lack of rigor. Klein's mode of exposition was largely due to the influence of the great Bernhard Riemann. Klein worshipped Riemann, whereas Weierstrass, the fanatic advocate of rigor (modern mathematics largely owes its spirit and style to him), assailed Riemann and his friend Lejeune Dirichlet and considered many of their results unproved or even incorrect. This constructive criticism gave rise to the theory of real numbers and to many topological concepts. In this connection, an interesting incident was recounted (at second hand) by Arnold Sommerfeld (1868–1951), one of the greatest of twentieth-century physicists. Sommerfeld was Klein's pupil and staff member for many years; Klein engaged him as his assistant, just as he himself had once been appointed assistant in the physics department by Plücker.[59] Sommerfeld told how, in the early 1860s, Weierstrass and the outstanding German physicist, mathematician, biologist, and medical doctor Hermann Helmholtz (1821–1894) spent a summer together in the country. Weierstrass had taken along Riemann's famous work (the source of the entire modern theory of functions of a complex variable) in order to elaborate and analyze it in his free time—while the highly "physicsminded" Helmholz could not imagine what there was to elaborate (see A. Sommerfeld, "Klein, Riemann und die mathematische Physik" *Naturwissenschaft*, **7** (1919), 300–303).[60]

Klein's close friendship with Lie and the latter's significant scientific influence on him made up for the absence of fruitful scientific contacts with Weierstrass. We have already described Lie and Klein's joint trip to Paris, which played an important role in the careers of both mathematicians. After returning from France and recovering from typhus, Klein settled in Göttingen not far from Clebsch and Weber; this was an extremely productive time for him. However, before dealing in detail with the scientific achievements of Klein and Lie, it is necessary to describe briefly the scientists who laid the foundation for their successes.

CHAPTER 3

Nineteenth-Century Geometry: Projective Geometry

Science never develops evenly. It is marked by rising and ebbing tides, linked to external conditions and stimulating the progress of some trend or, on the contrary, holding it back. In ancient Greece, geometry[61] was the basic branch of mathematics; from that time on the word "geometer" has been often used interchangeably with "mathematician," something we encounter even in the not so distant past (see Note 24). However, subsequently, successes scored in mathematics by-passed geometry for a long time. The principal mathematical achievements of the Renaissance and the following period were in algebra (see Chapter 1), and even medieval Arab (or, to be more exact, Arabic-language) mathematics was oriented towards algebra rather than geometry (see Chapter 1 and Note 9).

The seventeenth century was marked by the development of the calculus, which for centuries afterwards was regarded as the principal branch of mathematics. This was reflected in the appearance of the term "higher mathematics" (or "höhere Mathematik" in German), meaning analytic geometry, the differential and integral calculus, and related fields (e.g., differential equations). At the present time this term sounds absurd (surely, probability theory or mathematical logic cannot now be regarded as branches of mathematics "lower" than the calculus) but is nevertheless widely used, especially in Russian- and German-speaking countries.

In addition to the calculus, the seventeenth and eighteenth centuries saw the "coming of age" of number theory and probability. In number theory, the great names were those of the Frenchman Pierre de Fermat (1601–1665) later L. Euler and J.L. Lagrange, and, at the very beginning of the nineteenth century, C.F. Gauss. In probability theory they were those of Fermat, B. Pascal, and the Dutchman Christian Huygens (1654–1695) followed by Jacob Bernoulli (1654–1705) then the French Huguenot and English scientist Abraham de Moivre (1667–1754) and, in the early nineteenth century, Gauss and Laplace. The late eighteenth century was marked by decisive successes in algebra due to Lagrange and Ruffini. However, these two centuries, so rich in outstanding scientists and brilliant results, were marked by only modest success in geometry, where one can only point to the works of the French architect and military engineer Gerard Desargues (1593–1661), obviously ahead of their time and soon forgotten, to their continuation and extension (also unrecognized at the time) by the great scientist, writer, moralist, and religious figure Blaise Pascal (1623–1662),[62] and to some results obtained by Euler, whose encyclopedic knowledge prevented him from completely avoiding geometry and who had to his credit some very substantial works in this area.[63]

But in the nineteenth century the situation changed fundamentally. This can be called the golden age of geometry. According to Nicolas Bourbaki (see his *Eléments d'histoire des mathématiques* (Paris, Hermann, 1974)) this period extends, roughly, from the publication of Monge's *Géometrie descriptive* (1795) to Klein's *Erlanger Programm* (1872). Spectacular in the instant flourishing of the entire spectrum of geometric fields, in the simultaneous appearance of numerous brilliant geometers with different creative approaches, and—alas—in the rapid decline and unexpected fall of interest in the most ancient stream of mathematical research,[64] that age changed the face of geometry,

Gaspard Monge

which emerged from it a completely different science from the one that had entered the nineteenth century.

The founders of nineteenth-century French mathematics were the precursors of geometry's coming successes. They included one of the main organizers of French science and education, Gaspard Monge (1746–1816), who was at one time naval minister in the French revolutionary government, as well as Monge's pupil Lazare Nicolas Marguerite Carnot (1735–1823),[65] scientist and politician, famed organizer of the French revolutionary victory, influential member of the Public Salvation Committee and, in effect, revolutionary France's war minister.

Gaspard Monge was born in a small town in Burgundy, into the family of a shopkeeper. His father was an almost completely uneducated man who, however, profoundly respected knowledge and sought to give his sons the best education he could; and he was rewarded—Gaspard and his two younger brothers subsequently became professors of mathematics, an incredible success for the family of a poor provincial in pre-revolutionary France. An officer who happened to be travelling through Monge's home town saw a map of the town and its outskirts very ably compiled by the young Gaspard and arranged for him to be enrolled in a military school (or military academy, as we would say today; see Note 22) in Mezières in the Ardennes. This was one of the oldest and best higher military academies in France. Monge was accepted in the auxiliary department, which trained technical personnel for the army, because only people of noble origin were permitted to study at the department which trained officers. However, having presented the best possible solution of a problem concerning the layout of fortifications—he applied the ideas of descriptive geometry that he had by then worked out and would subsequently describe in the works of the École Polytechnique as the art of depicting three-dimensional objects on a flat piece of paper—,[66] Monge was granted the privilege of teaching mathematics and, soon thereafter, physics at Mezières. His outstanding pedagogical, scientific, and administrative career began at that point.

When Monge entered the school in Mezières he was 18; at 19 he became a teacher and assistant to the mathematics professor Charles Bossut (1730–1814); at 24 he was appointed full professor of mathematics and physics; and at 28 he was elected corresponding member of the French Academy of Sciences on a motion by Bossut and the famous d'Alembert (whose name will be encountered below) and Antoine Nicolas de Condorcet (1743–1794). In 1780 Monge became an academician. At first he combined teaching in Mezières with long stays in Paris, where he took part in the Academy's sessions, and later he moved permanently to Paris.

As a commoner Monge was elated by the revolution of 1789. His short term as naval minister was not particularly successful; yet his organizational activities in the revolutionary years were quite important. He took part in organizing the manufacture of gunpowder and the casting of cannons; he also played a major role in the commission which supervised France's transition to the metric system. But his greatest achievement was his participation in the founding of a new type of college, soon to be called the *École Polytechnique.*[67]

Monge's École Polytechnique was set up to train highly qualified engineers. It offered a three-year course, after which the graduates could continue their education in more specialized advanced institutions or military academies. The curriculum at the École Polytechnique was restricted to general fields, i.e., mathematics, theoretical mechanics, and physics; however, the level of instruction was extremely high because of a brilliant staff of lecturers and teachers, to whose selection Monge assigned top priority. Entrance examinations were selective and were held simultaneously in many places throughout the country. The examinees were supposed to solve a number of problems that they would select from a large list offered to them. (Of course, the list was the same for all of them: the examinees would open the envelopes with the problems simultaneously in all the places where the examination was held; the solution of each problem was worth a certain number of points.)

Solution of difficult problems continued to play an important role in the teaching process at the school, and the points gained during the studies were taken into account upon graduation. The subsequent careers of the graduates were heavily dependent on the places they had earned at graduation. Initially the school "headmaster" was changed every month, and, on Monge's suggestion, Lagrange was appointed the first headmaster; Monge himself was the second. Later, the inconvenient system of monthly changes in administration was abandoned, and Monge was the sole head for many years.

The École Polytechnique played a major role in nineteenth-century European science. In particular, it set an example for Klein in his activities at Göttingen University in Germany.[68] Undoubtedly, the example of the École Polytechnique was taken into account in the setting up of the German *Technische Hochschulen* in the late nineteenth and early twentieth century (in Zurich, Munich, Prague, and elsewhere); it was also instrumental in the founding of the famous American technological institutes (MIT, Cal Tech) and the Moscow Physics and Technology Institute in the town of Dolgoprudnaya near Moscow. The lectures delivered by Monge at the École Polytechnique, and in part at the École Normale, served as the basis for two textbooks in the new fields of geometry which he had in effect founded: *Géométrie descriptive* (1795), mentioned above, and *Application de l'analyse à la géométrie* (1795, 2nd ed. 1801; 3rd ed. 1807; and 4th ed. 1809).[69]

Monge's scientific, pedagogical, and organizational activities were held in high esteem by Napoleon, who bestowed numerous distinctions upon him. He was the first civilian to receive the Legion of Honor award instituted by Napoleon; he was made a senator and given the title of count. For his part, Monge was completely loyal to Napoleon. Monge's support of Napoleon during the hundred days when the latter attempted to regain power following the restoration of the Bourbons[70] was held against him when the Bourbons returned to power a second time: he was expelled from the Academy of Sciences and stripped of all titles, the École Polytechnique was temporarily closed, and Monge was no longer on its staff after it reopened. All this had a telling effect on the old professor. Monge was in a state of depression from the time of Napoleon's second defeat in 1815 until his own death in 1818. The pupils of the École Polytechnique were strictly forbidden to attend Monge's funeral; this, however, did not prevent them from collecting money to buy flowers which they lavished on his tomb the first Sunday after his burial.

We have called Monge (and Carnot) the precursors of the flourishing of geometry in the nineteenth century. Probably the first truly great geometer of that century was another French officer, Monge's pupil at the École Polytechnique, Jean Victor Poncelet (1788–1867).

Poncelet was an officer in Napoleon's army and was taken prisoner during the Russian campaign of 1812, spending two years as a prisoner of war in a village near Saratov on the Volga. However, living conditions were not particularly harsh for French officers, and, in order to while away the time, Poncelet began to lecture on geometry to a group of his fellow officers, mostly graduates of the École Polytechnique and like himself, former pupils of Monge. On returning home, the young officer read the existing literature and discovered that the ideas he set forth in the Saratov lectures were quite original and could serve as the basis for a completely new branch of geometry that Poncelet called *projective geometry*.

Poncelet summed up the results obtained while a prisoner of war in his large *Traité des propriétés projectives des figures* (1822) which brought its author fame. In later years, for example when publishing a new edition of the *Traité* (1864–1866), Poncelet—now a general—would complain bitterly of that early renown. The success of the book put out by the young officer in 1822 launched his administrative career. Poncelet attained extremely high military and scientific posts, including membership in the National Defence Committee and direction of the famous École Polytechnique, where he had once studied so fruitfully; he also took a leading part in organizing the London (1851) and Paris (1855) international expositions. But the duties attached to the high titles and ranks almost completely separated Poncelet from the science he loved so much[71]—thus, for example, the triumphs of projective geometry which followed the appearance of his *Traité* were achieved almost completely without his participation, a fact he bitterly deplored in the last years of his life. The old conflict between *vita activa* and *vita contemplativa* introduced a certain dissonance to those final years, as Klein pointed out in his biography.

Poncelet started from Monge's lectures on the descriptive geometry which he had created while studying the representation of three-dimensional figures in the plane (say, on a piece of paper). Since it is impossible to put a three-dimensional figure F in a plane, one must use the representation obtained by projecting all of its points on an image plane. Monge preferred orthogonal projection, which sends every point A of F to its orthogonal projection A' in the image plane (A' is the foot of the perpendicular AA' from A to the plane; see Fig. 4(a)). (The Monge method, now widely used in descriptive geometry,

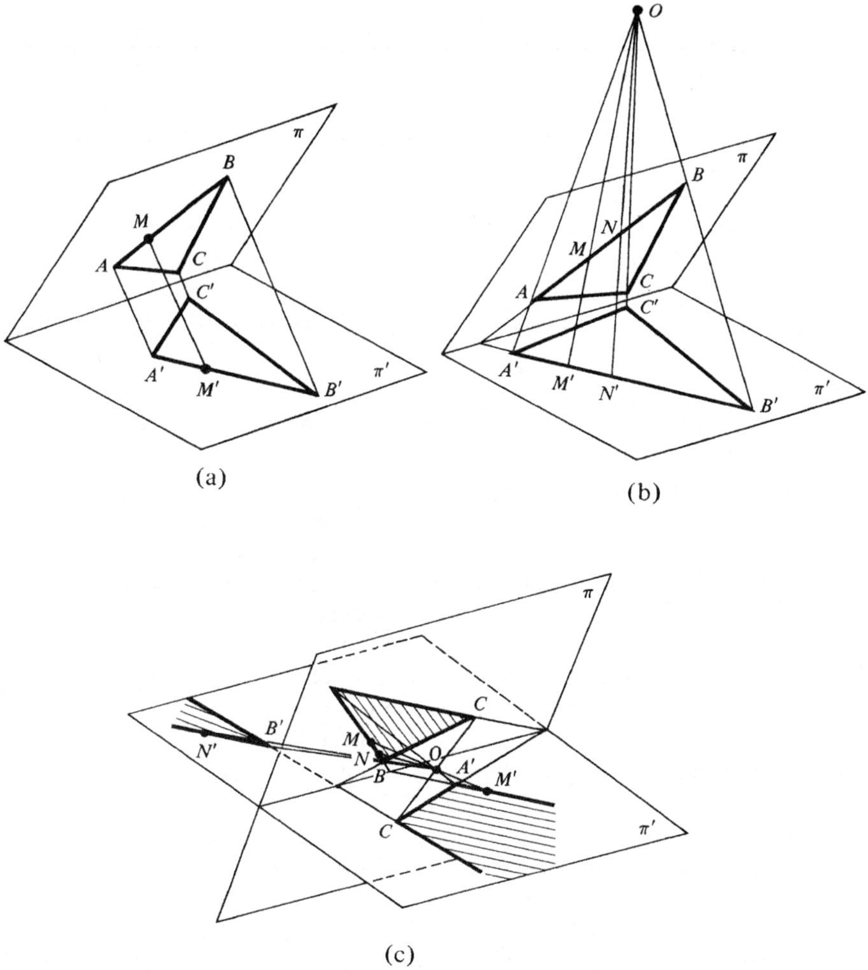

FIGURE 4

consists in replacing the three-dimensional figure F by its orthogonal projections F_1, F_2, and F_3 on three pairwise perpendicular planes. It is clear that the three-dimensional figure entirely determines the (plane) figures F_1, F_2, and F_3 and that the figures F_1, F_2, and F_3 also allow us to recover the figure F.)

Poncelet, on the other hand, was interested in the relationship between a (plane or space) figure F and its central projection F' consisting of the intersection points A' of the image plane π with all the lines OA, where O is the fixed center of projection and $A \in F$ (Fig. 4(b)).

It is clear that a parallel projection λ (sending each point A of the figure F to the point $A' = \lambda(A)$ such that $A' \in \pi$ and $AA' \| \ell$, where ℓ is a fixed line (if $\ell \perp \pi$, then λ is called an orthogonal projection)) does not change the figure

F very much. Thus if F is a triangle (see Fig. 4(c)), then $F' = \lambda(F)$ is also a triangle, though, in general, F' is different from F. A central projection ζ can alter a figure much more: thus the projection ζ in Fig. 4(b) sends the triangle $F \equiv ABC$ into the figure $F' = \zeta(F)$ consisting of two vertical angles at A' in one of which the tip is replaced by a segment $(B'C')$. Poncelet called the properties of figures preserved under central projection *projective properties* and the science studying these properties *projective geometry.*[72]

What are the basic notions of projective geometry? At this point it is convenient to compare projective geometry to affine geometry (originating with Euler; see Note 63) which studies the properties of figures preserved under parallel projections.[73] It is clear that a parallel projection λ sends every line a (regarded as the set of its points) into a new line $a_1 = \lambda(a)$; that it sends parallel lines a and a_1 into parallel lines $a' = \lambda(a)$ and $a'_1 = \lambda(a_1)$; and that it preserves the "simple ratio" $(A, B; C) = AC/BC$ of three collinear points A, B and C (in the notations of Fig. 4(a), $A'M'/B'M' = AM/BM$). Thus the notions of a line (and of a point A belonging to a line a), of lines a and a' being parallel $(a \| a_1)$ and of the simple ratio $(A, B; C) = AC/BC$ of three points A, B, $C \in n$ (three points of a line n) are all meaningful in affine geometry. On the other hand, parallel projection can transform a circle into an ellipse (see Fig. 5(a), where the circle S' in the plane π' is sent by parallel projection into the ellipse S of the plane π), and, therefore, the notion of a circle is meaningless in affine geometry, while its role is played, in a certain sense, by the ellipse. Similarly, a central projection ζ transforms a line a into a new line $a' = \zeta(a)$ and preserves the so-called *cross ratio* $(A, B; C, D) = (A, B; C)/(A, B; D) = (AC/BC)(AD/BD)$ (the quotient of two simple ratios) of four collinear points; if B, C, M, $N \in n$ and $B' = \zeta(B)$, $C' = \zeta(C)$, $M' = \zeta(M)$, $N' = \zeta(N)$ (see, for example, Fig. 4(b), then $(B', C'; M', N') = (B, C; M, N)$; that is why, in projective geometry, it is possible to speak of a line; of a point A belonging to a line a; and of the cross ratio $(A, B; C, D)$ of four collinear points. On the other hand, a central projection ζ may transform parallel lines into intersecting lines (see Fig. 6, where $AP \| BQ$, while $A'P' \cap B'Q' = C'$, so that the "halfstrip" $PABQ$ becomes the triangle $A'B'C'$); for that reason the notion of parallel lines does not exist in projective geometry. In projective geometry the role of circles is played by *conic sections*, i.e., ellipses, parabolas, and hyperbolas—in other words, by curves which can be obtained from a circle by means of central projection. Thus in Fig. 5(b) the central projection with center O transforms the circle ζ in the plane π' into the ellipse ζ in the plane π_1, or into the parabola ζ in the plane π_2, or into the hyperbola ζ in the plane π_3.

Figure 6 sheds light on another important fact: that a central projection ζ does not establish a one-to-one correspondence between the points of a given plane π and the image plane π'—no point $C' \in \pi'$ corresponds to the point $C \in \pi$ since $OC' \| \pi$. Thus neither π nor π' in the mapping $\zeta\colon \pi \to \pi'$ can be thought of as the usual (Euclidean or affine) planes, i.e., the very notion of a plane in projective geometry must be somewhat modified. Namely, we suppose that the image of the point C' under the projection $\zeta\colon \pi' \to \pi$ shown in Fig. 6

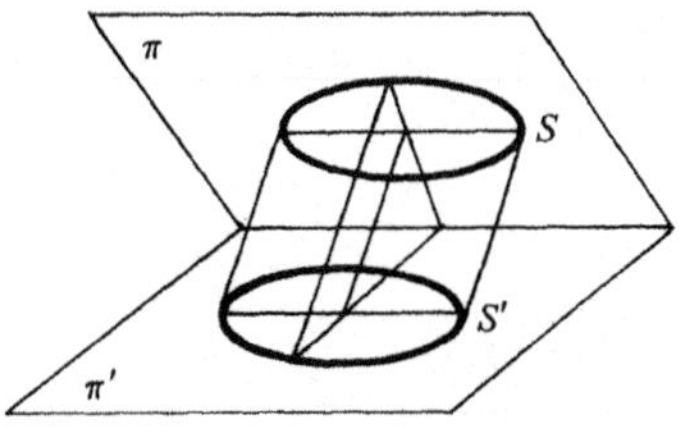

(a)

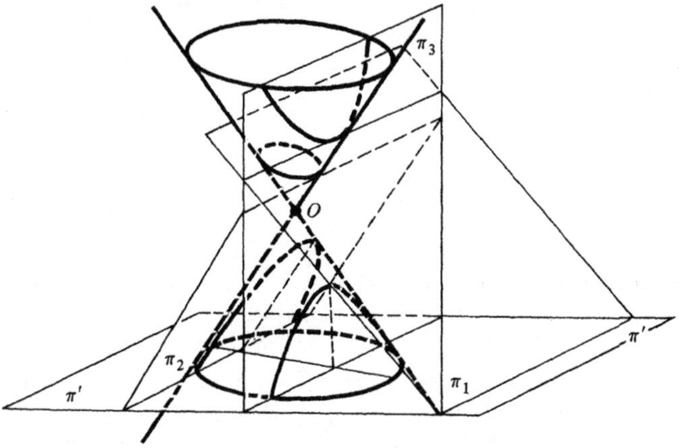

(b)

FIGURE 5

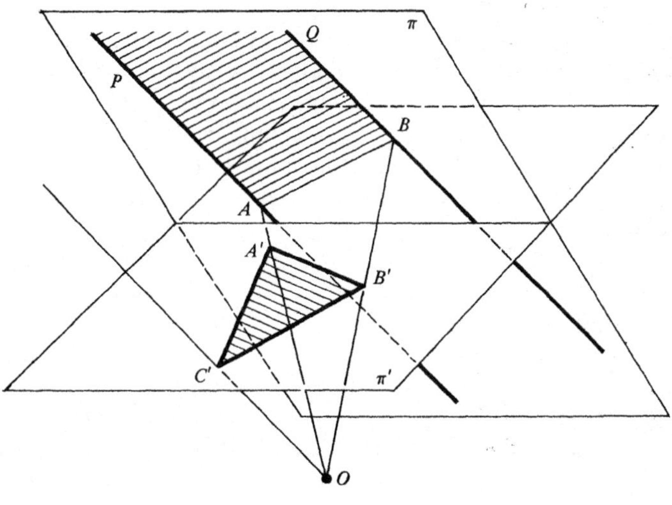

FIGURE 6

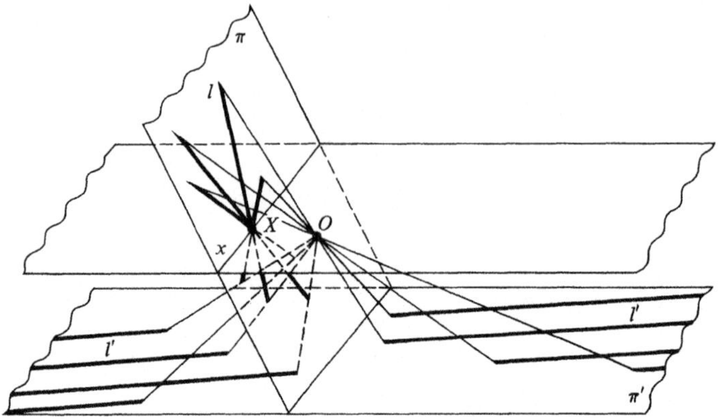

FIGURE 7

is the "point at infinity" C in which the lines AP and BQ intersect (although they do not intersect in the ordinary plane!), i.e., that $\zeta\colon C \to C'$ or $\zeta(C) = C'$. Now, since a sheaf of lines intersecting at the point M and contained in the plane π can be transformed by the central projection $\zeta\colon \pi \to \pi'$ into a sheaf of parallel lines in the plane π' (Fig. 7), it is convenient to assume that, from the point of view of projective geometry, all the lines of the (Euclidean or affine) plane parallel to some definite line in that plane converge to a single "point at infinity." (Thus, "from the affine point of view," we may assume that "points at infinity" are simply sheaves of parallel lines, just as an ordinary point M may be identified with the sheaf of lines intersecting at M.) It is useful to assume that all the "points at infinity" of the projective plane π belong to the same "line at infinity" o (whose image under the central projection $\zeta\colon \pi \to \pi'$ with center O is the line o' in which the plane π' intersects the plane ω passing through O parallel to π). The "Euclidean differences" between ellipses, parabolas, and hyperbolas can now be explained by saying that the ellipse contains no points at infinity, the parabola contains exactly one such point (corresponding to the direction of its axis of symmetry, along which the parabola "goes to infinity"), while the hyperbola contains two points at infinity (corresponding to the directions of its asymptotes—it is in these directions that the hyperbola "goes to infinity").

Thus the projective plane, the "domain of action" of projective geometry, differs from the ordinary Euclidean plane in that to each line of the latter one must add a point (the point at infinity, which does not exist in Euclidean geometry) so that all the points at infinity of a sheaf of parallel lines coincide and the set of all the points at infinity of the projective plane constitutes a single "line at infinity." Of course, such a description of the projective plane is based on our ordinary (Euclidean) ideas, which are meaningless in projective geometry; actually, in projective geometry the "points at infinity" are indistinguishable from other points, since central projection can transform any ordinary point into a point at infinity and conversely. The relationship between the notions of a projective plane and of a central projection leads us to the following "model" of the projective plane, which often turns out to be useful: the projective plane is the sheaf of all lines and planes of space passing through a fixed point O ("the center of projection"); here, the lines passing through O are called "points" of the projective plane π and the planes passing through O are called "lines."

Note that any two lines a and b of the projective plane necessarily intersect in a unique point (an ordinary point if $a \nparallel b$ in the Euclidean or affine plane π_0 from which the projective plane π was obtained by adding the line at infinity o, or a point at infinity if $a \| b$ or if one of the lines a or b is o); this fact determines a complete analogy between the properties of points and lines in the projective plane. Indeed, in the Euclidean plane, the analogy between the properties of points and lines breaks down because there exist nonintersecting (parallel) lines, while any two points belong to precisely one line; in the projective plane, however, there are no parallel lines. The analogy between the properties of points and lines may be stated in the form of the so-called *duality principle* of projective geometry, which states that in any theorem of projective geometry one can replace the word "point" by the word "line" and conversely, the expression "lies on" by the expression "passes through" and conversely, and the new theorem thus obtained (called *dual* to the given theorem) will be true precisely if the original theorem was true. The duality principle is one of the cornerstones of projective geometry. Unfortunately, its discovery was marred by a rather unpleasant argument over authorship between Poncelet and another French geometer, Joseph Diaz Gergonne (1771–1859), both of whom claimed priority.[74]

The nineteenth century was the golden age of projective geometry, which was undoubtedly the leading branch of geometry in that period.[75] Besides Poncelet and Gergonne, Michel Chasles (1793–1880), another prominent French geometer and for many years a professor at the École Polytechnique (where a geometry department was specially created for him in 1846), played an active role in the development of projective geometry. Chasles was only four years younger than Poncelet, but their scientific activities belonged to different epochs: whereas Poncelet did mathematical research only in his youth, Chasles realized his scientific potential only in later years. His first research publication appeared when he was a student at the École Polytechnique; but after completing his course of study Chasles was in no hurry to begin working in science, believing that he should first provide for himself, in material terms, by other means. He settled in his native city of Chartres, where he quickly gained fame as an entrepreneur and highly successful financial expert. Having become rich as a result of his banking activities, he switched to geometry. His first publication was the long *Aperçu historique sur l'origine et le développement des méthodes de la géométrie* (1837).[76] That review served as the starting point of research for many geometers, and above all, for Chasles himself. His other publications followed in quick succession. Chasles's courses

August Ferdinand Möbius

on higher geometry[77] at the École Polytechnique were particularly influential. Chasles was fully committed to the "analytic trend" which cultivated coordinate methods in geometry (about which more will be said below); and he possessed a unique analytic intuition that enabled him to obtain great numbers of impressive geometric facts from the store of faceless formulas. Unfortunately, Chasles's last years were tainted by a scandal which had wide repercussions in Paris. It turned out that the famous scientist and professor at France's leading college, who had been an avid collector of manuscripts for many years, inexplicably fell prey to a swindler who supplied him with obviously forged documents, such as Cleopatra's letters to Julius Caesar. That incident lent Chasles the dubious honor of being the prototype of the hero in Alphonse Daudet's *The Immortal* (Daudet, however, did not make his protagonist a member, like Chasles, of *L'Institut* but rather a member of the French Academy, to which only men of letters are elected (see Note 176 below)).

The analytic approach to geometry cultivated by Chasles was based on the research of the German geometer August Ferdinand Möbius (1790–1868). Möbius in turn readily acknowledged his indebtedness to the Frenchmen Poncelet and Gergonne; the development of science is always international, and the export of scientific ideas beyond state frontiers has never been regarded as smuggling. Möbius and Chasles who headed the analytic trend in French geometry were scientific equals. On the other hand, in purely human terms, the two were very different and, it must be said, Möbius's was a far more attractive personality.[78]

August Möbius was born in Saxony, in the royal school in Schulpfort not far from Leipzig. His father was a dance teacher at the court. (Readers who like to ponder problems of heredity may be interested in the fact that August Möbius's son later became a celebrated neurologist and the author of a notorious book about the physiological weakness of women.) Möbius finished secondary school in Schulpfort and entered Leipzig University in 1809 where he first studied law and then physics and mathematics. In 1813–14, the young Möbius was studying at Göttingen under Gauss who, however, only prepared him for a career as an astronomer and failed to discover his student's outstanding mathematical abilities.[79] Nevertheless, his studies at Göttingen made a lasting impression on Möbius and throughout his life he considered himself a pupil of Gauss, each of whose letters was the object of childlike pride on Möbius's part. In 1814, Möbius returned to Leipzig University. After graduation he accepted a post at the astronomical observatory in Pleisenburg, a suburb of Leipzig. He worked there for more than fifty years, until his death in 1868; he rose from the rank of a lower staff member to the position of observatory director (he combined the latter job at the end of his life with professorial duties at Leipzig University). Möbius's whole life passed within the walls of the Pleisenburg Observatory; his study, the flat in which he lived with his wife and children, and the hall where he was always glad to lecture, were all in this building. It was characteristic of Möbius that he took his observatory duties very seriously. He wrote a number of works on practical astronomy, including investigations concerning the improvement of the optical systems of telescopes, and his manuals on astronomical observations were still very popular in Germany in the 1920s (the last edition was dated 1916).

As a person, Möbius was the epitome of the absentminded professor. He was shy and unsociable, timid with unfamiliar people, and so absorbed in his thoughts that he was forced to work out a whole system of mnemonic rules (which did not always work) so as not to forget his keys or his inseparable umbrella and handkerchief when he set out from home for a walk or for the university. His entire life passed in one city and in one building. His study in Göttingen and two or three short excursions through Germany in his youth were his principal "adventures." A complete picture of his life can be gained from the scientific diary Möbius wrote every night and by which we can trace the evolution of his views, interests, and ideas, the only things which changed in that fully regulated life. It is paradoxical that modesty and even shyness in everyday life combined in that impressive figure with boldness, fantasy and inventiveness in science, profound thoughts, and outstanding teaching abilities. All of Möbius's works, including two of his long books *Der barycentrische Calcul* (1827) and the two-volume *Lehrbuch der Statik* (1837) are distinguished not only by innovative thinking and deep insights but also by crispness of style, clarity of narrative and excellence of structure. The mathematical talent of most mathematicians diminishes with age (Poncelet and Klein are relevant examples). But time did not diminish Möbius's gifts. What was perhaps his most impressive discovery—that of one-sided surfaces such as the famous "Möbius strip"[80]—was made when he was almost seventy, and all the works found among his papers after his death show the same excellence of form and profundity of thought.

Möbius's attainments stand out even against the background of the outstanding achievements that marked nineteenth-century geometry. Unfortunately, these achievements were not obtained in a spirit of cooperation but in a context of unending quarrels and bitter rivalry. The advocates of "purely geometric" (synthetic) methods attacked the "analysts," members of the French school warred against the Germans, and so on. These arguments and conflicts involved Poncelet, Chasles, Steiner and Plücker. The modest Möbius (like C. von Staudt, similar to him in character and temperament) remained aloof from any discussions not of a purely scientific nature. Moreover, in his works Möbius introduced a very impressive note of reserve into the debates: his works succeeded in uniting the analytic and synthetic approaches in geometry and served as the basis for many of its subsequent triumphs. At the same time, as pointed out above, Möbius readily conceded the priority of French mathematicians in the creation of projective geometry—that remarkable scientist was free not only of personal arrogance but of national prejudice as well.

The key idea in Möbius's view of projective geometry was, above all, the idea of projective coordinates, which assign to every point of the projective plane a system of numbers—the coordinates of this point. In modern expositions these coordinates are usually introduced as ordinary ("affine") coordinates in three-dimensional space $\mathbb{R}^3$ in which sheaves of lines and planes (with center at the origin O of the coordinate system) form a model of the projective plane. From this description it is clear that each point of the projective plane (i.e., each of the lines of three-dimensional space passing through O) is described by three coordinates x, y, z or x_0, x_1, x_2, at least one of which does not vanish, while proportional coordinate triples (x_0, x_1, x_2) or $(\lambda x_0, \lambda x_1, \lambda x_2)$ (where $\lambda \neq 0$) describe the same point of the projective plane (for if $M_1 = (x_0, x_1, x_2)$ and $M_2 = (\lambda x_0, \lambda x_1, \lambda x_2)$, then OM_1 and OM_2 denote the same line of the sheaf).

Möbius introduced coordinates in the projective plane in a different way. He considered an arbitrary fixed triangle $A_0A_1A_2$ of the plane π and the center of gravity of a system of masses m_0, m_1, m_2 placed at the points A_0, A_1, A_2. If we assume that the masses are also allowed to be "negative" (i.e., the corresponding "weights" may be directed not only vertically downward, but also upward), then it is easy to verify that for each point M of the plane we can choose a system of numbers m_0, m_1, m_2 (where $m_0 + m_1 + m_2 \neq 0$) such that the center of gravity of the masses $A_0(m_0)$, $A_1(m_1)$, and $A_2(m_2)$ coincides with M; it is these numbers m_0, m_1, m_2 or x_0, x_1, x_2 which Möbius called the *barycentric coordinates* of the point M (coordinates related to the notion of center of gravity or "barycenter"). The numbers m_0, m_1, m_2 for which $m_0 + m_1 + m_2 = 0$ are the barycentric coordinates of the "points at infinity" of the projective plane. Clearly, the (barycentric or projective) coordinates x_0, x_1, x_2 of a point are only determined up to multiplication by a common factor $\lambda \neq 0$, i.e., it is only the ratio of these coordinates which is relevant. This dictates the suggestive notation $M(x_0\!:\!x_1\!:\!x_2)$. Möbius now defined lines by means of linear homogeneous equations $a_0x_0 + a_1x_1 + a_2x_2 = 0$ which relate the (barycentric or projective) coordinates x_0, x_1, x_2 of points on a line, and defined conic sections as "second-order curves", i.e., by means of homogeneous equations of the form $a_{00}x_0^2 + a_{11}x_1^2 + a_{22}x_2^2 + 2a_{01}x_0x_1 + 2a_{02}x_0x_2 + 2a_{12}x_1x_2$, etc.[81]

It should be noted that another branch of geometry, which developed significantly in the nineteenth century, originated with Möbius. This is the so-called *circle geometry* or *inversive geometry*, which studies the properties of figures invariant under inversions of the plane. An inversion i with center O and degree k is defined as the map $i\colon A \leftrightarrow A'$ which sends each point A of the plane into the point A', such that A' belongs to the line OA and $OA \cdot OA' = k$. A characteristic property of inversions is the fact that they transform circles (to which, in this context, it is convenient to add "circles of infinite radius"—straight lines) into circles; inversions can transform straight lines into circles. Thus in circle geometry the notion of a circle (of finite or infinite radius) is meaningful, while the notion of a straight line is meaningless.[82] Finally, by his discovery of one-sided surfaces similar to the Möbius strip or the heptahedron (see Note 80) Möbius also made a substantial contribution to topology.

The development of projective geometry in Germany in the nineteenth century was quite rapid: the first volume of Crelle's Journal (mentioned above) which contained Steiner's early papers appeared in 1826; Möbius's *Barycentric calculus* was published in 1827; the first volume of Plücker's *Analytisch-geometrische Entwicklungen* appeared in 1828; the first (and, unfortunately, also last) volume of Steiner's *Systematische Entwicklung der Abhängigkeit geometrischer Gestalten von einander* appeared in 1832 and *Geometrie der Lage* by Staudt, which, in a certain sense, completed the evolution of projective geometry, appeared in 1847. Steiner headed the synthetic trend based on direct inferences from geometric axioms (usually not stated clearly at the time). Plücker headed the analytic trend with emphasis on coordinates (recall the title of Plücker's book). The two engaged in an endless feud.

The most colorful figure and the most brilliant geometric talent, in nineteenth-century mathematics was a sometime Swiss shepherd named Jacob Steiner (1796–1863). Steiner was born into a poor peasant family, far from the centers of science and culture, and received no education in his childhood. Later, he liked to recall that he could hardly write at eighteen, although he had gained some knowledge on his own in mathematics and in astronomy, which he particularly liked in his youth. The young shepherd's knowledge and

Jacob Steiner

interest amazed a colleague of the outstanding Swiss teacher Johann Heinrich Pestalozzi (1746–1827), who happened to meet Steiner, and, with some effort, persuaded the youth's father to let the badly needed agricultural hand go to Pestalozzi's school. There Steiner first studied and then taught mathematics. In 1818 he left Pestalozzi's school for the nearest major university center, Heidelberg, in Germany. Forced, however, to give too many private lessons—his only source of income—Steiner failed to graduate from the university and his stay in Heidelberg, though he attended several university courses, was relatively unprofitable. In 1821, having heard about an opening for a mathematics teacher in a Berlin gymnasium, Steiner moved to Berlin, where he was to stay until his death. Since he had no diploma, he was required to pass an examination. He demonstrated extensive knowledge of geometry, only a modest knowledge of algebra and trigonometry, and complete ignorance of the calculus; only by virtue of the laudatory recommendations he presented and his striking geometric abilities was he allowed to teach mathematics in all the classes of the gymnasium except the final one. Steiner taught in that secondary school until 1835. Only rarely, when he could no longer bear it, would he leave his regular job to earn a living (as he did in his youth) by giving private lessons to pupils lagging behind in mathematics (no great pleasure either!). Steiner was not very proficient as a schoolmaster, because he was oriented only to the most gifted pupils; the others only irritated him. One of the fortunate events in Steiner's life during those years was his acquaintance with the rich engineer and mathematics lover A. Crelle (see the reference to Crelle and Abel above). Crelle believed in Steiner from the first time he met him and supported him in all possible ways: the Crelle journal founded in 1826 became the rostrum from which the modest schoolmaster could proclaim his geometric ideas *urbi et orbi*. This golden opportunity was not lost by Steiner, who possessed exceptional persistence and ability to work. Because of his outstanding scientific work, he was elected in 1834 to the Berlin Academy of Sciences, and in 1835 he left the gymnasium for good to work permanently at Berlin University. It is curious that whereas Steiner found teaching school painful, his university lectures were a tremendous success from the very start. This would prove to some extent harmful to geometry: Steiner's courses were so influential that, even today, projective geometry courses in many universities are based on his outlines, and use an archaic terminology introduced by the former shepherd who never received formal instruction.[83]

Another schoolmaster who played a major part in nineteenth-century geometry was Christian von Staudt (1796–1868), who was Steiner's opposite in all other respects. Staudt came from an aristocratic Franconian family. In his youth he studied at Göttingen under Gauss, who, however, failed to detect the young man's abilities. After graduation, Staudt taught for many years in a gymnasium and a polytechnic school (similar to the modern technical college). It was only in 1835 that he became a professor at Erlangen University, where Klein later taught. Staudt worked there until his death, hardly communicating with people, and working unhurriedly on his books. His style was very strict and formal, corresponding to the twentieth rather than the nineteenth century. Since Staudt lacked all pedagogical abilities, the form of his works made them difficult to understand, and so they were not immediately acknowledged.[84]

Julius Plücker (1801–1868) came from a family of Rhine industrialists but was more closely linked to French and British scientists than to any of the contemporary German geometers. Plücker studied at the universities of Bonn and Paris, but his position in German mathematics was rather precarious, due to his conflict with the extremely influential Steiner (perhaps related to differences in the social status of these two outstanding scientists). In the book mentioned in Note 68, Klein points out that Plücker, who alternately devoted himself to mathematics and physics, abandoned geometry for a long period, only returning to it when he learned of Steiner's death. Another unexpected fact is that Plücker combined (what is very rare) the gift of dealing with the most abstract mathematics of his time and the skills of an experimental physicist. Some of his physical discoveries (for example, cathode rays[85]) are of interest even today.

Of course it would be impossible to describe in any detail the basic scientific achievements of Steiner, Staudt, and Plücker. Steiner's chief accomplishment in projective geometry was perhaps the (synthetic) theory of conic sections.[86]

Julius Plücker

An inveterate opponent of analytic methods, Steiner rejected the definition of conic sections by means of quadratic equations (see above) and defined them geometrically instead; the definition associated with Fig. 5(b) relates the concept of a conic section to the concept of circle (which does not exist in projective geometry[87]). An even greater triumph of synthetic geometry was the purely geometric introduction by Staudt of projective "coordinates" of points and lines (involving geometric constructions rather than numbers) and the cross ratio of collinear points (or concurrent lines). Staudt's refined constructions[88] provided the finishing touch for the proof that projective geometry was not self-contradictory and could be developed without appealing to Euclidean geometry.

Finally, an important achievement of Plücker was his line geometry, which made a profound impression both on Klein (who prepared for publication the second volume of his teacher's work in this field) and on Lie. It follows from the duality principle of plane projective geometry that if we take the straight line as the basic element (a kind of point) of plane projective geometry, we arrive at a geometric system in no way different from the initial one. Things are quite different in the case of projective geometry in space: here the duality principle asserts that the geometry of planes (and not lines!) in projective space (the geometry in which the plane is taken as its basic element) does not differ

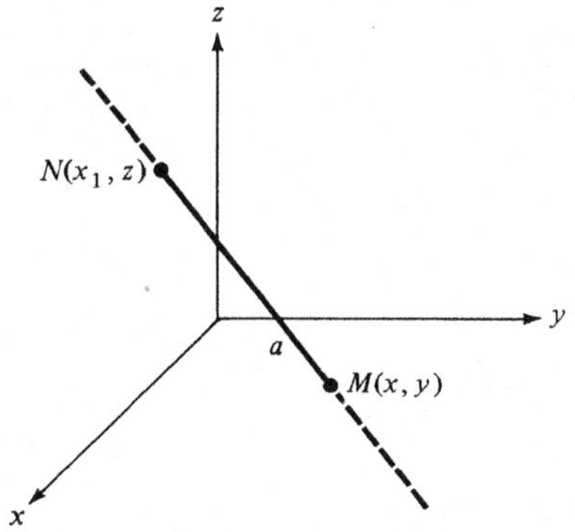

FIGURE 8

from ordinary projective space geometry. However, the geometry of lines in (projective) space is something quite new, if only for the reason that the set of lines in space is four-dimensional, i.e., the position of a line is determined by means of four parameters (coordinates); one can take as such coordinates, say, the coordinates (x, y) of the point M in which the line l intersects the horizontal plane xOy and the coordinates (x_1, z) of the point N of intersection of l with the plane xOz (Fig. 8). Thus Plücker's new geometry of space, in which the line is the basic element, is a four-dimensional geometry. Plücker developed that geometry with great depth and skill, using the aptly introduced and extremely convenient (extra) coordinates of the line known today as Plücker coordinates.[89]

CHAPTER 4

Nineteenth-Century Geometry: Non-Euclidean Geometries

In our review of the development of geometry in the nineteenth century we have so far completely left out an extremely important achievement—the discovery of the first non-Euclidean geometry, known as *Lobachevskian* or *hyperbolic geometry* and, in this connection, the refutation of the belief that there is, in principle, only one geometric system capable of "modelling" (to use a purely mathematical term) the real (physical) space surrounding us. Of course, Euler's *affine geometry* and Poncelet's *projective geometry* already were non-Euclidean systems, differing basically from the traditional geometry of Euclid studied at school; but it was the depth of the differences between affine (or projective) geometry and "classical" (or school) geometry which made it difficult to compare them and to realize that there are many possible geometric systems which deserve attention (similar in variety to the algebraic systems studied by mathematicians—groups, rings, fields, lattices, etc.). *Spherical geometry* (geometry on the surface of the Euclidean sphere) is much closer to Euclidean geometry and was well known in antiquity.[90] But spherical geometry was regarded as a mere chapter of Euclidean space geometry, studying spheres in Euclidean three-space (much as circles are studied in plane geometry). Riemann (or perhaps Lambert; see below) was the first to point out the independent significance of this chapter of geometry. The first truly non-Euclidean geometric system was *hyperbolic geometry*,[91] developed almost simultaneously and independently by the German Carl Friedrich Gauss (1777–1855), the Hungarian János Bolyai (1802–1860), and the Russian Nikolai Lobachevsky (1792–1856) of Kazan.[92]

Quite a large body of literature is devoted to the rise of non-Euclidean geometry.[93] Our own account will be cursory. We begin with the list of (rather unexpected) postulates given in most versions of Euclid's *Elements* that have come down to us. The list includes only five statements, of strikingly varied character. Postulates I–III assert that *it is possible to draw a straight line through any two given points, to draw a circle with a given center and a given radius*, and *to produce any line segment indefinitely*.[94] These statements explain Euclid's understanding of geometric constructions (mathematical rather than physical, i.e., having to do with theory rather than with actual drawing); such

Carl Friedrich Gauss

Nikolai Lobachevsky

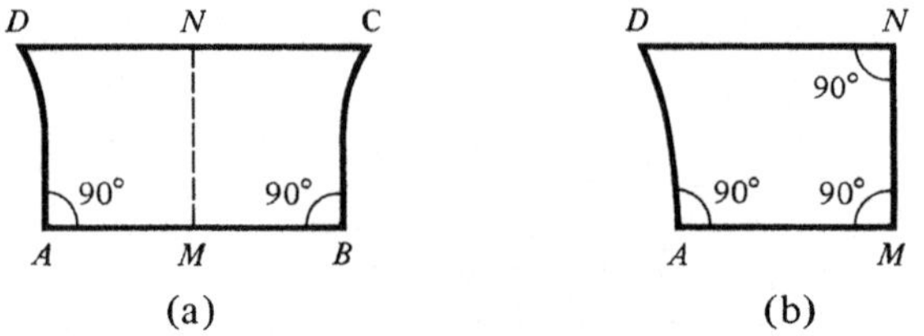

FIGURE 9

is the essential meaning of the "construction axiom" to which any construction problem must be reduced (such a reduction is called the solution of the problem).[94] Postulate IV, asserting *the equality of all right angles*, is not an axiom but a theorem: this proposition is easily proved.[95] Finally, there is the famous Postulate V: *If a straight line falling on two straight lines make the interior angles on the same side less than two right angles, the two straight lines, if produced indefinitely, meet on that side on which the angles are together less than the two right angles.* This clumsy and rather complicated statement[96] attracts one's attention—one wonders whether it is possible to do without this involved postulate.

Attempts to prove Euclid's fifth postulate began in antiquity and continued for centuries.[97] The Italian Jesuit Girolamo Saccheri (1667–1773) and, independently, the Alsatian Johann Heinrich Lambert (1728–1777),[98] one of the leading mathematicians of the eighteenth century, who worked in Munich and Berlin, probably came closest to discovering hyperbolic geometry. Saccheri's and Lambert's trains of thought were quite similar: Saccheri considered a symmetric "birectangle" $ABCD$ with $\angle A = \angle B = 90°$ and $AD = BC$ (Fig. 9(a)); Lambert considered a "trirectangle" $AMND$ with $\angle A = \angle M = \angle N = 90°$ (Fig. 9(b)); it is easy to see that the axis of symmetry MN of a Saccheri quadrilateral divides it into two Lambert quadrilaterals.[99] Saccheri showed that Euclid's fifth postulate is equivalent to the assertion that the (obviously equal) angles C and D in a Saccheri's quadrilateral are right angles; similarly, Lambert found that Euclid's fifth postulate holds if and only if angle D in a Lambert quadrilateral is a right angle.

Then Saccheri and Lambert considered three logical possibilities: *the right-angle hypothesis* (the assumption that D is a right angle, leading to Euclidean geometry); *the obtuse-angle hypothesis* (the assumption that the angle D is obtuse); and *the acute-angle hypothesis* (the assumption that the angle D is acute). The obtuse-angle hypothesis was rejected because, as Saccheri easily found, it contradicts the basic axioms of geometry and therefore "disproves itself." Lambert paid more attention to the hypothesis of the obtuse angle, noting that it is satisfied in *spherical geometry*. He even showed how certain facts known in spherical geometry follow from this hypothesis; two such facts are that the angle sum in every triangle PQR is more then π and the difference $\angle P + \angle Q + \angle R - \pi$ (the *angular excess* of a triangle) is proportional to the triangle's area. In view of the elegance and inner logic of the consequences derived by Lambert from the obtuse-angle hypothesis, as well as the existence of a simple and well known model—spherical geometry—in which both the hypothesis and all its consequences hold, Lambert was reluctant to reject this hypothesis out of hand; in fact, he dealt at some length with the "geometry" obtained from these assumptions.

Saccheri and Lambert next considered the acute-angle hypothesis (it should be recalled that their research was completely independent—Lambert was not familiar with Saccheri's results, published fifty years earlier[100]). Both obtained a number of profound geometric consequences from that hypothesis—theorems of hyperbolic geometry, as we would call them today. Saccheri rejected the hypothesis as implying that *two coplanar lines either intersect, or have a common perpendicular on both sides of which they diverge indefinitely, or else diverge indefinitely on one side of any transversal and converge indefinitely on the other* becoming "tangent to each other at infinity." This, in his opinion, contradicts the nature of a straight line. However, probing researcher that he was, Saccheri could not abstain from making a comparison: there is (he said) a difference between the refutations of the two hypotheses. In the obtuse-angle hypothesis everything is "as clear as the light of God's day," while the refutation of the acute-angle hypothesis seemed much less convincing.

Lambert went even further in the study of the geometry arising from the acute-angle hypothesis. He noted with surprise the absence of contradictions and the elegance of the consequences obtained, as well as the "converse similarity" of these consequences to those which arise from the obtuse-angle hypothesis and appear in spherical geometry: thus, here, the sum $\angle P + \angle Q + \angle R$ of the angles of any triangle PQR is less than π and its area is proportional to the difference $\pi - \angle P - \angle Q - \angle R$—the triangle's *angular defect*.[101] Still unable to disprove the hypothesis of the acute angle, Lambert made a truly prophetic statement: "I have almost reached the conclusion that the third hypothesis holds on some imaginary sphere—there must be something which makes it difficult to disprove it in the plane for so long." In what follows we will repeatedly recall Lambert's words concerning the imaginary sphere (the sphere of imaginary radius) on which the acute-angle hypothesis holds.

But both Lambert and Saccheri were convinced that Euclidean geometry is the only possible geometry and that the hypothesis of the acute angle does not hold.[102] The first person to state in writing that a geometric system differing from the traditional geometry of Euclid is possible was Ferdinand Karl Schweikart (1780–1859). Schweikart was not a mathematician but a lawyer, at that time professor of jurisprudence at Kharkov University in Russia (actually in the Ukraine). Schweikart had no formal mathematical education; this may explain the fact that he was not shackled by traditional ideas on geometry and on the nature of space.[103] In any case, in 1818 Schweikart gave Gauss's friend, the astronomer Christian Ludwig Gerling (1788–1864), a note for Gauss in which he asserted that there exist two geometries: the usual Euclidean geometry and an "astral" geometry which Schweikart supposed to hold on some distant stars. Schweikart took the sum of the angles of some triangle in astral geometry to be different from π. Proceeding on this assumption, he rigorously proved that this sum is less than π for all triangles, and that the larger the area of the triangle, the smaller this sum is. In addition, Schweikart established the existence of a "natural" (or "geometric") unit of length in astral geometry which he called *the constant* and defined as the limit of the height of a right isoceles triangle as its sides increase indefinitely. At the time everything Schweikart had written was familiar to Gauss. While he failed to support Schweikart or even to write to him (to say nothing of recommending the remarkable note he had received for publication), Gauss nevertheless wrote to Gerling: "Professor Schweikart's note caused me no end of joy, and please convey as many kind words to him as possible in my name. Almost all this has been copied from my soul."

The first printed expositions of elements of hyperbolic geometry were two booklets published at the expense of Schweikart's nephew Franz Adolf Taurinus (1794–1874), who was strongly influenced by his uncle: *Theorie der Parallellinien* (Cologne, 1825) and *Geometriae prima elementa* (Cologne, 1826). The first booklet in effect set forth the assumptions made in Schweikart's note. Taurinus emphasized the possibility of the existence of a large number of "astral" geometries corresponding to different values of Schweikart's constant, which he called the *parameter* (Taurinus tended to regard the multiplicity of values of Schweikart's constant as a serious shortcoming of the new geometry). In the second booklet Taurinus developed elements of "astral trigonometry" and pointed out that one can obtain the relevant formulas from those of spherical trigonometry by replacing the radius of a sphere by a pure imaginary number (recall Lambert's imaginary sphere, on which the acute angle hypothesis holds!); in astral trigonometry the role of trigonometric functions is played by the so-called hyperbolic functions:[104]

$$\cosh x = \cos ix \quad \left(=\frac{e^x + e^{-x}}{2}\right), \qquad \sinh x = -i \sin ix \quad \left(=\frac{e^x - e^{-x}}{2}\right).$$

Unfortunately, Taurinus's insistence that Gauss also publish his views on the question, expressed in particular in the introduction to the first booklet, only angered Gauss, who had previously given a favorable answer to Taurinus's letter. Gauss stopped writing to Taurinus who, receiving no answers to his letters was driven to despair, ceased his geometric studies, bought up the booklets he had published, and burned them.

Though Schweikart and Taurinus knew about the existence of a new non-Euclidean geometry (which Schweikart called *Astralgeometrie*), they are not usually regarded as being among the founders of hyperbolic geometry. Indeed, the jurist Schweikart never published anything on non-Euclidean geometry; it also seems that Schweikart regarded the very existence of his constant (Taurinus's "parameter") as a certain refutation of the new geometric

János Bolyai

system. Taurinus, on the other hand, apparently repudiated his uncle's ideas: he burned all the booklets he could get his hands on and never mentioned them again.

It is striking that the new geometric system was discovered almost simultaneously and independently by three researchers (Gauss, Lobachevsky, Bolyai) differing in scientific training and psychological traits (below we will tell about the differences between Lobachevsky and Bolyai in the approach to their discovery). The momentous words of Farkas Bolyai (1775–1856), who wrote to his brilliant son János that "when the time comes, scientific ideas are conceived by different people at the same time, like violets blossoming wherever the sun shines," point out, but fail to explain, the coincidence. Gauss repeatedly referred to the similarity between his own thoughts and the constructions of Lobachevsky and Bolyai (and, earlier, those of Schweikart and Taurinus) as "miraculous." For a long time János Bolyai refused to concede that he had not been the only person to reach his conclusions on geometry. He believed that Gauss had learned of the new geometry from works sent to him by János's father Farkas, had plagiarized these, and had concealed this fact by publishing an account of non-Euclidean geometry under the pen name "Nicolaus Lobatschevsky aus Kasan". Much later, Felix Klein argued against the simultaneous and independent discovery of hyperbolic geometry by three authors. In the first mimeographed variant of his book on non-Euclidean geometry,[105] he claimed that Gauss should be considered the only discoverer of the new geometric system: János Bolyai undoubtedly learned about it from his father, who had been Gauss's friend during his student years, while Lobachevsky learned about it from his teacher Johann Bartels (1769–1836), who had once taught Gauss in secondary school and was Gauss's bosom friend. When one compares Klein's different books, it is obvious how stubbornly he refused to abandon his mistaken idea—which was completely disproved by two of Lie's pupils, Paul Stäckel, 1862–1916, and Friedrich Engel, 1861–1941, who wrote an extensive history of non-Euclidean geometry.[106] In Klein's book on the history of mathematics mentioned in Note 68 the same assertion on Gauss's priority is made in less categorical form, so that

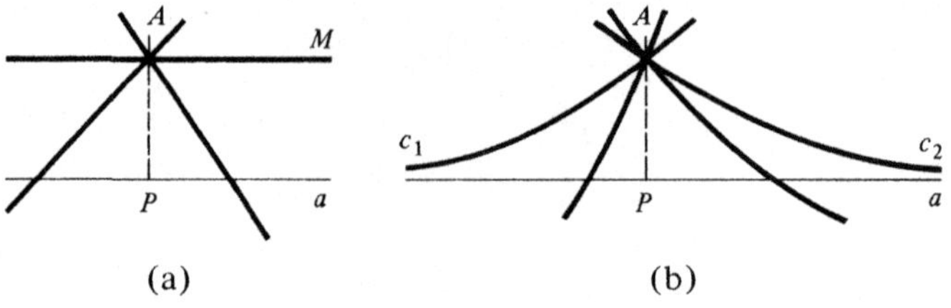

FIGURE 10

the ways in which Lobachevsky and Bolyai learned about Gauss's views are made to seem rather mysterious. In the printed edition of Klein's *Lectures on Non-Euclidean Geometry*[107] issued after his death, Gauss is still accorded priority among the discoverers of non-Euclidean geometry, but there is no trace of the assertion that Lobachevsky and Bolyai borrowed from Gauss!

Turning to the ways in which Gauss, Lobachevsky, and Bolyai came to hyperbolic geometry, it should be pointed out that there was an important difference in the basic assumptions of Lobachevsky and Gauss on the one hand, and Bolyai on the other. The starting point for all three scientists was apparently an attempt to prove Euclid's fifth postulate (or, what is the same thing, Playfair's axiom on parallel lines); at one time Lobachevsky even believed that he had a proof. However, having taken as the starting point the assertion that *through a point A not on a line a one can pass two lines not intersecting the line a* and using the *reductio ad absurdum* method of proof, all three authors gradually reached the conclusion that there were no contradictions in the geometric system which follows from these assumptions. They saw that the theorems obtained were unusual and strange but free of contradictions; in fact, they had, collectively, a certain elegance and perfection characteristic of all errorless mathematical theories. (It is clear that the assertion on the existence of two straight lines c_1 and c_2 passing through A and not intersecting a is equivalent to the assertion that an infinite set of such lines exists—here all the lines passing through A and enclosed within one of the pairs of vertical angles formed by c_1 and c_2 do not intersect a—see Fig. 10(b)). Of course it was not easy to take the step which Saccheri and Lambert had not dared to take in their time, i.e., to acknowledge that there is not one but there are two true geometric systems, equally valid in a certain sense—but Gauss, Bolyai, and Lobachevsky took that step, which required considerable scientific courage.

Gauss had become interested in the theory of parallel lines at the close of the eighteenth century. In 1799 he wrote to his friend Farkas Bolyai about his attempts to prove the fifth postulate: "It is true that I have achieved a great deal which, for most, would be sufficient as proof, but as I see it this proves absolutely nothing. For example, if someone were able to prove that a triangle whose area is greater than any given area is possible, I would be able to give a rigorous proof of all of geometry. Most would regard this as an axiom, I do not. Thus it is possible that the area of a triangle will always be less than some limit, however far the triangle's vertices may be from one another. I have many such propositions, but I do not find a single one satisfactory for the foundations of geometry." For many years Gauss apparently hoped to prove the fifth postulate, but in the mid-1810s[108] he finally concluded that two equally valid geometric systems exist. This fact completely contradicted not only all the views developed during previous centuries but also the beliefs (formerly fully shared by Gauss!) of the leading philosopher of the time, Immanuel Kant (1724–1804). According to Kant, the concepts of space and time are given to us *a priori*, and since geometry is the doctrine of space, it is necessarily *unique*. A letter, extremely important in this respect, was sent by Gauss in 1817 to his older friend, the astronomer Wilhelm Olbers (1758–1840). "I have increasingly come to believe," he wrote, "that the necessity of our geometry cannot be proved—at least not by human reason and for human reason. Perhaps in another life we will have different views on the nature of space which are inaccessible to us here. So far geometry has to be regarded as being on a par not with arithmetic, which exists *a priori*, but rather with mechanics." Thus here Gauss assigns geometry not to "mathematical" or "logical" fields like arithmetic (the theory of numbers with which he was absorbed) but to the natural (experimental) sciences like physics and mechanics. If this view were correct, then whether *non-Euclidean geometry* (as Gauss called his new geometric system) was true or false could be determined experimentally—and Gauss the geodesist attempted to measure, with utmost precision, the sum of the angles of a large triangle, whose vertices were the observer (Gauss himself) and the summits of two distant mountains. In this non-Euclidean geometry the sum of a triangle's angles is less than π—and the greater the area of the triangle, the more the sum differs from π. Had Gauss found that the sum of the angles of the triangle he considered was less than 180°, that would have meant that *non-Euclidean* rather than Euclidean geometry holds in the real world.[109] But Gauss failed to discover any significant difference between the sum of the angles in the triangle he examined and 180°, at least with the precision allowed by his measuring devices.

Lobachevsky regarded geometry much as Gauss did. In his general philosophical and methodological principles, if not in his views on non-Euclidean geometry and the nature of space, Lobachevsky may have been indirectly influenced by Gauss through the German professors teaching at Kazan, in particular by Bartels. Of course Lobachevsky was not a scientist of Gauss's stature,[110] but the general level of teaching in physics and mathematics was quite high in Kazan, despite its great distance from the main scientific centers—Lobachevsky's education had been reasonably thorough.[111] From the point of view of the foundations of geometry the geometry textbook written by Lobachevsky in 1823 on instructions from the superintendent of the Kazan educational district M.L. Magnitsky[112] is especially interesting. Owing to a very negative review by a well-known mathematician of the time, Euler's pupil and colleague, academician Nikolai Fuss (1755–1826), the textbook was never published in the author's lifetime.[113] The traditionalist Fuss disliked the book because its author tried to represent geometry not in Euclid's spirit, i.e., not as a series of scholastically interpreted deductions from stated axioms,[114] but as a science concerning (real) space, in which measurements of geometric magnitudes play a leading role and in which transformations ("motions") are freely used in proofs of theorems; transformations had been avoided by Euclid because of his metaphysical principles, originating with Zeno of Elea (c. 490–c. 430 B.C.) and Aristotle. Lobachevsky's principles were derived from the large entry "Geometry" in Diderot's and d'Alembert's famous *Encyclopedia*, written (as were all the other mathematical or natural science entries in that monumental work) by d'Alembert. In all likelihood, the contemporary textbook employed these principles as fully and consistently as did Lobachevsky's. Fuss's negative review was thus due not to real shortcomings, but rather to the merits of the textbook, to the fact that it differed in a positive way from all existing textbooks. This the conservative Fuss refused to acknowledge.[115] What is most interesting for us here is the author's "physical" (one can say "materialistic") approach to geometry, largely derived from d'Alembert, the brilliant figure of the French Enlightenment, which in some ways even contradicted Lobachevsky's traditional religious convictions.

It has already been mentioned that at first Lobachevsky believed that he had proved the fifth postulate. Later, in 1826, he came to firmly believe that the geometric system which denies the fifth postulate is true (and noncontradictory). Lobachevsky called that system "Imaginary Geometry" (a name subsequently criticized by Gauss), and later "Pangeometry" (by which name Lobachevsky sought to stress that Euclid's classical geometry may be obtained from his geometric system by means of a passage to the limit: by letting Schweikart's constant tend to infinity). In 1826 Lobachevsky delivered the first lecture on the new geometry at a session of the physics and mathematics department of Kazan University. The French text of the lecture, prepared for publication in the department's *Research Papers*, was sent to three university professors to be reviewed. But Lobachevsky's colleagues failed to understand his work. Since they did not want to write negative reviews they simply "lost" the text. As a result, the first publication on hyperbolic geometry was not this (apparently brief) report, but Lobachevsky's long two-part memoir "On the Elements of Geometry," published in the *Kazan Vestnik* (a Kazan university publication) in 1829–1830 and containing an advanced description of the foundations of the new geometry.[116] From that moment on, Lobachevsky's scientific and literary activities did not cease for several decades: he published a number of detailed works—papers and books—on the new geometry. In them he developed, among other things, *analytic* and *differential geometry* in non-Euclidean space, obtaining numerous specific results; compared to Gauss and Bolyai, Lobachevsky developed the new geometry furthest (see Note 125). His last presentation of hyperbolic geometry, published in 1855 in Russian and in 1856 in French under the concise title

Pangeometry, was dictated by Lobachevsky, blind by that time, to his pupils and colleagues; perhaps that was why the *Pangeometry* did not contain a single sketch, but numerous painstaking analytical passages apparently worked out by the author in his mind. Here—and not for the first time—Lobachevsky pondered which geometry—Euclid's or hyperbolic—operates in the space surrounding us; he concluded that the question could be answered by measuring the angle sum of a large triangle. Earlier Lobachevsky himself had attempted to find the sum of the angles of the triangle formed by the Earth and two fixed stars but, like Gauss, had failed to discover a significant difference from 180°.

János Bolyai's approach to hyperbolic geometry was somewhat different. Bolyai's attitude was that of a logician and not a physicist: he considered geometry as a purely logical system based on axioms, and not as a set of facts relating to real space. Accordingly, János would never have thought of actually measuring the angle sum of some triangle, since this would not prove anything (except the inadequacy of Euclid's geometry as a mathematical model of physical space). Characteristically, Bolyai devised for his exposition a symbolic language, to some extent comparable to the modern language of mathematical logic. This symbolic language, using a minimum of words, made it very difficult for Bolyai's contemporaries to read his great work. Bolyai's other works, which were not published during his lifetime, were also in the nature of logical treatises, in which utmost attention was paid to the language. It is not by chance that János Bolyai's only serious work not connected with non-Euclidean geometry contained a short but quite advanced account of the formal theory of complex numbers; this too was not correctly appraised during his lifetime.

Bolyai's idea of constructing an "absolute" geometry that does not depend on the parallel postulate and includes both some results from hyperbolic geometry and some of the theorems of Euclid's classical geometry was a brilliant and completely unexpected idea for its time.[117] Bolyai sought to formulate his definitions so that they would be relevant both to hyperbolic and to Euclidean geometry. Thus, for example, a *parallel* (in Bolyai's terminology, *asymptotic*) line passing through a point A not on a line a was defined by Bolyai as a line AM not intersecting a such that all the lines enclosed within the angle MAP, where P is the projection of A on a, intersect a. This definition is meaningful not only in Euclid's geometry, where just one parallel line passes through A, but in hyperbolic geometry as well, where there are two such lines (see Figs. 10(a) and (b)). Again, the sine theorem for triangles, say, is written by Bolyai in the following way (we have altered his symbols somewhat):[118]

$$\frac{\sin A}{s(a)} = \frac{\sin B}{s(b)} = \frac{\sin C}{s(c)} \tag{4.1}$$

where A, B, C are the angles of a triangle, a, b, c are respectively the sides opposite to them, and $s(x)$ is the length of the circle with radius x. In this form the theorem remains valid both in Euclidean and in hyperbolic geometry, despite the fact that the formulas for the circumference of a circle are different in the two cases.

From today's vantage point we must conclude that Bolyai's understanding of the logical structure of the new geometry was deeper than that of either Gauss or Lobachevsky. No wonder Bolyai was so upset by the absence of a complete proof of the fact that hyperbolic geometry is free of contradictions—he devoted many years of his life to finding such a proof. However, his mathematical training was obviously insufficient for the purpose. It should be noted that Lobachevsky nearly proved that the new geometry was free of contradictions. In effect his works contain what are now called the *Beltrami coordinates* of the hyperbolic plane or space; had he assigned to each point of the hyperbolic plane the Beltrami coordinates x and y of the point (x, y) in the Euclidean plane, he would have come to Beltrami's model,[119] and would have proved that hyperbolic geometry was free of contradictions. But, like Gauss, Lobachevsky was not greatly interested in the purely logical proof of the fact that the geometric scheme he had created was free of contradictions. Lobachevsky and Gauss were mostly concerned with that geometry's relation to physical space, a question far removed from any logical-axiomatic thinking. That was why Lobachevsky, having almost found a rigorous proof of the fact that hyperbolic geometry is free of contradictions, a proof Bolyai sought so avidly, failed to notice the treasure within his reach!

Now we can return to the history of non-Euclidean geometry. As we have already remarked, Gauss was the first to work out the new geometric system; however, he never published his results. Quite satisfied with his position as the world's leading mathematician, *mathematicorum princeps*,[120] Gauss readily shared his ideas with people who were close to him, but had no intention of making them public, fearing the "hornets that will rise above my head if I disturb their nest" (the phrase is from a letter to Gerling in 1818) or "the criticism of the Boeotians," as he wrote somewhat later to Friedrich Bessel (1784–1846). (Apparently by "hornets" and "Boeotians" Gauss meant the advocates of Kant's *a priori* philosophy.) Typically, at the end of a friendly and revealing letter to Taurinus, Gauss warned his correspondent not to make anything known about his views: when Taurinus urged Gauss to set forth his ideas in print, Gauss immediately broke off all relations with him.

Gauss's attitude had a tragic effect on the fates of Lobachevsky and Bolyai. Outwardly, Lobachevsky lived a life that could be called highly successful: in spite of the fact that he came from a poor family[121] he reached the rank of acting state councillor, corresponding to a general in the military bureaucracy;[122] this automatically gave him a hereditary title and a coat of arms. For many years he was rector of one of Russia's six universities and a leading figure in the Kazan educational district; he received many awards, including the highest ones. Nevertheless, he did not feel happy: he was respected only as an administrator, although he regarded himself as a scientist. Lobachevsky was no doubt an outstanding administrator in public education, and did a great deal of good for the university of his native Kazan. Those who surrounded him thought that such a distinguished person could be excused for his eccentricity—let him putter around with the imaginary geometry which no one needs and publish his incomprehensible works. Alas, not everyone thought so: in 1834 derogatory reviews of Lobachevsky's *On the Elements of Geometry* appeared in two St. Petersburg journals, signed with the initials S.S. and couched in coarse language. Subsequently, even the influential political author Nikolai Chernyshevsky[123] took part in the press campaign against Lobachevsky. Academician Victor Bunyakovsky (1804–1889), one of the most prominent mathematicians of the age, reviewed Lobachevsky's works for the Russian Academy of Sciences. His review was extremely negative. Bunyakovsky failed to understand Lobachevsky's idea; he was probably associated with the person hiding under the initials S.S. The book *Geometrische Untersuchungen zur Theorie der Parallellinien*, published in 1840 in Germany, received a very negative review which appeared in the same year in the German *Gersdorff's Repertorium*. Concerning this review, Gauss wrote in a letter to his friend Gerling that any competent person would immediately see that it was due to a completely uninformed author; and in another letter he called it "absurd." But Lobachevsky knew nothing about these words, nor did he learn about the numerous laudatory references to him by Gauss in his letters and conversations, and he knew nothing of the fact that Gauss studied Russian so as to be able to read Lobachevsky's works.[124] Thus, although Lobachevsky was honored by universities and scientific societies (he was professor *honoris causa* of Moscow and Kazan universities, member of the highly regarded Göttingen scientific society, to which he was elected on Gauss's initiative—but he knew nothing about Gauss's role in that event), Lobachevsky thought that it was not his scientific but his administrative achievement that was being marked. The recollections of Lobachevsky's son, as well as other memoirs, paint him in his last years as a sullen misanthrope, unhappy in family life and almost without friends.

Yet by contrast with János Bolyai's life, Lobachevsky's fate was idyllic! To begin with, Bolyai did not receive the education he deserved. When Farkas Bolyai, once Gauss's close friend, having described the interests and abilities of the schoolboy János to his old friend asked Gauss to lodge the young man in his house, Gauss did not even bother to answer that (rather tactless) letter. János was thus forced to abandon forever his dreams of attending Göttingen University. He entered one of the Hungarian military academies; there he obtained a respectable but limited education in physics and mathematics—but he could never forgive his father that he studied neither in a university nor at home under Farkas's tutorship. János was compelled to become an officer, although he had never been attracted to a military career. János had very bad relations with his fellow officers because he thought himself above the circle that surrounded him, a view that the other officers naturally were not inclined to accept. He was helped by his reputation as an excellent swordsman; it was not safe to quarrel with him. Nevertheless, the number of duels he fought was incredibly high: in one particular day he fought twelve(!), only asking permission to play the violin between bouts for a rest; and he won all twelve encounters. Difficult material circumstances prevented him from marrying; and he had hardly any friends, with the possible exception of the mathematician Karl Szász, with whom he had begun studying the theory of parallel lines. At the earliest opportunity, Bolyai left the military service he hated. Since he had no other source of income, he suffered from dire poverty for the rest of his life.

A full exposition of Bolyai's ideas on non-Euclidean geometry (brilliant, but extremely difficult to understand) was apparently ready in 1824. János could not afford to publish his work, but continued to improve it. Fortunately, his father agreed to publish the work (at his son's expense) as an *Appendix* to the first volume of his own book, known under the Latin title *Tentamen* (the first word in the long title of the older Bolyai's work). *Tentamen* appeared in print in 1831.[125] A copy was immediately sent to Gauss with an urgent request to review János's work: "my son relies on your review more than on the opinion of all of Europe," Farkas wrote to his old friend.

Apparently, Bolyai's *Appendix* made a profound impression on Gauss. On the very day after he received the book, Gauss wrote to Gerling that he had received a remarkable work from F. Bolyai, whose author (János Bolyai) was "a genius of the first magnitude." But Gauss wrote to Bolyai only a month later—and it was a gruelling month for János. What was even worse, the tone of the letter sent to Bolyai differed considerably from that of the letter addressed to Gerling. Gauss wrote to Farkas Bolyai that all of János's work was familiar to him, so he could not praise János because that would mean praising himself. (Actually Gauss's claim was not true: not only did János Bolyai's approach differ from that of Gauss, but also a number of concrete results contained in the *Appendix*—for example, the statement that in hyperbólic geometry there are solvable problems related to the quadrature of the circle—were undoubtedly new to Gauss).

We have already described how János responded to the letter and how offended he was by Lobachevsky's *Geometrische Untersuchungen*, to which Gauss drew his attention. And János could not long console himself with the thought of his priority: at the beginning of his book Lobachevsky wrote that a full exposition of the same ideas had been published by him in Russian in 1829–1830, i.e., before Bolyai's *Appendix* had appeared (Gauss had even written to Bolyai that since these publications were in Russian, they could be easily read by the Hungarian János; Gauss was under the mistaken impression that Hungarian belongs to the same language group as Russian). János wrote a profound, though in some ways biased, commentary on Lobachevsky's *Geometrische Untersuchungen*—still he had to acknowledge that Lobachevsky held priority. Bolyai was also offended by an account of Gauss's talk with another mathematician in which Gauss, referring to the theory of parallel lines, heaped praise on Lobachevsky's work (alas—Lobachevsky never knew about it) but failed even to mention Bolyai's *Appendix.*[126]

The subsequent years of János Bolyai's life were not productive. It has already been noted that his excellent memoir on the theory of complex numbers, written in János's laconic and cryptic style, was not appreciated. János sent it to a contest held by the Leipzig scientific society, but the sponsors failed to understand the work and it did not receive a prize. In the hope of eclipsing Gauss and Lobachevsky, Bolyai attempted to solve other problems but met with failure due to his lack of formal mathematical training. Thus he tried to prove that every algebraic equation can be solved in radicals and to obtain a general formula for the solution; he was unaware that by then Abel had already proved that no such formula exists. He attempted to prove that the integral of every algebraic function is expressible in terms of algebraic functions; this, as we know, is not true. He also tried to find a general formula for the nth prime number. In *Tentamen*, Farkas Bolyai proved for the first time that *any two polygons of equal area are equidecomposable*, i.e., can be cut up into smaller congruent ones and reassembled; János tried to extend that result to polyhedra in space—a problem that Gauss once studied but quickly abandoned as too difficult or perhaps incorrect (now we know that it is incorrect, but this was proved only in the twentieth century[127]). An even stranger impression is produced by Bolyai's founding of a "Science of Universal Welfare" which greatly concerned him for many years and to which he gave the German name "*Allheillehre*." Despite attempts to apply mathematics, his notes on the subject that have come down to us are closer to religion than to the natural sciences or the humanities. The genius János Bolyai, now celebrated as the glory of the Hungarian people,[128] died in a state of profound depression due to a serious mental illness. His life was also poisoned by many years of conflict with his father.

Thus in the first third of the nineteenth century, it seemed to have been finally established[129] that not only one but two equally valid geometries exist: *Euclid's geometry* and *the geometry of Lobachevsky–Bolyai–Gauss* (of course, at the time no one thought that Euler's *affine geometry*, Poncelet's *projective geometry*, and even Ptolemy's *spherical geometry* were comparable with Euclid's geometry or with non-Euclidean systems). However, the view that there are just two geometries, firmly held by Gauss, Lobachevsky, and Bolyai, was not to remain in force for long. The nineteenth century was a period of explosive development of different geometries. Further (and very considerable) progress of geometry was associated above all with the name of Georg Friedrich Bernhard Riemann (1826–1866), undoubtedly one of the greatest mathematicians in history and one of the two supreme mathematicians (with Gauss) of the nineteenth century.

The son of a poor clergyman, Riemann was sent by his father in 1846 to Göttingen, at the time the world's leading mathematical center (Gauss taught at Göttingen),[130] not to study mathematics but to study theology. Neverthe-

Georg Friedrich Bernhard Riemann

less, his mathematical interests soon prevailed and, despite family traditions, Riemann changed his course of study and devoted himself completely to mathematics. Riemann's short life in mathematics was entirely linked with Göttingen—at first as a student, then as a *Privatdozent*, then as an "extra-ordinary" professor and, finally, as an "ordinary" professor (the latter rank implied tenure).[131] Riemann's teaching career was not strewn with roses: shy and insecure, he was not a successful lecturer. Nor did he count on such success. In the autumn of 1854 he proudly wrote to his father that his course had attracted eight listeners. Only three students attended his remarkable course on functions of a complex variable, delivered in the winter of 1855/56 and in the summer of 1856. This course, incredibly rich in ideas, gave rise to the entire theory of functions of the nineteenth and early twentieth century, and many other fields of mathematics, for example topology, derived numerous ideas from it. Fortunately, in both cases the listeners included Richard Dedekind (1831–1916), the most gifted and dedicated of Riemann's pupils, whose contribution to orgainizing (mostly only posthumously) the publication of Riemann's works and of his books of lecture notes (which unfortunately are not very precise) is difficult to overestimate. Nevertheless, the modest position Riemann occupied at the university, where many much less talented colleagues treated him condescendingly, distinctly hurt Riemann, who was well aware of his own potential. The loyalty of the young Dedekind and a warm friendship with the Berlin professor P. Lejeune Dirichlet (1805–1859)[132] could not offset the slight to his self-esteem.

Riemann's position at the university improved with the return to Göttingen of the inflential Wilhelm Weber (1804–1891)[133]: Weber immediately recognized Riemann's talent and supported him in all possible ways. Weber also improved Riemann's formal status at the faculty, by inviting him to be an assistant in the department of experimental physics he headed—in view of Riemann's varied interests in physics, he was not at all troubled by having to perform the duties of an assistant professor in the practical seminar for physics students. Riemann began to feel even more sure of himself when, following Gauss's death, the latter's position was given to Riemann's friend Dirichlet. It was at this time that Riemann was raised to the rank of extraordinary professor, largely due to the efforts of Weber and Dirichlet, as well as to Dedekind, who had become quite influential by then. Upon Dirichlet's death, Riemann took over Gauss's chair. Now he could recover from the slights he had experienced in his youth—and even marry, something the shy Riemann had earlier deemed impossible. But unfortunately he had little time left. He married in 1862, but fell seriously ill in the same year. Three trips to Italy, arranged by Weber at the university's expense, failed to restore his health, and he died in Italy of tuberculosis at the age of 40.

Riemann's works greatly changed the face of modern mathematics. Klein's words that "No one had a more decisive impact on modern mathematics than Riemann" can hardly be regarded as outdated even today. The amazing thing is the scope of Riemann's scientific interests, extending to almost all the fields of the mathematics of his time (and sometimes even going beyond their bounds: thus Riemann may be regarded as the precursor of *topology*, which arose only in the twentieth century), to theoretical and applied physics, to the physiology of the sensory organs, and to the philosophy of natural science. He was a direct predecessor of Albert Einstein, whose "general theory of relativity" is wholly based on Riemann's ideas.

In the summer of 1854 Riemann was granted the opportunity to deliver a lecture on a subject of his own choice in the presence of the board members of Göttingen University. According to a custom existing at German universities, such lectures served as the basis for allowing the lecturer to assume his teaching duties (see below, page 129; in this case it concerned Riemann's appointment to the post of assistant professor[134]). Riemann offered a choice of completely different subjects in physics (which was quite natural for a colleague of Weber), analysis, and geometry. Apparently not without the influence of the elderly Gauss, the choice fell to a subject in geometry. The audience listened attentively to Riemann's lecture "Über die Hypothesen, welche der Geometrie zu Grunde liegen" but did not understand it. Riemann's brilliant ideas, creating a totally new and extremely profound concept of geometry, were so far ahead of their time that perhaps only Gauss could have understood them—and support of young talent, as we know, was not to be expected from the world's foremost mathematician.[135] Still, all those who were present noticed that Gauss left the lecture in deep thought; Dedekind even recalled that Gauss praised Riemann's work in a talk with Weber, with whom he returned from the session. Riemann was granted the right to teach at the university.

That lecture was first published by Dedekind in 1868, two years after Riemann's premature death; but Dedekind hardly realized the work's depth. In 1876 Dedekind issued for the first time the *Collected Works* of Riemann, which included the above-mentioned lecture. Subsequently, the *Collected Works* were republished many times and translated into other languages—now Riemann's lecture is available in practically all the European languages. But Riemann's ideas were truly appreciated only after they were revised by the outstanding twentieth-century mathematician Hermann Weyl and by Albert Einstein. In 1919 Weyl published a new edition of Riemann's lecture with penetrating comments[136] establishing the connection between Riemann's constructions (in the oral lecture they were, unavoidably, presented in very general form and almost without formulas)[137] and contemporary (tensor) approaches to what is now known as the theory of Riemannian spaces. On the other hand, Einstein's famous memoir of 1916, *Grundlagen der allgemeinen Relativitätstheorie*,[138] contained a very detailed—and remarkably clear and well written—examination of Riemann's ideas; possibly this reflected the influence of Weyl, Einstein's colleague at the department of the *Technische Hochschule* in Zürich in 1913–1914.[139]

Riemann's geometric ideas began with Gauss's remarkable memoir *Disquisitiones generales circa superficies curvas* (1828),[140] a work that continued and developed the trend in mathematics originating in ideas of Euler and Monge and now known as *differential geometry*. This subject is the study of local properties (i.e., properties only related to a small neighborhood of the selected point) of curves and surfaces and uses the apparatus of the differential calculus. Gauss's work, which arose from his practical activities (he was directed by the King of Hannover to do a detailed geodesic survey of the kingdom), developed the concept of the *intrinsic geometry* of an arbitrary (curved) surface Φ in three-dimensional space. The intrinsic geometry of Φ consists of those geometric properties of Φ which can be determined "without leaving Φ" i.e., with the help of measurements carried out "in Φ." Thus, for example, Gauss defines the distance between two points $A, B \in \Phi$ as the length of the shortest curve in Φ joining A and B. (A colorful description of the intrinsic geometry of a surface was proposed by the mathematician, naturalist, and medical doctor Hermann von Helmholtz (1821–1894), an outstanding German scientist with an incredibly wide range of interests. He started with the (purely speculative) assumption that the water bugs we see skimming over the surface of ponds possess a two-dimensional psyche, i.e., are just as incapable of imagining three-dimensional space as we are of imagining four-dimensional space. He then suggested that one define all the facts and theorems which such "two-dimensionally minded" creatures living on the given curved surface could discover as comprising the "inner geometry of the surface Φ." [141]) Gauss's main result was that it is possible to calculate the (intrinsic or Gaussian) *curvature*[142] at every point of a surface within the framework of its intrinsic geometry. If the curvature is identically zero, then the surface can be developed on a plane.

Instead of surfaces located in ordinary space (of dimension three), Riemann proposed to consider arbitrary "curved" *manifolds* of any dimension (more about this notion below) in which the "metric" is defined by a formula enabling one to measure the distance between any two points in the given manifold (today such manifolds with a metric are called *Riemannian spaces*). An important role in Riemann's constructions was played by the notion of curvature of space. Special types of manifolds with a metric are spaces (*homogeneous* and *isotropic*, i.e., such that all their points are equivalent and no direction at any point differs from any other) that have constant curvature: *Euclidean* space of zero curvature, *hyperbolic* space of negative curvature, and *elliptic* space of positive curvature. Take the case of two-dimensional manifolds. Here a manifold of zero curvature is exemplified by the Euclidean plane, the intrinsic geometry of the hyperbolic plane does not differ from that of the Lobachevsky plane, while the elliptic plane is "shaped" just like the surface of an ordinary Euclidean sphere.[143]

Thus three spaces occupy a special place among the infinite variety of curved spaces considered in Riemann's lecture; they are Euclidean space and two other spaces (which are in a certain sense just as "good" as the former)—the hyperbolic Lobachevskian space, and elliptic space, now often called *Riemann's non-Euclidean space.*[144] Of course, it is not these two spaces, in effect known earlier, which constituted Riemann's major contribution to non-Euclidean geometry. Much more important is the fact that he introduced in his lecture a wide class of non-Euclidean spaces with different curvatures. These spaces were to play an essential role in Einstein's subsequent attempts to include the distribution of mass (which brings about gravitational effects) directly into the geometry of the universe.

We know that the sphere is a two-dimensional space of constant positive curvature (two-dimensional elliptic space). However, by the *non-Euclidean Riemann plane* (*the elliptic plane*) one usually means a geometric entity differing somewhat from the sphere. The thing is that any pair of the sphere's *great circles* (the curves of the sphere's intersection with planes passing through its center), which play the role of straight lines in spherical geometry, *intersect* not in one but *in two points.* That circumstance creates a sharp but not fundamental difference between plane and spherical geometry, which may be eliminated if the sphere is viewed as the set of diametrically opposite pairs of points, and each such pair is taken to be the basic element of the geometry. Such a gluing of opposite points may be imagined as the result of removing

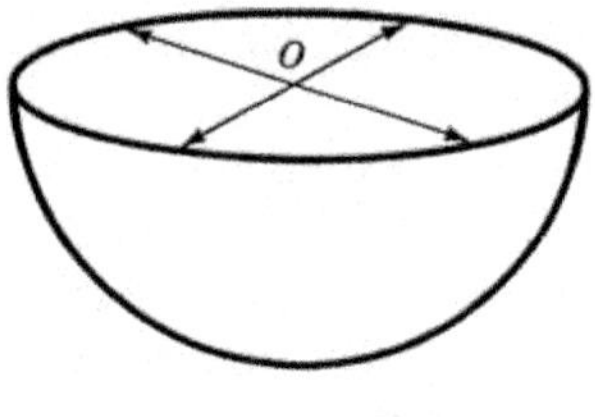

FIGURE 11

one (say the upper) hemisphere followed by the identification (gluing) of diametrically opposite points of the "equator" bounding the removed hemisphere (Fig. 11). It is this geometric entity (the hemisphere with glued boundary points or the sphere viewed as a set of pairs of antipodal points) that is most often called *Riemann's elliptic space.*[144]

It has already been mentioned that Riemann's lecture, which now appeals so strongly to our imagination, was not appreciated by mathematicians in its time because of its profound and unusual insights and the implicit nature of its content. Although Riemann's last years were marked by truly international recognition by academies and scientific societies,[145] his "Über die Hypothesen" was never mentioned. Thus when Riemann was appointed corresponding member of the Berlin (Prussian) academy of sciences in 1859 and (foreign[146]) member of the same academy in 1866, neither of the relevant recommendations, largely written by the respected Berlin mathematician Karl Weierstrass, made the slightest reference to that great address. (Weierstrass was as much a pure "logician" (or algebraist) as Riemann was a pure "physicist" (or geometer).) This is all the more noteworthy because Weierstrass's recommendations included references to Riemann's works on the theory of functions, much admired but at the same time severely criticized by Weierstrass. (This, incidentally, made itself felt in the slowness with which Riemann's achievements were recognized in the scientific community.) Many of Riemann's works were based on the *Dirichlet principle*, borrowed from one of the latter's lecture courses, a principle that seemed rather doubtful to Weierstrass.[147] Characteristically, the first acknowledgment of Riemann's lecture came from a man who was not regarded by others or by himself (wrongly, as we see it) as a mathematician. In 1868, Hermann Helmholtz responded to Riemann's great lecture by a revealing paper "Über die Thatsachen, die der Geometrie zu Grunde liegen."[148] Helmholtz's paper was based on his research into the physiology of vision,[149] and its title was all but a copy of the title of Riemann's lecture.[150] Helmholtz's idea was to describe spaces of constant curvature by their characteristic properties of *homogeneity* and *isotropy* (*free mobility* and *monodromy*, in Helmholtz's terminology) as the most important properties for physics.[151] Helmholtz's problem of describing spaces, which we would now, in the spirit of Chapter 1, describe as having *maximal symmetry*, was subsequently specified and somewhat reinterpreted by Sophus Lie[152] on the basis of his *theory of continuous groups*, to which we will return below; today that problem is called the Helmholtz–Lie problem.

Another attempt to expand Lobachevsky's geometry by including it in a system of new geometric structures,[153] was made by Felix Klein, one of the two protagonists of our story. Compared with Riemann's grandiose constructions, this attempt was more limited and, perhaps for that very reason, its publication evoked a much more lively response. It is conceivable that this attempt played a leading role in forming Klein's general views on the nature of geometry, views we will deal with in greater detail below.

Klein started from a work by the English algebraist Arthur Cayley. In 1854–1859, in six memoirs on "quantics" in the *London Philosophical Transactions*, Cayley considered *homogeneous algebraic polynomials* (*forms* or *quantics*, as he called them) of the second degree or higher. The methods developed by Cayley were purely algebraic, but the space of variables on which the polynomials depended could, of course, be interpreted as projective space (described in projective coordinates). This space is two-dimensional (the projective plane) or three-dimensional projective space, according as the number of independent variables is three or four. (Recall that a point in the projective plane has 3 homogeneous coordinates, while a point in projective space has 4 such coordinates.) From this geometric standpoint, the "Sixth Memoir upon Quantics,"[154] which appeared in 1859, could be regarded as an attempt to introduce a "metric" of sorts into projective space, enabling the distance between points in space or angles between lines to be measured by means of a quadratic form defined on that space. Depending on the type of the form, Cayley obtained different kinds of "projective metrics." In February 1870, during Klein's visit to Berlin, he delivered a report at Weierstrass's seminar on Cayley's work. In particular, he suggested that Cayley's work might be linked with Lobachevsky's non-Euclidean geometry (which Klein knew only very superficially at the time[155]). But Weierstrass, the purist and fanatical believer in mathematical rigor, did not take kindly to Klein's thought, which was still in the formative stage. Weierstrass could not abide hastily conceived ideas and recognized only completely finished and formally impeccable constructions in mathematics. He severely criticized Klein. For only a brief period of time, however, did Klein abandon the thought that the results of the algebraist Cayley and those of the geometer Lobachevsky are closely related. He asked his friend Stoltz (who was very widely educated in mathematics and had acquainted Klein with Staudt's research; see Note 84) to give him a detailed account of Lobachevsky's and Bolyai's results. Klein's talks with Stoltz resulted in a long paper by Klein called "Über die sogennante nichteuklidische Geometrie" (1871),[156] containing a broad interpretation of the projective metric systems in the plane and in space (today, they are called *Cayley–Klein geometries*).[157]

Only one of the geometric systems considered by Klein was classical *Euclidean geometry* in which (in the plane case) the distance d_{AA_1} between the points $A(x, y)$ and $A_1(x_1, y_1)$ is determined by the formula $d_{AA_1} =$

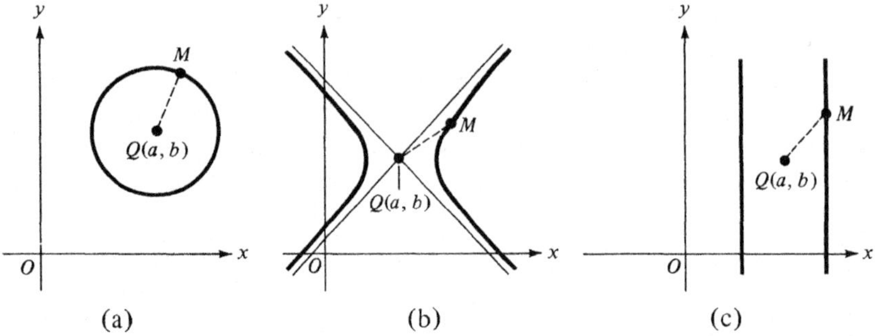

FIGURE 12

$\sqrt{(x_1 - x)^2 + (y_1 - y)^2}$ (see Fig. 12(a), showing the Euclidean *circle* with center $Q(a,b)$ and radius r—the set of points at a distance r from Q). In a certain sense, they are matched by the *pseudo-Euclidean* and *semi-Euclidean* geometries in which the distance d_{AA_1} between the points $A(x,y)$ and $A_1(x_1,y_1)$ is determined by the formulas $d_{AA_1} = \sqrt{(x_1 - x)^2 - (y_1 - y)^2}$, $d_{AA_1} = \sqrt{(x_1 - x)^2} = |x_1 - x|$ respectively (see Figs. 12(b) and (c), in which the pseudo-Euclidean and semi-Euclidean circles are shown). These two simple geometries possess significant mechanical interpretations: semi-Euclidean geometry is used to describe Newton's *classical mechanics*, while pseudo-Euclidean geometry was proposed by Hermann Minkowski (1864–1909) for the geometric interpretation of Einstein's (*special*) *theory of relativity*,[158] with which we cannot deal in greater detail here.[159]

Klein also singled out *Lobachevsky's hyperbolic geometry* and *Riemann's elliptic geometry*, which form part of the system of "projective measurement" under consideration. In particular, Klein interpreted Lobachevskian (plane) geometry as the geometry of the interior of a conic section (bounded, say, by the circle $\mathscr{K}$; see Fig. 13). Here, the points of the interior of the disk bounded by the circle $\mathscr{K}$ are called the "points of the Lobachevskian plane"; the (open, i.e., without ends) chords of the circle are the "straight lines"; the "distance" d_{AB} between the points A and B of the Lobachevskian plane is computed by means of the simple formula

$$d_{AB} = \log(A, B; U, V) = \log\left[\frac{AU}{BU}\bigg/\frac{AV}{BV}\right], \tag{4.2}$$

where U and V are the points in which the line $AB = m$ intersects the *absolute* $\mathscr{K}$ of our non-Euclidean plane, while the choice of a definite logarithmic base is equivalent to the choice of a unit of length (Fig. 13; it follows directly from formula (4.2) that the entire non-Euclidean line $UV = m$ or even the non-Euclidean half-lines AU and AV are infinite). It is easy to see from Fig. 13 that through a point M not on the line m there pass infinitely many lines (such as

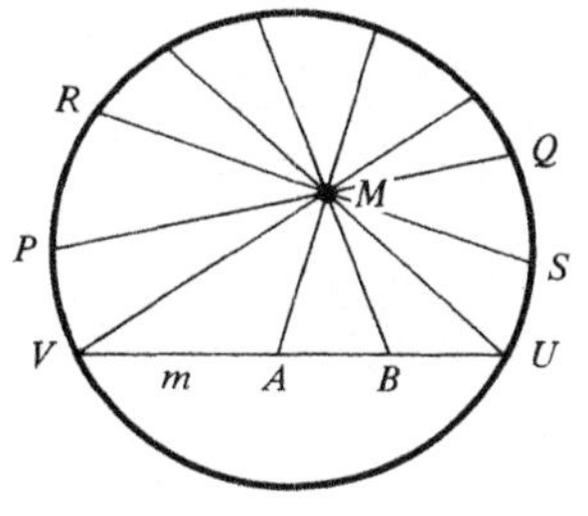

FIGURE 13

the line MA) that intersect m and infinitely many lines (such as PQ and RS) that do not intersect m; the lines MU and MV separating the lines of the first kind from those of the second kind are called *parallel to the line m in the sense of Lobachevsky* (or *Bolyai*). This "model" of Lobachevskian geometry is very easy to imagine and is very often used in the teaching of Lobachevskian geometry.[160]

It was due largely to the latter circumstance that Klein's works provoked wide repercussions. At first the discovery of a large class of new geometries, including the systems of Euclid, Lobachevsky, and Riemann as special cases, did not evoke particular enthusiasm among mathematicians. Some scientists (even Arthur Cayley, whose name is (fittingly) linked with the geometric systems described by Klein) never accepted that discovery, for they suspected the presence of contradictions in this theory. Perhaps this was an unconscious reaction against the overabundance of geometries.[161] In this connection we note that Klein himself, in his 1871 paper (as well as in the long book mentioned in Note 107), saw his main achievement not in the discovery of a large number of new geometric systems, but merely in the discovery of a new (and general) approach that linked the previously known hyperbolic and elliptic geometries with (traditional) Euclidean geometry and, at the same time, established their consistency.[162] It is striking that Klein did not even bother to count the number of geometries he had obtained. A complete classification of these geometries was first produced in 1910 by the English geometer Duncan Sommerville (1879–1934).[163] According to Sommerville, there are 9 Cayley–Klein geometries in the plane—Euclid's geometry and 8 non-Euclidean geometries[164]; 27 Cayley–Klein geometries in space, and 3^n n-dimensional Cayley–Klein geometries, when n is any natural number.

Thus the part of Klein's long 1871 memoir that was immediately accepted by many and became quite popular was his simple model of Lobachevsky's hyperbolic geometry presented in Fig. 13, which graphically illustrated the basic facts of that geometry and convincingly demonstrated its freedom from contradictions. However, at the time it was believed that the noncontradictory nature of Lobachevskian geometry had already been proved by the Italian mathematician Eugenio Beltrami (1835–1900),[165] who seemed to have shown

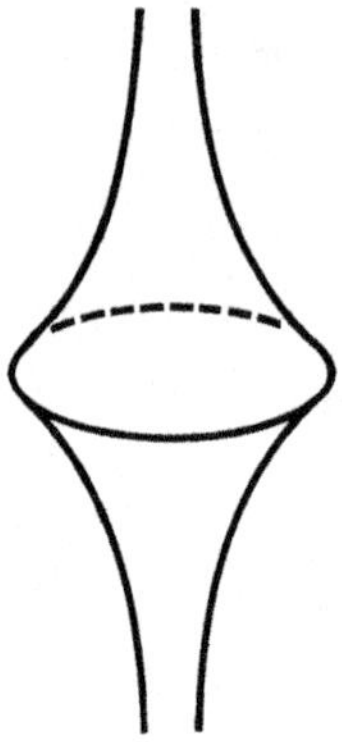

FIGURE 14

in 1868 that Lobachevskian (plane) geometry can be realized on a certain surface in Euclidean space (in the form of the "intrinsic geometry" of a surface with constant negative curvature,[166] for example on the so-called *pseudosphere* depicted in Fig. 14).[167] Since the pseudosphere can be determined by a simple equation in space,[168] Beltrami's result, if correct, implied that if plane Euclidean geometry does not lead to contradictions, then neither does plane Lobachevskian geometry. It was readily conceded that for teaching purposes Beltrami's model on the pseudosphere was less persuasive than Klein's very simple model, much better suited to give one an initial idea of hyperbolic geometry. But this is not the end of the story.

In 1903, Hilbert discovered fundamental and insurmountable flaws in Beltrami's pseudosphere model of Lobachevsky's hyperbolic geometry: he showed that the presence of the acute "edge" of the pseudosphere, seen in Fig. 14, does not allow us to establish a one-to-one correspondence between all the points in the Lobachevskian plane and all the points on the pseudosphere. Similarly, it is possible to map only part of the Lobachevskian plane on any other surface of constant negative curvature.[169] In view of this, the pseudosphere (or any surface of constant negative curvature) could not be considered as a model of Lobachevskian geometry as a whole—and today Beltrami's pseudosphere construction is no longer regarded as proving the logical consistency of the axioms of Lobachevskian geometry. However, following his striking (but, as Hilbert showed, improper) model on the pseudosphere, Beltrami (in the first 1868 paper mentioned in Note 167) briefly indicates another "model in a disk" of hyperbolic geometry, completely coincident with the one shown in Fig. 13 At the time, this conclusion of Beltrami's memoir did not attract sufficient attention, and Klein himself failed to notice it. At the present time, the model of the non-Euclidean Lobachevskian plane shown in Fig. 13 is appropriately called the *Beltrami–Klein model.*[170]

Finally, we point out another interpretation (or model) of hyperbolic geometry, this one due to the famous French mathematician and physicist Henri Poincaré.[171] That model prompts us once again to recall Lambert's prophetic words about an imaginary sphere on which Lobachevsky's non-Euclidean geometry would be realized if such a sphere were to exist (see page 49). We have already pointed out that spherical geometry may serve as a model of Riemann's *elliptic geometry*. Then *points* of the elliptic plane are by definition points of the sphere (or pairs of diametrically opposite points of the sphere, or points of a hemisphere with opposite points of the boundary equator identified; see Fig. 11 on page 64). The great circles of the sphere are the straight lines. Rotations of the sphere about its center are the isometries (in this case it is more convenient to refer to the set of pairs of diametrically opposite points than to the hemisphere model). Of course, it is impossible to consider any imaginary sphere or sphere with imaginary radius in Euclidean space.[172] That is why we pass from Euclidean to *pseudo-Euclidean* three-dimensional space with coordinates x, y, z and the metric

$$d^2 = (x_1 - x)^2 + (y_1 - y)^2 - (z_1 - z)^2, \tag{4.3}$$

where d is the distance between the points $M(x, y, z)$ and $M_1(x_1, y_1, z_1)$. Unlike Euclidean space, that space contains spheres of *real* and *imaginary* radius; for example, the sphere of radius i and center $O(0, 0, 0)$ is the set of points at an (imaginary!) distance i from the origin that satisfy the equation

$$x^2 + y^2 - z^2 = -1 \tag{4.3a}$$

in Euclidean space (x, y, z). This set is a surface known as a *two-sheeted hyperboloid*. One sheet (part) of the hyperboloid (Fig. 15), singled out, say, by the condition $z > 0$, may serve as the imaginary hemisphere, similar to the

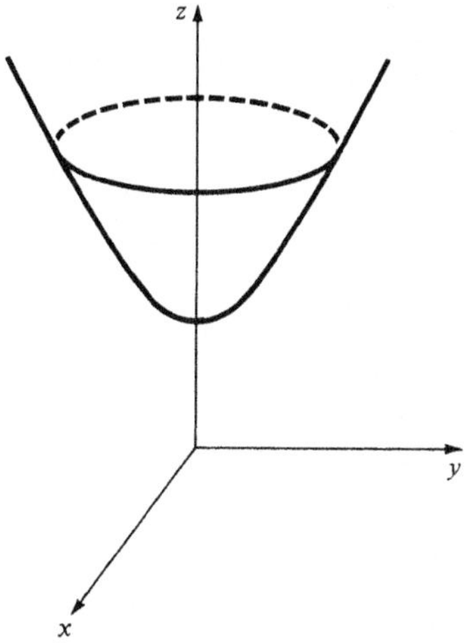

FIGURE 15

real (ordinary) hemisphere depicted in Fig. 11. If we take the points of this imaginary "hemisphere" to be the points of the new (non-Euclidean) plane, the sections of the "hemisphere" by planes passing through the origin (the center of this "sphere") to be the straight lines and the "pseudo-Euclidean rotations" of the sphere that preserve the pseudo-Euclidean distances between points to be the isometries, then we obtain (a model of) Lobachevsky's hyperbolic geometry.[173]

CHAPTER 5

Nineteenth-Century Geometry: Multidimensional Spaces; Vectors and (Hyper) Complex Numbers

The subject matter of this chapter is somewhat removed from the ideas of symmetry and their evolution through the 19th century, which form the core of the book, but is needed to understand Klein's and Lie's work.

In antiquity, mathematicians had already noticed the analogies between a segment of a straight line, a triangle in a plane, and a tetrahedron (a triangular pyramid) in space (Fig. 16(a)); pairs of points on a line, a circle in a plane, and a sphere in space (Fig. 16(b)); a closed interval on a line, a parallelogram in a plane, and a parallelepiped in space (Fig. 16(c)).[174]. Even before the introduction of coordinates, mathematicians could express the difference between the straight line, where figures possess one dimension—their length a; the plane, where figures can be characterized by two dimensions—length a and breadth b; and space, where every solid has three dimensions—length a, breadth b, and height c (see Fig. 17). Once coordinates are introduced, this difference can be conveniently expressed as follows: a point on a (one-dimensional) line is characterized by a single number—its single coordinate (the abscissa) x; a point in a two-dimensional plane has two coordinates (its abcissa x and its ordinate y); finally, in three-dimensional space the position of a point is determined by three numbers (coordinates)—the abscissa x, the ordinate y, and the applicate z (compare Fig. 16(c)). A segment on a line can be described as a set of points $M(x)$ with $0 \leqslant x \leqslant a$, a rectangle in the plane as a set of points $M(x, y)$ with $0 \leqslant x \leqslant a$, $0 \leqslant y \leqslant b$, and a parallelepiped in space as a set of points $M(x, y, z)$ with $0 \leqslant x \leqslant a$, $0 \leqslant y \leqslant b$, $0 \leqslant z \leqslant c$ (see Fig. 16(c)).

Rudiments of the idea of "multidimensional" spaces—spaces of four or more dimensions—turn up, in implicit, and sometimes semi-mystical form, in the works of many mathematicians and philosophers of different nationalities.[175] A clear understanding of the fact that the world we live in is four-dimensional, and that, therefore, every event is characterized by three "space" coordinates x, y, z, and time t, was expressed by the famous French mathematician and philosopher Jean le Rond d'Alembert (1717–1783).[176] D'Alembert and Denis Diderot (1713–1764) were the joint editors of the 35-volume *Encyclopédie* or *Dictionnaire raisonné des sciences, des arts and des métiers* (Paris, 1751–1780). The volume which appeared in 1764 contained an

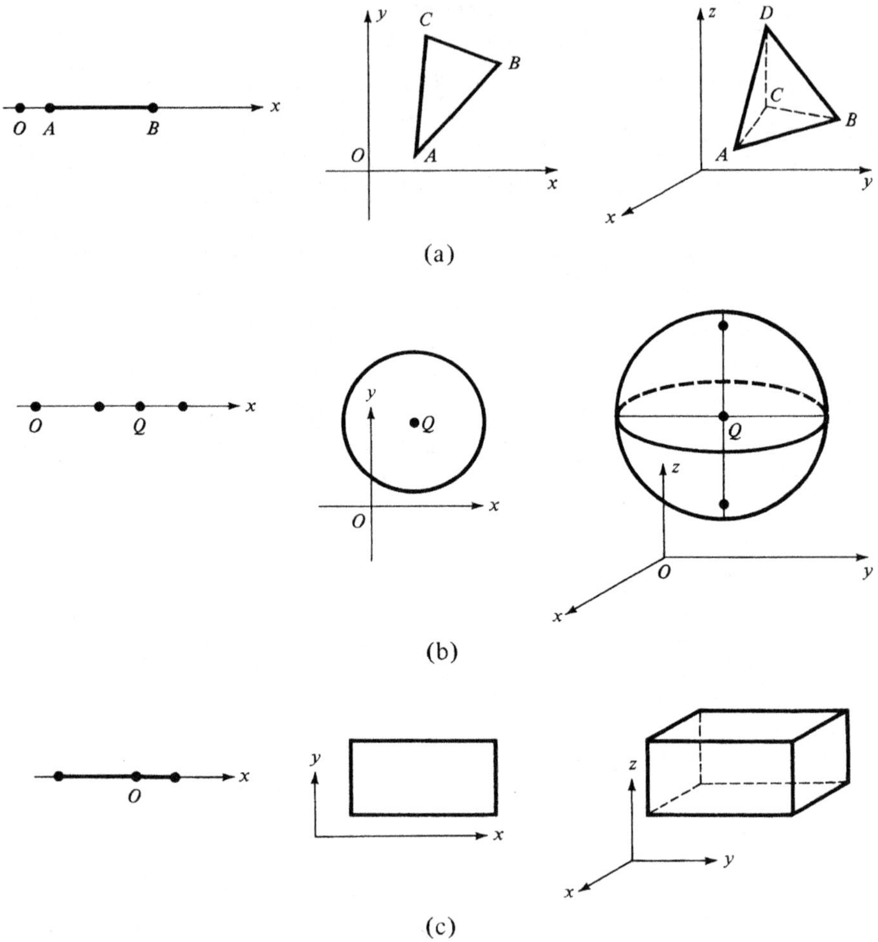

FIGURE 16

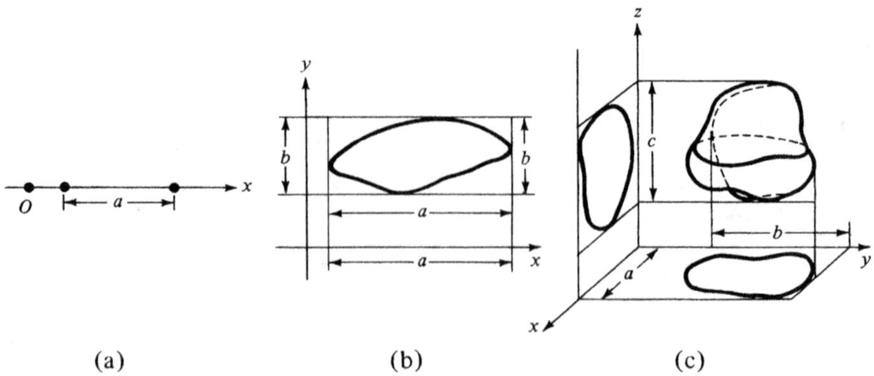

FIGURE 17

Arthur Cayley

entry ("dimension") remarkable for an eighteenth-century encyclopaedia, in which d'Alembert wrote that "One could consider time as a fourth dimension, so that the product of time by volume would, in a certain sense, be the product of four dimensions; this idea is perhaps debatable, but I feel that it has certain merits...."

It seems that the term "n-dimensional geometry" was first used by the outstanding English algebraist Arthur Cayley (1821–1895), repeatedly mentioned above, in the article "Chapters on the Analytical Geometry of n Dimensions."[177] Cayley came from a well-to-do English family. Owing to business affairs, Cayley's father lived in St. Petersburg, in Russia, where young Arthur spent his childhood. Cayley studied at Cambridge, where he became senior wrangler and the first winner of Smith's prize. His first scientific papers appeared in the year of his graduation. But since he felt that mathematics did not guarantee sufficient material rewards, he began to study law. Soon he became a well-known and successful London lawyer. However, in contrast to M. Chasles, who also postponed the pursuit of an academic career because of financial considerations (see page 37), Cayley never suspended his mathematical work during his time as a lawyer, combining his practice with intense and extremely fruitful research activity. In 1864, considering himself sufficiently well off, Cayley left the bar for a professorship at Cambridge University, where he remained until his death.

Paradoxically, Cayley's science combined profound creativity with caution and conservatism which often hampered his deeper understanding of the achievements of others as well as his own. Thus while he was practically the first important mathematician to comment in writing on Lobachevsky's non-Euclidean geometry (see above), he did not for a time understand its real significance. Similarly, he never acquired a deep appreciation of the geometric systems now known as projective metrics or Cayley geometries[178] (see page 67). Cayley's conservatism also appears in the paper "Chapters on the Analytical Geometry of n Dimensions." In his profound analysis of certain facts of $(n-1)$-dimensional projective geometry[177] and in his proofs of important theorems in this field, Cayley is very cautious in the use of geometric terminology (except for the title) and symbolism which was obviously appropriate but not yet fully established. Only at the end of the paper does he note that in the three-dimensional case (why only this one?) his algebraic results (algebraic only in form, not in content) are equivalent to meaningful theorems of geometry.[179] Similarly, in his famous "Memoirs upon Quantics," particularly in the Sixth Memoir where the foundations of "Cayley metrics or geometries" are set forth, he deals exclusively with second-order curves in the projective plane and with surfaces in three-dimensional space, although the extension of all these results to the general (n-dimensional) case is obvious. Incidentally, Klein's famous 1871 paper "On so-called non-Euclidean geometry" considers in detail only plane (two-dimensional) and space (three-dimensional) geometries, and mentions the possibility of extensions to n dimensions only in a single sentence at the end of the paper.

Despite all his cautiousness, it is hard to overestimate Cayley's role in the creation of the concept of n-dimensional space and his contribution, by no means limited to the *Chapters* article, to the development of "multi-dimensional intuition". Cayley's research, as well as that of his followers, friends, and sometime rivals (in particular the indomitable traveller, the Anglo-American James Joseph Sylvester (1814–1897) who moved from country to country and from one activity to another,[180] and the Irish mathematician and theologian George Salmon (1819–1904)),[181] was mainly devoted to the development of *linear algebra* (which, in its geometric aspect, reduces to the study of linear and quadratic loci planes and quadrics in multidimensional affine and projective space) and to the theory of invariants, arising in the affine and projective classification of quadrics in the plane and in space. And perhaps it is to this "invariant trio," as they were called by the famous French analyst Charles Hermite (1822–1901), that we owe the fact that by the end of the 1870s the notion of an n-dimensional vector space became accessible even to average students of mathematics. In connection with Cayley, we should also mention that he created the theory of matrices, i.e., tables of numbers (square at first and then rectangular when the theory developed), which provide an important tool in the theory of n-dimensional vector spaces as well as a meaningful example of an n^2-dimensional space (cf. page 99; see A. Cayley, "Memoir on the Theory of Matrices" (*Philos. Transactions*, 1858) which also appears in Cayley's *Collected Works*).

Hermann Günther Grassmann

And yet the true founder of the concept of *n*-dimensional space is probably the German Hermann Günther Grassmann (1809–1877), one of the most original (and for that reason unrecognized in his time) mathematicians of the nineteenth century. Grassmann's *Die lineale Ausdehnungslehre*[182] appeared in 1844, the same year as Cayley's *Chapters*, while the second, completely revised, text of the same work (*Die Ausdehnungslehre*) came out seventeen years later (in 1861). The vector calculus, closely related to *n*-dimensional geometry, was created simultaneously and independently by Grassman and by the famous Irishman William Rowan Hamilton (1805–1865). (This is yet another instance of a phenomenon that we encountered and described on a few occasions above.) Hamilton was an outstanding physicist (we recall that according to some modern views physics "belongs" to the right hemisphere of the brain) and Grassmann a leading linguist (everything having to do with linguistics is associated with the left hemisphere).

Hermann Grassmann[183] was born in Stettin, a provincial German town with which practically his entire life is connected. His family had long been known for its religious and scientific interests, and both aspects of intellectual life were always very close to Grassmann. His grandfather was a pastor; his father, who influenced Hermann greatly, was also a clergyman until his scientific interests prevailed and he became professor of physics and mathematics in the Stettin gymnasium from which his son graduated and where he taught for many years.

The senior Grassmann also wrote several works on physics, technology, and elementary mathematics. After graduating from the gymnasium, Hermann Grassmann studied for three years at Berlin University. He studied not mathematics but philosophy, psychology, philology, and theology; following family tradition, he thought seriously of becoming a clergyman, and only abandoned the idea completely while working on his *Ausdehnungslehre* (later he repeatedly expressed regret at not having become a pastor). After completing the university course, Grassmann passed an examination granting him the right to teach, and he taught at a secondary school in Berlin for one-and-a-half years. Then, after passing two comprehensive examinations in theology (at the time he still intended to become a pastor), and then a state examination for the right to teach in the higher forms of the gymnasium (all these examinations took place in Berlin), Grassmann received a document certifying that he was superbly qualified to teach mathematics, physics, mineralogy, and chemistry, as well as theology or religion. From 1836 onwards Grassmann worked exclusively in Stettin, where at first he taught at the Friedrich Wilhelm secondary school, and then, after his father's death, occupied the latter's post in the city gymnasium He moved to Stettin, where at different times he taught German, Latin, chemistry, mineralogy, physics, and mathematics. Most of the time his teaching load was very heavy—up to twenty hours a week of different subjects. It is remarkable then that Grassman found the time to do significant research in various fields—in addition to mathematics he obtained significant results in physics; in particular in the theory of electricity and colors (this work was highly valued by the famous Helmholtz). Grassmann studied music and its theory and the theory of vowels; he had a very keen ear, which in these fields served him well. For a long time he was also an editor or (with his younger brother Robert) a co-editor of, and an active contributor to the local newspaper, as well as a prominent Freemason and church figure. We will yet have occasion to discuss Grassmann's philological interests and contributions, which are regarded by some as the most important part of his intense intellectual life. Klein ends his highly sympathetic biographical essay on Grassmann with the following words: "It is not surprising that in view of such a variety of activities, there was one field Grassmann failed to master: he was a very poor teacher" (compare this with what we said about Jacob Steiner on pages 42–43). Mildmannered and friendly with everyone, Grassmann was unable to maintain the needed discipline in class. He only talked with a few of the most interested pupils, while the rest "had a good time".

Grassmann's "theory of extension" was certainly not his only achievement in pure mathematics,[184] but here it is appropriate to examine only his two basic books in the 1844 and 1861 editions. In effect both of these books present the theory of n-dimensional space (nonmetric "linear" or "affine" space in the 1844 book; n-dimensional Euclidean space in the 1861 book) and employ, respectively, the two basic methods adopted today. In *Die lineale Ausdehnungslehre*, the author, unfortunately, keeps his promise, given to the readers in the introduction, to set forth his work "proceeding from general philosophical notions without the help of any formulas." These notions correspond to a modern theory of linear (vector) spaces; more specifically, to their axiomatic, i.e., descriptive, development. Nowadays, all mathematics students (and even some secondary school pupils) know that a **vector space** is a set of (undefined) objects a, b, c, ..., called vectors (Hamilton's term; Grassmann talks about "extensive magnitudes") closed under two operations, namely addition and scalar multiplication, and satisfying the following axioms:

$$a + b = b + a \quad \text{(commutativity of addition);}$$

$$\lambda(\mu a) = (\lambda\mu)a;$$

$$(a + b) + c = a + (b + c) \quad \text{(associativity of addition);}$$

$$1a = a; \qquad a + 0 = a; \qquad a + (-a) = 0$$

$$(\lambda + \mu)a = \lambda a + \mu a, \quad \lambda(a + b) = \lambda a + \lambda b \quad \text{(distributive laws);}$$

here 0 is called the *zero vector*, and $(-a)$ the *additive inverse* of the vector a. (Dieudonné's *Linear Algebra and Elementary Geometry*, intended for secondary school teachers, opens with the axioms for a vector space; it seems that Dieudonné believes that the school geometry course should begin in that way).[185] However, in Grassmann's time, the reader was totally unprepared for such a manner of presentation and for such an approach to the essence of mathematics (regardless of the fact that it goes back to Pythagoras and Plato, or, at the very least, to Leibniz). Neither could Grassmann's general philosophical standpoint,[186] completely shared by Hankel (see Note 189) and George Boole (1815–1864),[187] be generally absorbed at that time (and perhaps not even today). Readers were unprepared for Grassmann's approach and for his idiosyncratic style, in particular, his use of many strange terms of his own invention; if the lawyer Cayley was excessively cautious in using new terms, the linguist Grassmann obviously enjoyed inventing them! Hence Grassmann's first book was ignored by mathematicians. There was not a single reference to it, nor one review; and, 20 years after its appearance, about 600 copies of the book (of the 900 which seem to have been printed) were disposed of as waste (the other unsold copies were distributed free to anyone who wanted one). Neither was the book supported by Gauss, to whom Grassmann sent a copy. Gauss responded as usual, thanking the author in a short letter; instead of appraising the book, he said: "The tendencies in your book partly intersect the roads along which I wandered for almost half a century." Gauss failed to respond to Grassmann's calculus (or algebra), the most important thing in *Die Ausdehnungslehre*. We deal with it below.

Grassmann was not at all so philosophical as to be indifferent to his failure. In the introduction to the book's second version (1861) he stressed that it "was completely revised and presented in the rigorous language of mathematical formulas." Indeed, the underlying approach here is the "arithmetic" (constructive) approach, i.e., all the constructions are based on "arithmetic" or "coordinate" space—a set of "points" defined by their coordinates $(x_1, x_2, \ldots, x_n)$, or, in modern terms, the set of n-tuples of (real) numbers $(x_1, x_2, \ldots, x_n)$. The respective operations of addition of points $x = (x_1, x_2, \ldots, x_n)$, and $y = (y_1, y_2, \ldots, y_n)$ and of multiplication of a point x by a number λ, are defined as follows:

$$x + y = (x_1 + y_1, x_2 + y_2, \ldots, x_n + y_n)$$

$$\lambda x = (\lambda x_1, \lambda x_2, \ldots, \lambda x_n).$$

Linear subspaces of the space are defined as the solution sets of one (or several) linear homogeneous equations relating the coordinates of points:

$$A_1 x_1 + A_2 x_2 + \cdots + A_n x_n = 0.$$

Grassmann introduces, for the first time, the crucial notion of linear dependence of points $a_1, a_2, \ldots, a_k$ (we would now say vectors; Grassmann talks of "extensive magnitudes" instead). The points in question are linearly dependent if there is a relation

$$\lambda_1 a_1 + \lambda_2 a_2 + \cdots + \lambda_k a_k = 0,$$

where $0 = (0, 0, \ldots, 0)$ is the zero point (vector) and not all the numbers $\lambda_1, \lambda_2, \ldots, \lambda_k$ are equal to zero. This notion enables him to define the dimension of the entire space (or of the associated linear space) as the largest possible number of linearly independent points. One consequence of that definition, is the simple Grassmann formula

$$\dim U + \dim V = \dim(U \cdot V) + \dim(U + V).$$

The notation is modern; dim means dimension, U and V are two linear subspaces, $U \cdot V = U \cap V$ is their intersection, while $U + V$ is their (vector) sum.[188] Another consequence is what is now called a Grassmann algebra (see below), described in a clear "formula form." In the second version of *Die lineale Ausdehnungslehre*, Grassmann introduces a metric defined by the expression $\sqrt{x_1^2 + x_2^2 + \cdots + x_n^2}$ into the manifold under consideration and thus transforms it into an (n-dimensional!) Euclidean space. (Riemann's famous lecture "Über die Hypothesen," where n-dimensional manifolds with much more general metrics were considered, had already been delivered at the time; however, Grassmann had no way of knowing about it, because he lived far away from scientific centers in total isolation from mathematical circles; not even the best-known research journals reached him.) Although the book's second version could be more easily understood by a persistent and favorably-inclined reader it too was presented in severely abstract form and abounded in new terms; it seems that it attracted even less attention than the 1844 edition. In his last years Grassmann prepared a new edition of the first version of his book; this came out in 1878, a year after the author's death. It should be noted that the awkward language, more philosophical than mathematical, of both versions of his book and the abundance of new terms, unfamiliar to the reader, served as an obstacle to Grassmann's being invited to teach at the university. Twice he applied for a university post and twice he was rejected, and once a negative review of his works (stressing the above-mentioned shortcomings) was written by the well-known mathematician Ernst Eduard Kummer (1810–1893).

It seems that the first mathematician who really appreciated Grassmann's achievements was W.R. Hamilton. In 1853 Hamilton wrote a number of letters to the Cambridge algebraist and logician Augustus de Morgan in which he explained and praised Grassmann's *Ausdehnungslehre.* Hamilton's long *Lectures on Quaternions* appeared in the same year. The author attached special importance to these lectures and in the introduction gave credit to his German colleague's accomplishments. Unfortunately, Grassmann never learned of the impression his works made on Hamilton. Still, the established

German mathematician Hermann Hankel (1839–1873), who learned of the *Ausdehnungslehre* from Hamilton's lectures, sent Grassmann an enthusiastic letter[189] in 1866. In 1871 Grassmann was elected a corresponding member of the Göttingen scientific society, but this belated (and far from adequate) recognition came at a time when Grassmann had already largely forsaken mathematics for the purely philological studies to which he had always been inclined. As early as 1843, Grassmann and his brother Robert,[190] put out an elementary textbook of the German language with exercises ("mit zahlreichen Übungen").[191] Its fourth edition appeared in 1876. Grassmann's extensive work on German names of plants was much more substantial,[191] as was his study of German folklore and, in particular, his collections of German folk songs. In his last years his scientific interests focussed on his study (now regarded as a classic) of the famous literary and religious work of ancient India known as the *Rig-Veda*. In 1873, Brockhaus, a Leipzig publishing house, issued Grassmann's extensive dictionary for the *Rig-Veda* (written in Sanskrit) and in 1876–1877 (Grassmann died in 1877) the same publisher issued his two-volume translation of this outstanding work into German. These were the only scientific works for which Grassmann was awarded a degree: the remarkable scholar was made doctor of philosophy *honoris causa* by Tübingen University. Recognition of Grassmann's mathematical achievements came only after his death, and was due to the high praise bestowed on his work by Felix Klein and Sophus Lie. In particular, much space was devoted to Grassmann in Klein's history of nineteenth-century mathematics.[192] In 1894–1911, largely on Lie's initiative, B.G. Teubner publishers of Leipzig (with whom Lie was closely associated) issued *Gesammelte mathematische und physikalische Werke von Hermann Grassmann* in three volumes (six substantial books). An active part in the publication was taken by Lie's closest pupils Georg Scheffers and, especially, Friedrich Engel, who was the editor of the whole work. The last of the six books contained a biography of Grassmann, written by Engel (see Note 183).

Surprisingly, in view of the few contemporary comments on Grassmann's n-dimensional geometry, the relevant ideas very quickly became common knowledge. In 1851, the Swiss Ludwig Schläfli (1814–1895), a professor at Bern University, presented a large work, *Theorie der vielfachen Kontinuität*, to the Vienna Academy of Sciences; however, it was published only fifty years later (Basel, 1901). In his introduction (which he may have written without knowing Grassmann's work) Schläfli states that his book "sets out to describe the foundations of a new branch of analysis, a sort of analytical geometry in n dimensions, containing ordinary analytic geometry of the surface and of space when $n = 2$ and 3." Schläfli's terminology and notations were very close to the modern ones. He also solved a number of specific problems of n-dimensional (Euclidean) geometry, such as extension to n dimensions of the famous Euler formula[193] linking the number of vertices, edges, and faces of an arbitrary (convex) polyhedron, and drawing up a list of regular polyhedra in n-dimensional Euclidean space.[194] Despite their delayed publication,

Schläfli's results became known to mathematicians from his published papers (and perhaps from manuscripts circulating in Vienna and Bern). Of the two specific problems mentioned above, Euler's n-dimensional formula (although it seems to have been rediscovered by Poincaré, who makes no reference to Schläfli) is today associated with the Schläfli's name. The classification of regular polyhedra in n-dimensional space is, unfortunately, often attributed in the literature to later authors.[195] In any case, Klein points out in his history of nineteenth-century mathematics that in the 1870s the notion of n-dimensional space was widely known. In his 1872 *Erlangen program* Klein himself considered only two- and three-dimensional geometries but noted (in a single sentence in the memoir *On the so-called non-Euclidean geometry*) that his main ideas obviously hold in the general (n-dimensional) case. Finally, our account would be incomplete if we failed to mention our "old acquaintance" Camille Jordan who, in a long memoir *Essai sur la géomètrie à n dimensions*[196] offered a substantial presentation, quite modern in form, of n-dimensional Euclidean geometry, including solutions of the problem of finding the angles between two linear subspaces (known as stationary angles—their number depends on the dimensions of the subspaces) and of the shortest distance between them.[197] Jordan also developed the differential geometry of (smooth) curves in n-dimensional Euclidean space in two short notes in 1874 (both were included in Vol. 3 of the *Works*[196]), and laid the basis for differential geometry of m-dimensional surfaces in n-dimensional space (where $m \leqslant n - 1$; the simplest cases are $m = 1$ (curves) and $m = n - 1$ (hypersurfaces)).

Let us now turn to the second subject cited in this Chapter's title and closely related to the topic of n-dimensional space, namely the rise of vector calculus. Gottfried Leibniz (1646–1716) had dreamed of a geometric calculus dealing directly with geometric objects rather than with numbers. Monge's pupil Lazare Carnot (see Note 65) outlined a rather inadequate model of such a calculus in his *Géométrie de Position* (Paris, 1803). Leibniz's dream was realized by Grassmann and Hamilton, who came to their respective calculi from completely different directions.

In accordance with his philosophical and general scientific interests, Grassmann had a very respectful attitude toward Leibniz. Grassmann devoted a special work to the explanation of Leibniz's geometric ideas: *Geometrische Analyse geknüpft an die von Leibniz erfundene geometrische Charakteristik* (Leipzig, 1847). Incidentally, this was the only work of Grassmann that mathematicians noticed: it received an award at an all-German contest of scientific works and was published in Leipzig (not in Stettin, an extremely provincial town in terms of science) with an introductory "explanatory essay" (*mit einer erläuternden Abhandlung*) by Möbius. This term, apparently originating with Möbius, shows clearly that Möbius, a master of lucid exposition, hardly sympathized with Grassmann's awkward "philosophemas". Grassmann regarded his calculus, in which purely geometric objects were added, subtracted, multiplied and divided (and admitted scalar multiplication) as a precise realization of Leibniz's program.

It is very easy to explain *Grassmann's algebra* in modern terms. Proceeding from n-dimensional space (Euclidean, although much of the theory does not require a Euclidean metric), Grassmann introduces n (linearly independent!) basis vectors or units $e_1, e_2, \ldots, e_n$ and examines the formal sums of these units and their products subject to the rule that products of units are anticommutative,

$$[e_i, e_j] = -[e_j, e_i] \quad \text{when} \quad i \neq j, \quad [e_i, e_i] = [e_i^2] = 0,$$

and associative. We thus have products of units and "products of products" $[e_{i_1} e_{i_2} \cdots e_{i_p}] \cdot [e_{j_1} e_{j_2} \cdots e_{j_q}]$, which are nonzero only if all the indices i_1, $i_2, \ldots, i_p$ are different, all the indices $j_1, j_2, \ldots, j_q$ are different, and no i_s $(s = 1, \ldots, p)$ is equal to a j_t $(t = 1, \ldots, q)$. If this is the case, then such a product can be written in the form

$$\pm e_{k_1} e_{k_2} \cdots e_{k_r}, \quad \text{where} \quad 1 \leqslant k_1 < k_2 < \cdots < k_r \leqslant n, \quad r = p + q.$$

Thus the Grassmann algebra has to do with "ordinary" (n-dimensional) vectors

$$x_1 e_1 + x_2 e_2 + \cdots + x_n e_n$$

where the x_i are numbers, as well as with more complicated sums

$$x_0 + \sum x_i e_i + \sum x_{ij} e_{ij} + \sum x_{ijk} e_{ijk} + \cdots + x_{12\cdots n} e_{12\cdots n}, \tag{5.1}$$

where the x's still denote numbers and, for example,

$$e_{ij} = [e_i e_j], \quad i < j, \qquad e_{ijk} = [e_i e_j e_k], \quad i < j < k.$$

It is precisely such expressions that Grassmann called "extensive magnitudes".

The great importance of Grassmann's "exterior algebra" for mathematical analysis (without going into details, we point out that in any multiple integral the expression under the integral sign contains the exterior product of differentials of the independent variables), for geometry and for topology (which was in its infancy in Grassmann's time) could not have been anticipated at the time. Nor could it have been anticipated by the school of fanatical Grassmannites which arose soon after his death (see pages 88–89). The importance of this algebraic system was fully appreciated only in the twentieth century by Henri Poincaré, by Elie Cartan,[198] one of the leading figures in, among other areas, 20th-century geometry, and by the Swiss Georges de Rham, who carried out the synthesis of Cartan's algebra and geometry with Poincaré's analysis and topology.

In addition to the antisymmetric "exterior" product of units in which $[e_i e_j] = -[e_j e_i]$, Grassmann also considered the symmetric "inner" product subject to the rules

$$(e_i e_j) = 0 \quad \text{when} \quad i \neq j, \quad (e_i e_i) = (e_i^2) = 1. \tag{5.2}$$

The inner product of a vector by itself, i.e., the inner or scalar square of a vector, was related by Grassmann to its length (in the Euclidean metric).

Similarly, Grassmann defined the inner product of arbitrary extensive magnitudes or expressions. In particular, he related "extensive expressions of the second order" $\sum x_{ij}e_{ij}$ $(=\sum x_{ij}[e_i e_j])$ to surface elements[199] (these constructions were carried out in verbal geometric form in the first version of the *Ausdehnungslehre*) and the "inner square" of such an expression to the area of the surface element.

It is instructive to look at the Grassmann algebra of vectors in the simplest cases of two- and three-dimensional vector spaces. In both cases, the inner product operation assigns to each pair of vectors $a = x_1 e_1 + x_2 e_2 (+x_3 e_3)$ and $b = y_1 e_1 + y_2 e_2 (+y_3 e_3)$ the number (see formula (2))[200]

$$a \cdot b = (a, b) = x_1 y_1 + x_2 y_2 \, (+x_3 y_3). \tag{5.3}$$

In the two-dimensional case, the outer or exterior product assigns to each pair of vectors a and b the number

$$a \times b = [a, b] = (x_1 e_1 + x_2 e_2) \times (y_1 e_1 + y_2 e_2) = (x_1 y_2 - x_2 y_1) e_{12}; \tag{5.4a}$$

we say "number" advisedly since the "vector factor" e_{12} is the same for all pairs a and b and may therefore be omitted. In the three-dimensional case the outer product of two vectors is also a vector

$$\begin{aligned} a \times b = [a, b] &= (x_1 e_1 + x_2 e_2 + x_3 e_3) \times (y_1 e_1 + y_2 e_2 + y_3 e_3) \\ &= (x_1 y_2 - x_2 y_2) e_{12} + (x_2 y_3 - x_3 y_2) e_{23} + (x_3 y_1 - x_1 y_3) e_{31} \\ &= X_{12} e_{12} + X_{23} e_{23} + X_{31} e_{31}, \end{aligned} \tag{5.4b}$$

in the three-dimensional space with the basis $\{e_{12}, e_{23}, e_{31}\}$;[201] in the Euclidean case this three-dimensional space can be identified with the original space by putting $e_{12} = e_3$, $e_{23} = e_1$, and $e_{31} = e_2$. $S^2 = (a \times b) \cdot (a \times b)$, "the inner square" of the outer product $a \times b$, is related to the area S of the parallelogram spanned by the vectors a and b. Specifically, in the two-dimensional case $S^2 = (x_1 y_2 - x_2 y_1)^2$, and in the three-dimensional case $S^2 = X_{12}^2 + X_{23}^2 + X_{31}^2$, that is, $S^2 = (a \times b) \cdot (a \times b)$. Actually, in the two-dimensional case one usually writes

$$S = x_1 y_2 - x_2 y_1,$$

where S stands for the so-called "oriented area", i.e., the area of the parallelogram spanned by the vectors a and b taken with a plus or minus sign.

Another way of arriving at the vector calculus involved the geometric interpretation of complex numbers $z = x + iy$, where $i^2 = -1$, widely used by Gauss. Gauss[202] assigned to each such number the point of the plane with Cartesian (rectangular) coordinates (x, y) and polar coordinates $\langle r, \varphi \rangle$, where $r = |z| = \sqrt{x^2 + y^2}$ and $\varphi = \operatorname{Arg} z$, i.e., $\cos\varphi = x/r$, $\sin\varphi = y/r$, while $\tan\varphi = y/x$ (Fig. 18). Here the sum

$$z_1 + z = (x_1 + iy_1) + (x + iy) = (x_1 + x) + i(y_1 + y)$$

is computed according to the parallelogram rule $(x, y) + (x_1, y_1) = (x + x_1, y + y_1)$ (Fig. 19(a)) while multiplication $z_1 z = \langle r_1 \varphi_1 \rangle \cdot \langle r, \varphi \rangle = \langle r_1 r, \varphi_1 + \varphi \rangle$

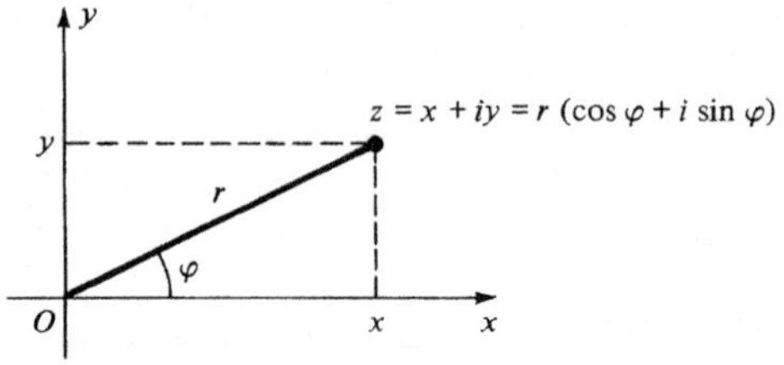

FIGURE 18

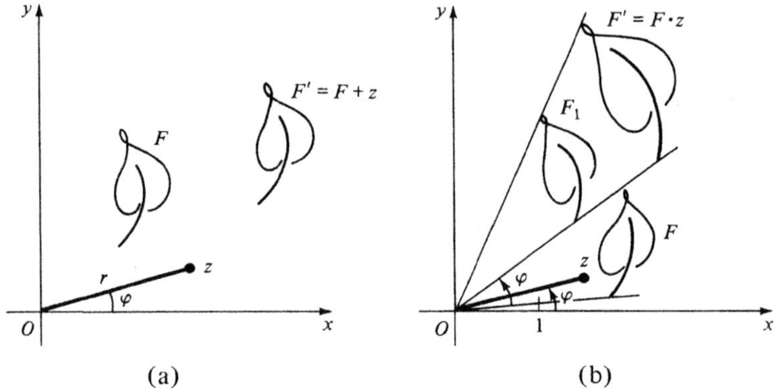

FIGURE 19

"rotates" each point z_1 through the angle φ and "stretches" it r times (see Fig. 19(b); note that the operations of "increasing" by z and multiplying by z are applied to the entire figure formed by the points z_1). Thus it can be said that the addition operation ("the adding of z") geometrically expresses a translation of the plane, while multiplication by z is a rotation followed by a stretching; in particular, if $|z| = 1$, then this operation is a pure rotation through the angle $\varphi = \operatorname{Arg} z$).[203]

By "crossing" Gauss's complex numbers and the algebra of Grassmann numbers ("extensive expressions") we get the *Clifford algebra* generated by n "units" $e_1, e_2, \ldots, e_n$ and all their possible products, where the products of units are anticommutative ($e_i \cdot e_j = -e_j \cdot e_i$) and associative, but the square of each unit $e_i \cdot e_j = e_i^2$ is not zero, as in Grassmann's case, but -1, as in the algebra of complex numbers (see W.K. Clifford, "Application of Grassmann's extensive algebra," *Amer. Journ. of Maths.*, 1879). As in the case of Grassmann numbers, the product of any number of "products of units" $(e_{i_1} e_{i_2} \cdots e_{i_p})$ and $(e_{j_1} e_{j_2} \cdots e_{j_q})$ can be written in the form $\pm e_{k_1} e_{k_2} \cdots e_{k_r}$, where $1 \leqslant k_1 < k_2 < \cdots < k_r \leqslant n$, by using the associativity of multiplication of units and the main rules $e_i e_j = -e_j e_i$ and $e_i^2 = -1$. Thus the Clifford algebra involves the same formal sums (1) as the Grassmann algebra, but uses somewhat modified rules of operation. The Grassmann algebra (called "exterior algebra" because it is based on Grassmann's exterior multiplication) plays a fundamental role in modern mathematics. The main "users" of the Clifford algebra are probably physicists rather than mathematicians;[204] the former, however, have also expressed interest in Grassmann numbers.

An important role in the history of mathematics was played by attempts to extend ordinary complex numbers from the plane to space by creating new numbers with three units instead of two. Gauss had considered this problem and arrived at the conclusion that a system of complex numbers of the form $u = xe_1 + ye_2 + ze_3$ was not viable but did not publish his arguments. Attempts to solve the problem continued on the other side of the Channel—they were undertaken by Augustus de Morgan (1806–1871), one of the leaders of the Cambridge "formal" school, and Charles Graves (1810–1860), an Irish mathematician. In a voluminous treatise called *On the Foundations of Algebra*, de Morgan presents, in particular, an axiomatic approach to what he calls "the triple algebra."[205] A different form of essentially the same algebra appears in Graves's article[206] "On Algebraic Triplets," published the same year (1847) as de Morgan's book. Graves writes his triplets as $u = x + ye + ze^2$, multiplies them "the natural way" subject to the rule $e^3 = 1$, and assigns to each triplet the point (x, y, z) in space; he also gives a neat geometric interpretation of his multiplication.[207]

The greatest progress in the field we are now concerned with was achieved in another attempt to generalize complex numbers and is linked with the name of one of the most brilliant nineteenth-century mathematician and physicists William Rowan Hamilton (1805–1865). Hamilton's mathematical work was not directly related to n-dimensional spaces, but there is a deep connection between these problems and his research in mechanics. On the other hand, the vector calculus (both vector algebra and vector analysis) was entirely created by Hamilton, although not quite in the form in which it is presented in modern textbooks.

Hamilton's extraordinary abilities were manifested at an early age. At five, he knew, in addition to English, five languages—French, Italian, Latin, Greek, and Hebrew; by twelve he had added seven more, including such exotics as Arabic, Persian, Sanskrit, and Malay. Hamilton attended (protestant) Trinity College in Dublin, which we had occasion to mention before, in connection with another alumnus of (and teacher at) this institution (see Note 181). There he was so highly regarded as a student that he received the rank of a college astronomy professor in 1827 even before he had completed the full college course! Soon afterwards, Hamilton occupied the high (and well paid) post of Ireland's Astronomer-Royal, which he retained till his death, and he was appointed observatory director at Dunsink (on the outskirts of Dublin). Besides his outstanding abilities in physics, mathematics, and linguistics, Hamilton was also talented in the arts. He was a prolific poet, whose verses were appreciated by his friend William Wordsworth (1770–1850).

Unfortunately, the early and rapid development of Hamilton's outstanding talents, and his meteoric career in science and administration, had their tragic side. Later in life Hamilton experienced bouts of depression. He attempted to overcome these, and simultaneously to improve his ability to work, with the help of alcohol. Persistent drinking reduced his creative capacities still further, and undermined his intellectual powers; even such an outstanding mind could not withstand the consequences of alcohol abuse. Hamilton died at sixty, but his last years were truly tragic. He was broken in mind and in spirit.

While still a student at Trinity College, and right after his graduation, Hamilton carried out superb research on geometric optics, and in 1834–1835 he published the basic works on mechanics that were the high point of his scientific career. Today, Hamiltonian mechanics forms the foundation of quantum mechanics and of many other modern mathematical constructions,[208] and is learned by all students of mathematics and physics. Despite their overall mathematical importance, Hamilton's later works, which he regarded as the chief accomplishment of his life and which are of most interest for our story, are not as profound as his works in mechanics.

In 1837 the *Transactions of the Irish Academy* published Hamilton's long paper (linked to the research of the "Cambridge formal school" mentioned above; see page 85 and Note 205), "Theory of Conjugate Functions or Algebraic Couples," with a "Preliminary and Elementary Essay on Algebra as the Science of Pure Time." The second part of the title refers to the following idea, which goes back to Kant: geometry, as the science of space, should be contrasted with algebra conceived as the science of time based on the notion of "ordinal" numbers (of course, not only natural numbers and integers!). It turns out that a number of ideas in this paper were fairly close to those contained in Grassmann's *Lehrbuch der Arithmetik* (see Note 184), a book which Hamilton did not know at that time. As to the first part of the paper's title, it refers above all to Hamilton's interpretation (actually proposed earlier by János Bolyai; see page 59) of complex numbers $z = x + iy$ as pairs (x, y) of real numbers with the following rules of multiplication and addition

$$(x, y) + (x_1, y_1) = (x + x_1, y + y_1); \quad (x, y)\cdot(x_1, y_1) = (xx_1 - yy_1, xy_1 + yx_1).$$

What especially pleased Hamilton was that he managed to do away with the mysterious symbol $\sqrt{-1}$, a permanent source of difficulty for people taught to accept as revealed truth that only positive numbers have square roots. Of course, Hamilton kept in mind the usual geometric interpretation of complex numbers (as points in the plane); his own interpretation of complex numbers (as pairs of real numbers) is essentially the same, since points in the plane can be described by their coordinates.

In subsequent years Hamilton spent much time in several attempts to construct a reasonable system of "spatial" complex numbers with three units. This led to Hamilton's friendly contacts and correspondence with the Cambridge formalists. Here he expressed, among other things, his high opinion of Grassmann's work (see page 79). After much thought and considerable computation, Hamilton concluded that it is not possible to construct a system of complex numbers with three units in which division is possible.[209]

He then passed to systems with four complex units and almost immediately discovered the **quaternions.**[210] Already in his attempts to construct numbers with three units Hamilton conceived the idea of assigning to each complex unit a rotation in space, just as in the plane case multiplication by the complex unit $i(=1(\cos 90° + i \sin 90°))$ is equivalent to a rotation of the plane through 90° (see Fig. 19(b)). (Before Hamilton, and apparently unknown to him, the same idea was expressed by the Danish geodesist Caspar Vessel (see Note 202)). Since rotations in space do not, in general, commute (they form a noncommutative group; see page 102), Hamilton was forced to give up the requirement of commutativity of multiplication of his complex units.

We shall not dwell here on the operator theory of quaternions, in which each quaternion is viewed as a certain operator in space. This theory is sketched in Klein's book mentioned in Note 68, and developed in greater detail in Klein's book cited in Note 201. Here we come directly to the final results.

Hamilton's quaternions are complex numbers of the form

$$q = x_0 + x_1 i + x_2 j + x_3 k,$$

where x_0, x_1, x_2, x_3 are real numbers while i, j, k are complex numbers with the following rules of multiplication:

$$i^2 = j^2 = k^2 = -1; \qquad ij = -ji = k, \qquad jk = -kj = i, \qquad ki = -ik = j.$$

It was essential for Hamilton to split each quaternion q into two summands: the number $x_0 = Sq$, which he called the scalar or the scalar part of the quaternion (from the Latin word *scala*, meaning ladder), and the expression $x_i i + x_2 j + x_3 k = Vq$, which he called the vector or vector part of the quaternion (from the Latin word *vector*, meaning carrier).[211] In the case when

$$v = xi + yj + zk \quad \text{and} \quad v_1 = x_1 i + y_1 j + z_1 k$$

are two vectors, their product vv_1 is a "general" quaternion

$$vv_1 = S(vv_1) + V(vv_1),$$

where, as can be easily checked,

$$\begin{aligned} S(vv_1) &= -xx_1 - yy_1 - zz_1; \\ V(vv_1) &= (yz_1 - y_1 z)i + (zx_1 - z_1 x)j + (xy_1 - x_1 y)k. \end{aligned} \tag{5.5}$$

Following Hamilton, the expressions $S(vv_1)$ and $V(vv_1)$ (a number and a vector!) are called, respectively, the scalar and vector products of the vectors v and v_1 (more precisely, the scalar and vector parts of the "quaternion" product of the two vectors). Comparing relations (5.5) with relations (5.3) and (5.4b) due to Grassmann, we are struck by the remarkable similarity between Hamilton's scalar and vector products and Grassmann's inner and exterior products. This similarity has become increasingly clear today, since in the further evolution of the vector calculus, Hamilton's scalar product was stripped of its minus sign so that

$$vv_1 = xx_1 + yy_1 + zz_1.$$

Hamilton's rule for dividing quaternions is very similar to the one for dividing complex numbers. For each complex number $z = x + iy$ one can define the "conjugate" number $\bar{z} = x - iy$ (obviously, the map $z \to \bar{z}$ is geometrically equivalent to a reflection in the real axis Ox). We then have

$$z\bar{z} = x^2 + y^2 (= |z|^2).$$

Now if the number z is not zero and $z_1 = x_1 + iy_1$, then,

$$\begin{aligned} z_1/z &= (z_1\bar{z})/(z\bar{z}) = (x_1 + iy_1)(x - iy)/(x^2 + y^2) \\ &= (xx_1 + yy_1)/|z|^2 + [(xy_1 - x_1y)/|z|^2]i. \end{aligned}$$

Similarly, if $q = s + xi + yi + zk = s + v$, so that $v = xi + yj + zk$, and $\bar{q} = s - v = s - xi - yj - zk$, (the map $q \mapsto \bar{q}$ is obviously equivalent to changing the sign of the vector part of the quaternion, or reflecting the vector v in the origin), then

$$q\bar{q} = s^2 + x^2 + y^2 + z^2 \qquad (= |q|^2).$$

Now if $q \neq 0$ and $q_1 = s_1 + x_1 i + y_1 j + z_1 k$, then

$$q_1/q = (q_1\bar{q})/(q\bar{q}) = (q_1\bar{q})/|q|^2,$$

which can easily be written as an ordinary quaternion $\sigma + \xi i + \eta j + \zeta k$.

The similarity between quaternions and complex numbers inspired Hamilton. He decided to tackle the grandiose problem of carrying over all the results of the theory of functions of a complex variable to the theory of quaternions. He devoted twenty years of his life to this problem. Hamilton discovered quaternions in 1843; his first publication on the topic is dated 1844. His two decades of labor produced two lengthy books, *Lectures on Quaternions* (Dublin, 1853) and *Elements of Quaternions* (Dublin, 1866) the latter published posthumously; its has been recently republished (N.Y., Dover, 1969).[212]

Hamilton's achievements in the theory of quaternions were considerable, yet he himself exaggerated their importance. Quaternions play a far less important role in mathematics than do complex numbers. Even more than their creator, members of the fanatical school of "quaternionists" or "Hamiltonians," which arose around Hamilton in Dublin and eventually spread its influence over England, exaggerated the importance of quaternions. During the same period, in continental Europe, particularly in Germany, these excesses gave rise to a suspicious attitude to the topic. In 1895 the "quaternionists" even created a "World Union for the promotion of quaternions." They felt that the theory of functions of a quaternion variable is not only comparable to the theory of functions of a complex variable, but even superior to it. Also, whereas the relations between Hamilton and Grassmann were marked by mutual respect, the relations between the quaternionists and the equally fanatical school of Grassmannites, which arose in Germany soon after Grassmann's death, immediately became very hostile.

In modern mathematics Hamilton's quaternions undoubtedly occupy an important, but not a central place. It is easy to see that they are actually "second-order Clifford numbers" with principal units i and j and compound unit $k = i \cdot j$, whence it follows by anticommutativity and associativity that

$$k^2 = (ij)(ij) = i(ji)j = i(-k)\cdot j = -i(kj) = -i\cdot(-i) = i^2 = -1.$$

However, compared to the general "Clifford numbers" (for which, incidentally, we also have the operation of conjugation and which have a norm similar to the quaternion norm $q\cdot\bar{q} = |q|^2$) they have a number of merits, e.g., only quaternions admit of a well-defined operation of division by any nonzero number. Arbitrary motions (isometries) in space can be conveniently expressed in terms of quaternions. Thus a translation in space can be written as

$$v' = v + a,$$

where v, v' are variable vector quaternions (radius vectors of points) and a is a constant vector—the translation operator, and a rotation about the origin can be written as

$$v' = qvq^{-1}, \tag{5.6}$$

where q is a fixed quaternion, and q^{-1} is its inverse, i.e., $qq^{-1} = q^{-1}q = 1$ $(= 1 + 0i + 0j + 0k)$. It is remarkable that formula (5.6) (written in the language of numerical quadruples, which he did not call quaternions) was already known to Gauss, while P. Stäckel, in his book *Gauss als Geometer*, referred to in Note 106, states that it was already known to Euler.

We have already pointed out that of all the "reasonable" systems of "complex" numbers only the numbers commonly so called, and the quaternions, admit a well-defined operation of division by nonzero numbers. Now we discuss this statement in more detail. The founders of the theory of general complex numbers (or hypercomplex numbers, as they were later called) were W.R. Hamilton (who began his monograph *Lectures on Quaternions* with this notion) and H. Hankel in the book *Theorie der complexen Zahlensysteme* (Leipzig, Voss, 1867). Hankel here stated the general problems of symbolic or axiomatic algebra[213] in great detail and gave a rigorous definition of arbitrary systems of general complex numbers, exemplified by Hamilton's quaternions. It was Hankel's theory of general complex numbers that inspired Hamilton's admiration for Grassmann's *Ausdehnungslehre*. Such numbers are understood to be systems of symbols

$$u = x_0 + x_1e_1 + x_2e_2 + \cdots + x_ne_n, \tag{5.7}$$

where $x_0, x_1, \ldots, x_n$ are arbitrary real numbers and $e_1, e_2, \ldots, e_n$ are complex units. The number $0 + 0e_1 + \cdots + 0e_n$ is usually denoted by the digit 0 (or the symbol **0**) while the number $1 + 0e_1 + \cdots + 0e_n$ is denoted by the digit 1 (or the symbol **1**). Addition and subtraction of general complex numbers are defined in the natural way: if u and $v = y_0 + y_1e_1 + \cdots + y_ne_n$ are two (hyper) complex numbers, then

$$u \pm v = (x_0 \pm y_0) + (x_1 \pm y_1)e_1 + (x_2 \pm y_2)e_2 + \cdots + (x_n \pm y_n)e_n.$$

The product of u and v was introduced by use of the usual rules for eliminating brackets in multiplication; these are based on the distributive laws of multiplication with respect to addition, always assumed true for hypercomplex systems:

$$\begin{aligned} uv &= (x_0 + x_1e_1 + x_2e_2 + \cdots + x_ne_n)(y_0 + y_1e_1 + y_2e_2 + \cdots + y_ne_n) \\ &= x_0y_0 + x_0y_1e_1 + x_0y_2e_2 + \cdots + x_0y_ne_n + x_1y_0e_1 + x_1y_1(e_1)^2 \\ &\quad + x_1y_2(e_1e_2) + \cdots + x_ny_n(e_n)^2. \end{aligned} \tag{5.8}$$

Now in order to be able to view this expression as a number of the same nature as u and v, it is necessary to define the pairwise products of units:

$$e_1e_1 = e_1^2, e_1e_2, e_1e_3, \ldots, e_1e_n, e_2e_1, e_2^2, \ldots, e_{n-1}e_n, e_n^2.$$

Thus, for example, if $n = 1$ and the multiplication table of units reduces to the (obviously unique) rule $e_1^2 = -1$, then we get the ordinary complex numbers. If $e_1^2 = 1$, then we obtain the double Clifford numbers and if $e_1^2 = 0$ then we obtain the dual Clifford numbers (see Note 203). For Hamilton's quaternions n equals three, and the multiplication table for the units $e_1 = i$, $e_2 = j$ and $e_3 = k$ is

	e_1	e_2	e_3
e_1	-1	e_3	$-e_2$
e_2	$-e_3$	-1	e_1
e_3	e_2	$-e_1$	-1

where the first factor is in the column on the left and the second is in the top row (so that, for example, $e_1e_3 = -e_2$ and $e_2e_3 = e_1$).[214]

Under addition (and subtraction, which is the operation inverse to addition), hypercomplex numbers form a $(n + 1)$-dimensional vector space. The new properties of the space have to do with the multiplication of numbers (see the next chapter, where we return to hypercomplex numbers and even give this notion a new name). It is clear that the multiplication of numbers is commutative if and only if the multiplication of any two complex units is commutative. Thus, for example, for Graves's "triplets"

$$u = x_0 + x_1e + x_2e^2$$

(see Notes 206 and 207) we always have $uv = vu$, since in that case the units e and e^2 satisfy $e^ie^j = e^je^i = e^{i+j}$, where $i, j = 1$ or 2 and $e^3 = 1$. Multiplication is associative (i.e., for any three numbers u, v and w we have $(uv)w = u(vw)$) if the multiplication of any three units e_i, e_j and e_k, where i, j and $k = 1, 2, \ldots$ or n (here the three indices need not be distinct), is associative: $(e_ie_j)e_k = e_i(e_je_k)$. (Usually the term "hypercomplex number" already *implies* associativity.) Thus the multiplication of ordinary complex numbers is associative (since for the unique unit satisfying $e^2 = -1$ we have $(e^2)e = e(e^2) = -e$), and the same is true of the multiplication of dual numbers, double numbers and (more importantly) of the multiplication of quaternions (check this!). Multiplication of Grassmann numbers is also associative, and so too is multiplication of Clifford numbers (why?).

If a system of numbers is not commutative, then we have two "quotients" u/v, namely the two numbers t_1 and t_2 satisfying $u = t_1 v$ and $u = vt_2$; and these two numbers may turn out to be distinct.[214] Hence the notation u/v is inappropriate here and the following more convenient notations are used: $t_1 = uv^{-1}$ and $t_2 = v^{-1}u$.[215] Of course, in the general case the quotients uv^{-1} and $v^{-1}u$ of two given (hyper)complex numbers u and v may not exist. If the two numbers $t_1(=uv^{-1})$ and $t_2(=v^{-1}u)$ do exist, and are unique for each two numbers u and v in our system, where v differs from zero ($v \neq 0$), then we say that the (hyper)complex numbers under consideration constitute a numerical system with division. (Division by zero is impossible in any system of hypercomplex numbers (why?).)

As we have already pointed out, the general notion of hypercomplex number is due to the Irishman Hamilton and the German Hankel (and also to the Americans B. Peirce and C. Peirce, who will be mentioned in the next chapter). However, the main achievements here are due to the algebraic-geometric German school, whose acknowledged leader was Sophus Lie (we shall discuss the connection between the (hyper)complex topics and Lie's main scientific interests below) and which included E. Study, G. Scheffers,[216] and somewhat later Friedrich Heinrich Schur (1856–1932)[217] and Theodor Eduard Molin (1861–1941),[218] and to the Berlin arithmetical school headed by Weierstrass. A member of both German schools—which were headed by Lie and Weierstrass respectively—was the Berliner Georg Ferdinand Frobenius (1849–1917), one of the leading algebraists of the time. Frobenius proved in 1878 the theorem, now considered classical, that any associative system of hypercomplex numbers with division is either the real numbers (the case $n = 0$ in the representation (5.7) of hypercomplex numbers), the ordinary complex numbers, or Hamilton's quaternions. The same theorem was proved independently by the American Charles Peirce (who published his result two years later, however). Thus associative systems of hypercomplex numbers (or algebras, as Peirce preferred to call them; see Chapter 6) with division can have only $1(=2^0)$, $2(=2^1)$, or $4(=2^2)$ complex units, which must include the real number 1.[219] Moreover, for each of the above-mentioned numbers of units (1, 2, 4) there exists a unique associative system of numbers with division.[220]

The Frobenius theorem does not contradict the existence of the so-called *Cayley numbers*, discovered by the latter in 1845 (see A. Cayley, "On Jacobi's Elliptic functions and on quaternions," *London-Edinburgh-Dublin Phil. Magazine* (3); V. 26, 1845, pp. 208–211). Another term, more often used for these numbers, is "octaves," due to John Thomas Graves (1806–1870), brother of the Charles Graves mentioned earlier. J.T. Graves discovered these numbers independently of Cayley and indeed somewhat

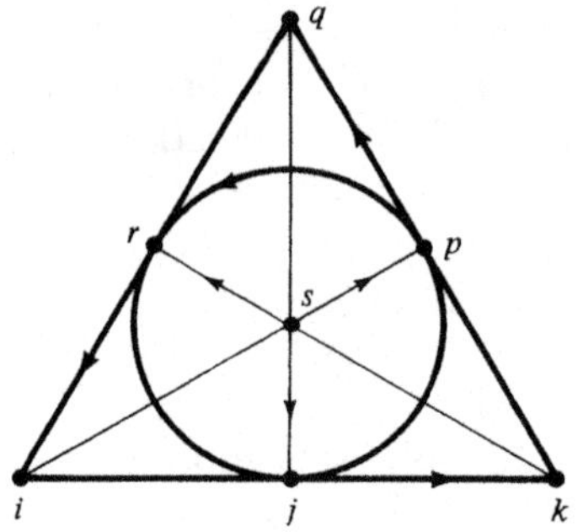

FIGURE 20

earlier (in 1843). However his paper was not published and mathematicians learned about it from an article by Hamilton (published only in 1848, i.e., later than Cayley's article) in which Hamilton related the achievements of his friend Graves. Octaves or Cayley numbers are numbers of the form

$$u = a_0 + a_1 i + a_2 j + a_3 k + a_4 p + a_5 q + a_6 r + a_7 s, \tag{5.9}$$

where $i^2 = j^2 = k^2 = p^2 = q^2 = r^2 = s^2 = -1$ while the product of any two units is equal to a third unit taken with a plus or minus sign. Here, as in the case of quaternions, the octave units anticommute: the sign of the product changes when the units are interchanged (e.g., $ij = -ji = k$; $pq = -qp = k$, etc.). A complete multiplication table for octaves is easy to set up by means of the "Freudenthal diagram" (named after the Dutch mathematician Hans Freudenthal, born in 1905) shown in Fig. 20. This diagram depicts a "finite projective plane" with seven points i, j, k, p, q, r, s and seven lines—a line being a triple of collinear or concyclic points, where any two points belong to a unique line, any two lines intersect in a unique point, each line contains three points, and three lines pass through every point. The product of any two units represented by neighboring points on a line is always equal to the third point of this line, taken with a plus sign if the arrow on the line leads from the first factor to the second one and with a minus sign otherwise. Further, the formulas for multiplication of triples of "points" on one line are obtained from one another by a cyclic permutation (so that for example the relation $ij = k$ implies $jk = i$ and $ki = j$) and the "complex units" $i, j, k, \ldots$ anticommute (so that, say, $qr = i$ implies $rq = -i$). Thus the system of octaves contains seven "quaternion lines":

$$a_0 + a_1 i + a_2 j + a_3 k, \quad a_0 + a_3 k + a_4 p + a_5 q, \quad a_0 + a_5 q + a_6 r + a_1 i,$$

$$a_0 + a_1 i + a_7 s + a_4 p, \quad a_0 + a_3 k + a_7 s + a_6 r, \quad a_0 + a_5 q + a_7 s + a_2 j,$$

$$a_0 + a_2 j + a_4 p + a_6 r.$$

There exist other, possibly simpler, descriptions of the system of octaves. Quaternions may be described as "Clifford numbers of the second order," i.e., as the system of numbers with two principal units i and j satisfying $i^2 = j^2 = -1$ and also the anticommutativity and associativity conditions for the multiplication of units. Putting $ij = k$, this already implies that $ji = -k$ (anticommutativity!), $k^2 = -1$ (compare page 89), $kj = (ij)j = i(jj) = i(-1) = -i$, etc. (here we are constantly using the anticommutativity and associativity of multiplication of units). Similarly, octaves may be defined as a system of hypercomplex numbers with three principal units i, j, and s satisfying $i^2 = j^2 = s^2 = -1$ and anticommutativity and associativity for the multiplication of units. Then, letting $ij = k$, $is = p$, $si = q$ and $ijs = ks = r$, it is easy to show that $k^2 = p^2 = q^2 = r^2 = -1$ and, say, $kp = (ij)(is) = -(ji)(is) = -j(ii)s = js = q$ etc., i.e., we again obtain the multiplication table for the eight octave units (including the number 1).[221]

To any octave (5.9) we can associate the conjugate number

$$\bar{u} = a_0 - a_1 i - a_2 j - a_3 k - a_4 p - a_5 q - a_6 r - a_7 s;$$

so that the product of two conjugate numbers will be a real positive number:

$$u\bar{u} = a_0^2 + a_1^2 + a_2^2 + a_3^2 + a_4^2 + a_5^2 + a_6^2 + a_7^2,$$

which it is natural to denote by $|u|^2$ or $\|u\|$ and to call the norm or the square of the absolute value of the octave. (Here we have assumed that the octave u differs from 0; in addition, we define $|0| = 0$.) Therefore for every octave $v \neq 0$ one can define the inverse octave $v^{-1} = \bar{v}/|v|^2$ (satisfying $vv^{-1} = v^{-1}v = 1$) and hence also the quotient of any two octaves u and $v \neq 0$ (to be more precise, two quotients uv^{-1} and $v^{-1}u$).[222]

The existence of octaves does not contradict the Frobenius theorem because the octaves constitute a nonassociative system of hypercomplex numbers. Thus, for example, $(ij)s = ks = r$ while $i(js) = i(-q) = -iq = -r$. Incidentally, "nonassociativity" does not imply that for any three octaves u, v, w we have the inequality $(uv)w \neq u(vw)$. It only means that such an inequality holds for at least one triple of octaves, while for many other triples we do have $(uv)w = u(vw)$. Thus, for example, we always have

$$[(\alpha u)(\beta u)](\gamma u) = (\alpha u)[(\beta u)(\gamma u)] = (\alpha\beta\gamma)(uuu) = (\alpha\beta\gamma)u^3,$$

where α, β, γ are three arbitrary real numbers and $u^3 = (uu)u = u(uu)$. Indeed, each octave u generates an "octave ray" λu (where λ is an arbitrary real number); in this way any three octaves constitute an associative triple (which is obvious). It is much more interesting to note that any two octaves u, v generate an octave plane, the set of octaves $\alpha u + \beta v$ where α and β are arbitrary real numbers; in this plane we also have the associativity condition for any three octaves

$$[(\alpha u + \beta v)(\alpha_1 u + \beta_1 v)](\alpha_2 u + \beta_2 v) = (\alpha u + \beta v)[(\alpha_1 u + \beta_1 v)(\alpha_2 u + \beta_2 v)]. \quad (5.10)$$

Thus, for example, for any two octaves u and v we always have

$$(uv)v = u(vv) \quad \text{and} \quad (vv)u = v(vu) \quad (5.10a)$$

(it is easy to see that (5.10a) and (5.10) are equivalent).

Systems of hypercomplex numbers satisfying (5.10) (or equivalently (5.10a)) are called *"alternative" number systems*. Thus alternativity is a kind of generalization of associativity. The Frobenius theorem can be extended to the following statement: any alternative system of hypercomplex numbers with division is either the (associative and commutative) system of real numbers or of ordinary complex numbers, or the (associative but not commutative) system of quaternions, or, finally, the (noncommutative and nonassociative, but alternative) system of octaves.[223] (It is clear that the associativity requirement is stronger than that of alternativity.)[224] The proof of this generalized Frobenius theorem resembles that of the theorem about associative systems of hypercomplex numbers with division. For this reason the theorem about alternative systems is often referred to as Frobenius's theorem, although historically this is not quite appropriate. The first proof of the theorem on alternative systems with division is apparently due to the American algebraist Adrian Albert (1905–1972). Thus the list of all the alternative hypercomplex systems with division consists of four remarkable numerical systems with $1(=2^0)$, $2=(2^1)$, $4(=2^2)$ and $8(=2^3)$ complex units.

It should be noted that N. Bourbaki in his *Elements of the History of Mathematics* (cf. Note 117) does not rate the alternative system of octaves very highly, indicating that the negation of associativity used by Graves and Cayley, the creators of the system of Cayley numbers, did not open up any interesting new directions. But Bourbaki's view seems refuted by an "octave boom" now occurring both in mathematics, where octaves undoubtedly provide the key—which unfortunately we do not yet know how to fully use!—to the puzzle of the "singular semisimple Lie groups" (see pages 105–106), and in physics, where many researchers pounced on octaves, with varying success, in trying to decipher the remarkable properties of the elementary particles that build the Universe.[225]

One other elegant trait which singles out, among all possible systems of hypercomplex numbers, the four listed above—the real numbers x_0, ordinary complex numbers $x_0 + x_1e_1$, quaternions $x_0 + x_1e_1 + x_2e_2 + x_3e_3$ and octaves $x_0 + x_1e_1 + x_2e_2 + x_3e_3 + x_4e_4 + x_5e_5 + x_6e_6 + x_7e_7$—was pointed out by the outstanding German analyst Adolf Hurwitz (1859–1919), a friend of the David Hilbert so frequently mentioned in this book. This property is the following: for these numbers, and only for them, each number u has a "norm" (the square of the absolute value) $|u|^2$ or $\|u\|$ that is a quadratic function in the "coefficients" x_i of the number u satisfying $|uv| = |u|\cdot|v|$ and $\|u\| \geqslant 0$, where $\|u\| = 0$ only if $u = 0$.[226] Actually, if we did not require that the norm be positive, and that it vanish only for vanishing numbers, then these conditions would be satisfied by one system of real numbers; three systems of complex numbers (ordinary complex numbers, double numbers, dual numbers) with the respective norms $x_0^2 + x_1^2$; $x_0^2 - x_1^2$ and x_0^2; five systems of quaternions with the respective norms

$$x_0^2 + x_1^2 + x_2^2 + x_3^2; \quad x_0^2 + x_1^2 - x_2^2 - x_3^2, \quad x_0^2 + x_1^2, \quad x_0^2 - x_1^2, \ x_0^2;$$

and seven systems of octaves with the respective norms

$$x_0^2 + x_1^2 + x_2^2 + x_3^2 + x_4^2 + x_5^2 + x_6^2 + x_7^2, \quad x_0^2 + x_1^2 + x_2^2 + x_3^2 - x_4^2 - x_5^2 - x_6^2 - x_7^2,$$
$$x_0^2 + x_1^2 + x_2^2 + x_3^2, \quad x_0^2 + x_1^2 - x_2^2 - x_3^2, \quad x_0^2 + x_1^2, \quad x_0^2 - x_1^2, \ x_0^2.$$

All these $1 + 3 + 5 + 7 = 16$ systems of (hyper)complex numbers are written out on pages 24–25 and 221–223 of the book referred to in Note 203: I.M. Yaglom, *Complex Numbers in Geometry*.[227]

CHAPTER 6

Sophus Lie and Continuous Groups

After this unavoidably long digression on to the progress of geometry in the nineteenth-century (see Chapters 3–5), we can return to our protagonists, Sophus Marius Lie and Felix Christian Klein, whom we left after their return from a period of study in Paris with Camille Jordan, whose main research interests at the time centered around the theory of groups. Jordan was firmly convinced that the theory of groups was destined to play an outstanding role in the future development of mathematics and he had imparted this conviction to Lie and Klein. To be sure, even Jordan underestimated the impact that group theory would have—though in the late 1860s and early 1870s this was hardly forseeable.[228] Credit for introducing group-theoretic concepts into literally all branches of mathematics is mainly due to Lie and Klein.

Jordan believed that in geometry the main role of groups would be played by *geometric transformation groups*,[229] such as the group of isometries (or the group of similitudes) of the Euclidean plane, the group of affine transformations (without a metric) of the affine plane or the group of projective transformations of the projective plane, as well as groups such as the group (1.5) of symmetries of the square, consisting of only eight elements (see Chapter 1). Jordan drew a clear distinction between discrete groups, such as the group of isometries of the square whose elements are "separate", and continuous groups, such as the group $\mathfrak{J}$ of (direct) *isometries* of the Euclidean plane

$$\begin{aligned} x' &= x\cos\alpha + y\sin\alpha + p, \\ y' &= -x\cos\alpha + y\cos\alpha + q, \end{aligned} \tag{6.1}$$

and the more general group $\mathfrak{A}$ of *affine transformations*[230] of the (affine) plane xOy:

$$\begin{aligned} x' &= ax + by + p, \\ y' &= cx + dy + q, \\ \Delta &= \begin{vmatrix} a & b \\ c & d \end{vmatrix} = ad - bc \neq 0. \end{aligned} \tag{6.2}$$

Thus the group (6.1) may be described as the "symmetry group" of the Euclidean plane (the group of transformations of this plane which do not change any of its properties). In the next chapter we shall discuss in more detail the relationship between groups of geometric transformations and the notion of symmetry. Here we shall merely indicate that since the group (6.2) of affine transformations is richer in elements than the group (6.1) of isometries (indeed, each element of the latter also belongs to the former, but the converse is obviously false), it would seem that we can claim that the affine plane is "more symmetric" than the Euclidean plane, just as the square, with a (numerically) larger group of symmetries, is "more symmetric" than the equilateral trapezoid (see Figs. 2(a), (b)). The resulting connections between Euclidean geometry and affine geometry will also be discussed in more detail in Chapter 7.

Discrete groups such as the group (1.5) of symmetries of a square are now known as *crystallographic* groups, because the symmetry groups of crystals are of that type. The importance of such groups in the study of crystals was fully accepted in the second half of the nineteenth century.[231] The adjective "continuous" in the name of groups such as (6.1) and (6.2) underscores that the transformations belonging to the group can be changed continuously by slight alteration of the parameters determining a particular element of the group. Thus in the case of the group of isometries $\mathfrak{I}$, where every isometry $\delta \in \mathfrak{I}$ is determined by the angle α of the rotation ν and the vector $\mathbf{t} = (p, q)$ that determines the translation τ (see Fig. 21, where $A^* = \nu(A)$ and $A' = \tau(A^*)$, so that $A' = \tau\nu(A) = \delta(A)$), a slight alteration of the parameters α, p and q (the coefficients of $\cos\alpha$, $\sin\alpha$, p and q in formula (1)) will change the transformation δ "slightly", i.e., if we replace them by slightly different magnitudes α_1, p_1, q_1, we shall obtain the transformation $\delta_1 = \delta(\alpha_1, p_1, q_1)$, which is close to the

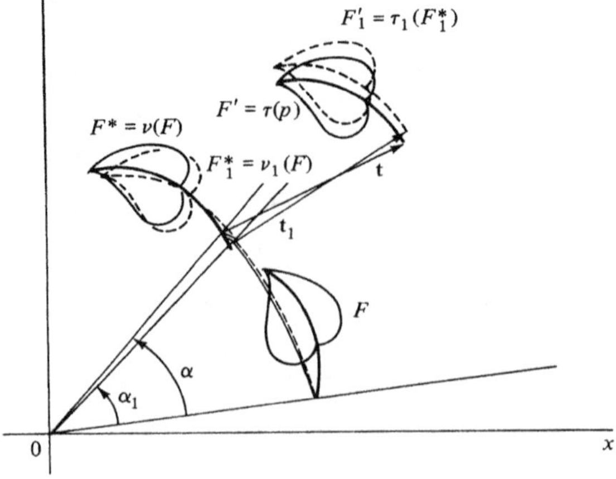

FIGURE 21

transformation $\delta = \delta(\alpha, p, q)$ (see the same Fig. 21, where $F' = \delta(F)$ and $F_1 = \delta_1(F)$).

Lie's and Klein's research was to a certain extent inspired by their deep interest in the theory of groups and in various aspects of the notion of symmetry. However, after the initial period of joint studies their areas of scientific work diverged somewhat. Lie devoted his entire life to the *theory of continuous groups* (such groups are now usually known as **Lie groups**[232]) and was the sole creator of this extensive theory, while Klein was more concerned with discrete transformation groups.

Lie's theory rested on his discovery of the intimate connection between continuous groups and specific algebraic systems now known as *Lie algebras.* The term "algebra" used here is not new. It appeared under another name in Chapter 5.

When discussing general complex and hypercomplex numbers in Chapter 5, we gave a relatively detailed description of the relevant contributions of the English school (the Englishmen A. de Morgan, A. Cayley, and especially W.K. Clifford; the Irishmen C. Graves, J.T. Graves, and especially W.R. Hamilton) and of the German school (the Stettin mathematician G. Grassmann; the Leipzig group of E. Study, G. Sheffers, T. Molin, and especially G. Hankel and S.M. Lie; the Berliners K. Weierstrass and G. Frobenius[233]). But we merely mentioned the Americans B. and C.S. Pierce. We shall fill this gap now.

Benjamin Pierce (1809–1890) was an influential scientist at Harvard who specialized in astronomy. He compiled detailed tables of the motion of the moon and of Neptune, calculated the orbits of numerous comets, and wrote certain theoretical studies (such as the theory of the possible structure and equilibrium of Saturn's rings); he also concerned himself with philosophy and mathematics. His main mathematical achievements came later in his life and were to a certain extent inspired by the research interests of his son Charles Sanders Peirce (1839–1914). Some of the mathematical researches of the two Peirces were carried out jointly. We mentioned C. Peirce before (in Chapter 5) as the independent discoverer of Frobenius's theorem on systems of hypercomplex numbers with division. Conversations with W.K. Clifford, which took place while B. Peirce was in London in 1871, helped maintain his active interest in mathematics in the last years of his life. (Later, both participants readily and frequently recalled these conversations.) Actually one of the reasons for Peirce's visit to London was his desire to report on his algebraic results (already obtained at the time) to the London mathematical society, of which he was a member. It should be noted that the astronomical works of C.S. Peirce (whose range of scientific interests was very wide) were to a great extent due to to his father's influence.

B. Peirce is the author of the posthumously published fundamental work "Linear Associative Algebras" (*Amer. Journ. of Math.*, **4**, 1881, p. 97–221), in which the term "algebra" is used in the sense in which the term hypercomplex numbers was then used in Europe. That is, Peirce calls a (linear) algebra any finite-dimensional vector space with basis $\mathbf{e}_1, \mathbf{e}_2, \ldots, \mathbf{e}_n$ whose elements are of

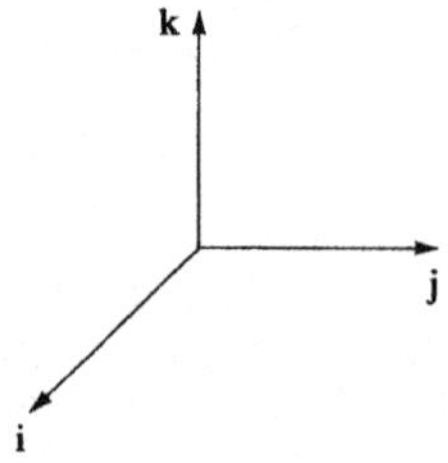

FIGURE 22

the form

$$\begin{aligned} \mathbf{u} &= x_1\mathbf{e}_1 + x_2\mathbf{e}_2 + \cdots + x_n\mathbf{e}_n, \\ \mathbf{v} &= y_1\mathbf{e}_1 + y_2\mathbf{e}_2 + \cdots + y_n\mathbf{e}_n \end{aligned} \tag{6.3}$$

(compare (5.7); here $x_1, x_2, \ldots, x_n; y_1, y_2, \ldots, y_n$ are real or complex numbers) and are multiplied according to the usual rules (i.e., the distributive law holds for algebras) subject to the "*structural equations*"

$$\mathbf{e}_i\mathbf{e}_j = c_{ij}^1\mathbf{e}_1 + c_{ij}^2\mathbf{e}_2 + \cdots + c_{ij}^n\mathbf{e}_n \quad \left(= \sum_{t=1}^{n} c_{ij}^t\mathbf{e}_t\right); \tag{6.4}$$

the constants c_{ij}^t (for which $i, j, t = 1, 2, \ldots, n$) are called the *structure constants* of the algebra. The associativity requirement eliminates from our present considerations such algebras as the algebra of *octaves* (the *Cayley numbers*; see Chapter 5 above) or the simpler (and, for us, more important) example of the anticommutative algebra of three-dimensional vectors, with the operation of vector multiplication, due to Hamilton.[234] Indeed, the algebra in this last example is not associative: if **i**, **j**, and **k** are three perpendicular unit vectors in ordinary space that form a right-hand triple of vectors (see Fig. 22), then (compare the multiplication table of quaternion units in Chapter 7) we have

$$\mathbf{i} \circ \mathbf{i} = \mathbf{i}^2 = \mathbf{0}, \quad \mathbf{j}^2 = \mathbf{0}, \quad \mathbf{k}^2 = \mathbf{0}, \quad \mathbf{i} \circ \mathbf{j} = -\mathbf{j} \circ \mathbf{i} = \mathbf{k},$$
$$\mathbf{j} \circ \mathbf{k} = -\mathbf{k} \circ \mathbf{j} = \mathbf{i}, \quad \mathbf{k} \circ \mathbf{i} = -\mathbf{i} \circ \mathbf{k} = \mathbf{j},$$

where $\mathbf{a} \circ \mathbf{b}$ is the vector product of vectors **a** and **b** so that, for example,

$$(\mathbf{i} \circ \mathbf{i}) \circ \mathbf{j}) = \mathbf{0} \circ \mathbf{j} = \mathbf{0},$$

while

$$\mathbf{i} \circ (\mathbf{i} \circ \mathbf{j}) = \mathbf{i} \circ \mathbf{k} = -\mathbf{j} \neq (\mathbf{i} \circ \mathbf{i}) \circ \mathbf{j}.$$

B. and C. Peirce were fully aware of the conditional nature of requirements such as associativity and commutativity of multiplication in an algebra. Multiplication in an algebra is *commutative*, i.e., $\mathbf{u} \cdot \mathbf{v} = \mathbf{v} \cdot \mathbf{u}$ for any two elements of the algebra, if and only if any two basis elements commute:

$$\mathbf{e}_i \cdot \mathbf{e}_j = \mathbf{e}_j \cdot \mathbf{e}_i \quad \text{or} \quad c_{ij}^t = c_{ji}^t \qquad \text{for all } i, j, t;$$

similarly, multiplication is *anticommutative*, i.e., $\mathbf{u} \cdot \mathbf{v} = -\mathbf{v} \cdot \mathbf{u}$ for all $\mathbf{u}$ and $\mathbf{v}$, if any two basis elements anticommute:

$$\mathbf{e}_i \cdot \mathbf{e}_j = -\mathbf{e}_j \cdot \mathbf{e}_i \quad \text{or} \quad c_{ij}^t = -c_{ji}^t, \quad \text{for all } i, j, t.$$

Finally, multiplication is *associative* if and only if for any three (not necessarily distinct) elements i, j and k we have

$$(\mathbf{e}_i \cdot \mathbf{e}_j) \cdot \mathbf{e}_k = \mathbf{e}_i \cdot (\mathbf{e}_j \cdot \mathbf{e}_k) \quad \text{or} \quad \sum_{t=1}^{n} c_{ij}^t c_{tk}^s = \sum_{t=1}^{n} c_{it}^s c_{jk}^t \qquad \text{for all } i, j, k \text{ and } s.$$

If $\mathbf{e}_1$ is the multiplicative identity of the algebra (it is frequently denoted by the number 1; we did not write it down in the representation of the element $\mathbf{u}$ in (5.7)), then, obviously,

$$c_{1i}^j = \begin{cases} 0, & \text{if } i \neq j, \\ 1, & \text{if } i = j, \end{cases}$$

or, as mathematicians write this last condition,[235] $c_{1i}^j = \delta_i^j$. It is to the Peirces that we owe our understanding of the fact that the system of (square) *matrices* introduced by Cayley can be viewed as a certain (associative!) system of hypercomplex numbers (or an algebra) of dimension n^2, where n is the order of the matrix, since the matrix $A = (x_{ij})$ (where $i, j = 1, 2, \ldots, n$; here x_{ij} is the number in the ith row and jth column of the table (matrix)), may be represented as the sum

$$x_{11}\mathbf{e}_{11} + x_{12}\mathbf{e}_{12} + \cdots + x_n\mathbf{e}_{nn}$$

of n^2 summands corresponding to the n^2 units

$$\mathbf{e}_{ij} = \begin{pmatrix} 0 & \vdots & 0 \\ \cdots\cdots & 1 & \cdots\cdots \\ 0 & \vdots & 0 \end{pmatrix} \begin{matrix} \\ i\text{th row} \\ \\ \end{matrix}$$

(jth column above the 1)

of our algebra: matrices one element of which is 1 while the others are zero. The structural equations of such an algebra are of the form

$$\mathbf{e}_{ij} \cdot \mathbf{e}_{kl} = \begin{cases} 0, & \text{if } j \neq k, \\ \mathbf{e}_{il}, & \text{if } j = k, \end{cases}$$

i.e., here $c_{(ij)(kl)}^{(pq)} = 1$ whenever $j = k$, $p = i$, $q = l$, and zero and in all other cases; such a choice of structural constants yields Cayley's rule of matrix multiplication.[236]

B. Peirce introduced a number of notions which even today play important roles in the general theory of algebras, such as the notion of a *nilpotent* element $\mathbf{n}$, defined as an element some power r of which is zero,[237]

$$\mathbf{n}^r = 0,$$

and the notion of an *idempotent* element $\mathbf{e}$, defined as an element whose square coincides with the element itself:[237]

$$\mathbf{e}^2 = \mathbf{e}.$$

(Obviously, if $\mathbf{e}$ is an idempotent element of an algebra, then the set of all elements of the form $x\mathbf{e}$, where x is a real number, can be viewed as a representation of the real numbers). B. Peirce used the apparatus which he had developed for the classification of (associative complex) algebras of small dimension. Weierstrass had earlier developed a general theory of algebras (hypercomplex systems); this was the subject of a course of lectures which Weierstrass delivered in 1861. But Weierstrass's results were published only in 1884, in the paper "On the Theory of Complex Magnitudes Generated by n Principal Units" ("Zur Theorie der aus n Haupteinheiten gebildeten komplexen Grössen," *Gött. Nachrichten*, 1884; this was republished in the second volume of Weierstrass's *Mathematical Works* (*Mathematische Werke*)).

Now suppose $\mathfrak{A} = \{\mathbf{u}, \mathbf{v}, \mathbf{w}, \ldots\}$ is an arbitrary associative algebra, in which the multiplication, denoted by a dot "$\cdot$", is not assumed to be commutative or anticommutative. We can use our multiplication "$\cdot$" to construct two new "multiplications" of elements of our algebra, namely the symmetric multiplication

$$\mathbf{u} * \mathbf{v} = \mathbf{u} \cdot \mathbf{v} + \mathbf{v} \cdot \mathbf{u} \tag{6.5}$$

which has the advantage of being commutative (it is clear that $\mathbf{u} * \mathbf{v} = \mathbf{v} * \mathbf{u}$), and the skew symmetric multiplication

$$\mathbf{u} \circ \mathbf{v} = \mathbf{u} \cdot \mathbf{v} - \mathbf{v} \cdot \mathbf{u} \tag{6.6}$$

which has the advantage of being anticommutative (obviously, $\mathbf{u} \circ \mathbf{v} = -\mathbf{v} \circ \mathbf{u}$). Unfortunately, these new multiplications of the elements of our algebra are not, in general, associative. The connections between the structural constants $\overset{*}{c}{}^t_{ij}$ and $\overset{\circ}{c}{}^t_{ij}$ of the multiplications "$*$" and "$\circ$" and the structural constants c^t_{ij} of the (original) multiplication "$\cdot$" are given by the respective formulas

$$\overset{*}{c}{}^t_{ij} = c^t_{ij} + c^t_{ji} \quad \text{(a)}, \qquad \overset{\circ}{c}{}^t_{ij} = c^t_{ij} - c^t_{ji} \quad \text{(b)}. \tag{6.7}$$

However, we can say that the definitions (5) and (6) do not destroy associativity completely: in algebras with the respective operations (5) and (6), we have certain identities which can be viewed as weakened forms of associativity.[238] Thus we always have[239]

$$(\mathbf{u}^2 * \mathbf{v}) * \mathbf{u} = \mathbf{u}^2 * (\mathbf{v} * \mathbf{u}), \quad \text{where } \mathbf{u}^2 = \mathbf{u} * \mathbf{u}; \tag{6.8}$$

and (also for all elements $\mathbf{u}$, $\mathbf{v}$ and $\mathbf{w}$ of the algebra $\mathfrak{A}$[239])

$$(\mathbf{u} \circ \mathbf{v}) \circ \mathbf{w} + (\mathbf{v} \circ \mathbf{w}) \circ \mathbf{u} + (\mathbf{w} \circ \mathbf{u}) \circ \mathbf{v} = \mathbf{0}. \tag{6.9}$$

The identity (6.8) can be called the *Jordan identity*, since an algebra with commutative multiplication "$*$" satisfying condition (6.8) was first considered by the well-known German physicist Pascual Jordan (cf. Note 224 above). The identity (6.9) is known as the *Jacobi identity* after one of the leading

German mathematicians of the nineteenth century, Carl Gustav Jacob Jacobi (1804–1851).[240]

Algebras (hypercomplex systems) with commutative multiplication satisfying the Jordan condition (6.8) are known as *Jordan algebras.* Algebras with anticommutative multiplication satisfying the Jacobi condition (6.9) are called *Lie algebras.* We have already mentioned the interest and attention which Jordan algebras attract today. But at present they are not nearly as important as *Lie algebras* and *Lie groups*, which constitute two of the central notions of mathematical science.[241] A concept whose importance for science in general is comparable to that of a Lie algebra is not that of a Jordan algebra but that of a Euclidean space.

In Chapter 5 we saw that Hamilton and Grassmann, the founders of the theory of vectors, introduced two types of products of vectors in a vector space $\mathfrak{B} = \{\mathbf{a}, \mathbf{b}, \mathbf{c}, \ldots; \mathbf{0}\}$, which we shall assume, for the sake of simplicity, to be 3-dimensional. The *scalar* product (according to Hamilton) or *inner* product (in Grassmann's terminology) assigns to every two vectors **a** and **b** the scalar (i.e., number) $\mathbf{a} \cdot \mathbf{b}$ or $(\mathbf{a}, \mathbf{b})$, and the *vector* (*outer*) product, assigns to these two vectors the vector $\mathbf{a} \times \mathbf{b}$ or $[\mathbf{a}, \mathbf{b}]$. These products have the following properties, which we shall write next to each other for easy comparison:

scalar (inner) product		vector (outer) product[242]
$(\mathbf{a}, \mathbf{b}) = (\mathbf{b}, \mathbf{a})$ (commutativity)		$[\mathbf{a}, \mathbf{b}] = -[\mathbf{b}, \mathbf{a}]$ (anticommutativity)
$(\lambda\mathbf{a}, \mathbf{b}) = \lambda(\mathbf{a}, \mathbf{b})$	(associativity with respect to multiplication of vectors by numbers)	$[\lambda\mathbf{a}, \mathbf{b}] = \lambda[\mathbf{a}, \mathbf{b}]$
$(\mathbf{a}_1 + \mathbf{a}_2, \mathbf{b}) = (\mathbf{a}_1, \mathbf{b}) + (\mathbf{a}_2, \mathbf{b})$	(distributivity)	$[\mathbf{a}_1 + \mathbf{a}_2, \mathbf{b}] = [\mathbf{a}_1, \mathbf{b}] + [\mathbf{a}_2, \mathbf{b}]$
		$[[\mathbf{a}, \mathbf{b}], \mathbf{c}] + [[\mathbf{b}, \mathbf{c}], \mathbf{a}] + [[\mathbf{c}, \mathbf{a}], \mathbf{b}] = \mathbf{0}$ (Jacobi identity).

The two affine operations of vector algebra defined on a vector space are addition of vectors and multiplication of vectors by numbers. A vector space with an additional operation of scalar multiplication with the three properties listed above is called a **Euclidean space**. A vector space with a vector product possessing the four properties listed above is called a **Lie algebra**. We note that there is just one Euclidean space of given dimension with a positive definite scalar product, (i.e., such that $\mathbf{a} \cdot \mathbf{a} = \mathbf{a}^2 \geqslant 0$ and $\mathbf{a}^2 = 0$ only if $\mathbf{a} = \mathbf{0}$), and that even the problem of listing Euclidean spaces of given dimension, without supplementary requirements as to nondegeneracy[243] or positive definiteness of the scalar product is very simple.[244] On the other hand, the problem of classifying Lie algebras is extremely difficult, and at present there are no approaches that promise its complete solution.[245]

If we compare the properties of scalar and vector products, then it seems that the Jacobi identity is superfluous; it is the most complicated of the relations above. It is not difficult, however, to clarify its origin. We require that the scalar and vector products be associative with respect to the multiplication of a vector by a number: $(\lambda\mathbf{a})\circ\mathbf{b} = \lambda(\mathbf{a}\circ\mathbf{b})$, where the little circle "$\circ$" denotes either of the two products. However, there is no true associativity here. Indeed, for scalar multiplication, where the product $\mathbf{a}\circ\mathbf{b}$ is an object of a different nature than the factors $\mathbf{a}$ and $\mathbf{b}$ (it is a number, not a vector), ordinary associativity is meaningless. In the case of vector multiplication, it is easy to write down the associativity condition, but there is no basis for hoping that it holds. Indeed, we are used to the fact that the commutative operation $\mathbf{a}\circ\mathbf{b}$ "must" also be associative, i.e., that here we usually have $(\mathbf{a}\circ\mathbf{b})\circ\mathbf{c} = (\mathbf{b}\circ\mathbf{c})\circ\mathbf{a} = (\mathbf{c}\circ\mathbf{a})\circ\mathbf{b}$ for any $\mathbf{a}$, $\mathbf{b}$, $\mathbf{c}$; but if we replace commutativity $\mathbf{a}\circ\mathbf{b} = \mathbf{b}\circ\mathbf{a}$ by anticommutativity $\mathbf{a}\circ\mathbf{b} + \mathbf{b}\circ\mathbf{a} = \mathbf{0}$, as we have done in the case of vector multiplication (in the last relation $\mathbf{0}$ is the zero element of our arithmetic, i.e., the zero vector), then it is natural to replace associativity by "antiassociativity" or the Jacobi identity, $(\mathbf{a}\circ\mathbf{b})\circ\mathbf{c} + (\mathbf{b}\circ\mathbf{c})\circ\mathbf{a} + (\mathbf{c}\circ\mathbf{a})\circ\mathbf{b} = \mathbf{0}$.

One of the crucial points of Lie's studies was the possibility of assigning to each continuous group a much simpler algebraic object—its Lie algebra. The relationship between these two objects can be clarified using the example of the group $\mathfrak{V}$ of (direct) rotations of space with a fixed center of rotation O. All such isometries are rotations about axes passing through O and are characterized by a straight line l (the axis of rotation) and an angle φ (the angle of rotation); this allows us to split the group $\mathfrak{V}$ into a family of one-parameter subgroups, each of which consists of rotations about a fixed axis l; it is clear that these rotations are characterized by a single parameter, namely the angle of rotation φ. Each element of the group $\mathfrak{V}$ close to the identity transformation is characterized by an axis l (which indicates to which of the above-mentioned one-parameter subgroups it belongs) and by a (small!) angle of rotation $\Delta\varphi$. If we agree to assume that the transformations under consideration are carried out during a fixed (but very small) interval of time Δt, then we can replace the (small) angle $\Delta\varphi$ by the "angular velocity" $\omega = \Delta\varphi/\Delta t$ of the corresponding rotation. If, as is usually done in mechanics, we lay off the angular velocity vector $\mathbf{l}$ ($|\mathbf{l}| = \omega$) along the axis l so that the rotation by the angle $\Delta\varphi$ observed from the tip of the vector $\mathbf{l}$ takes place in the positive direction, then we can assign to the group $\mathfrak{V}$ the set of "angular velocity vectors" with origin O [246] (a three-dimensional vector space).

Further, an important role in Lie's theory is played by the specific characteristics of noncommutativity of a continuous group. Consider two rotations δ and δ_1, both very close to the identity transformation, or "infinitesimal" rotations, and characterized by the angular velocity vectors $\mathbf{l}$ and $\mathbf{l}_1$. The difference between the transformations $\delta\delta_1$ and $\delta_1\delta$ is characterized by the transformation $\kappa = (\delta\delta_1)^{-1}(\delta_1\delta) = \delta_1^{-1}\delta^{-1}\delta_1\delta$, called the *commutator* of the transformations δ and δ_1 and denoted by the symbol $[\delta\delta_1]$ (cf. Note 38). To the transformation κ there corresponds its own angular velocity vector $\mathbf{k}$ which Lie denotes by $[\mathbf{l}\mathbf{l}_1]$. Thus to every two vectors $\mathbf{l}$ and $\mathbf{l}_1$ of our three-dimensional vector space we have assigned a third vector $[\mathbf{l}\mathbf{l}_1] = \mathbf{k}$. A simple computation shows that in the case under consideration the vector $[\mathbf{l}\mathbf{l}_1]$ is none other than the *vector product* of the vectors $\mathbf{l}$ and $\mathbf{l}_1$, i.e.,

$$\mathbf{k} \perp \mathbf{l}, \qquad \mathbf{k} \perp \mathbf{l}_1, \qquad |\mathbf{k}| = |\mathbf{l}||\mathbf{l}_1|\sin(\angle \mathbf{l}_1, \mathbf{l}_2),$$

and that the vectors $\mathbf{l}$, $\mathbf{l}_1$, and $\mathbf{k}$ constitute a "positive triple": seen from the tip of $\mathbf{k}$ the rotation (by the smallest possible angle) from the vector $\mathbf{l}$ to the vector $\mathbf{l}_1$ takes place in the positive direction. Thus Lie assigns to the group $\mathfrak{B}$ the three-dimensional vector space $\mathbf{V}$ in which, together with the principal operations in any vector space (addition of vectors and multiplication of a vector by a number), there is an operation of *vector multiplication* that assigns to each two vectors $\mathbf{a}$ and $\mathbf{b}$ a new vector $[\mathbf{a}, \mathbf{b}]$. This operation satisfies all the requirements listed in the right-hand column in the table that compares the properties of scalar and vector products (see page 101).

Lie's main result is the proof that *it is always possible to assign to a continuous group (Lie group) a corresponding Lie algebra.* Lie also worked out the converse construction that allows one to assign to each Lie algebra (a vector space $\mathbf{V}$ with a vector multiplication operation: $\forall \mathbf{a}, \mathbf{b} \in \mathbf{V} \exists! \mathbf{c} \in \mathbf{V} | \mathbf{c} = [\mathbf{a}, \mathbf{b}]$, satisfying all the requirements listed above) a specific Lie group[247] corresponding to this algebra (which Lie considered only locally, i.e., simply as a domain (of the continuous group) consisting of elements close to the identity (unit) transformation ε[248]). Thus in the case of the commutative group $\mathfrak{T}$ of translations of space, we arrive at the trivial Lie algebra: the three-dimensional vector space $\mathbf{V}_T$, where $[\mathbf{a}, \mathbf{b}] = \mathbf{0}$ for all $\mathbf{a}, \mathbf{b} \in \mathbf{V}_T$. In the case of the group $\mathfrak{B}$ of (direct) rotations of space, the "Lie product" (or vector product) in three-dimensional space $\mathbf{V}_d$ with basis $\{\mathbf{e}_1, \mathbf{e}_2, \mathbf{e}_3\}$ corresponding to three (infinitesimal) rotations about three perpendicular axes Ox, Oy, Oz (see Fig. 23(a)) is given by the conditions

$$[\mathbf{e}_1, \mathbf{e}_2] = \mathbf{e}_3, \qquad [\mathbf{e}_2, \mathbf{e}_3] = \mathbf{e}_1, \qquad [\mathbf{e}_3, \mathbf{e}_1] = \mathbf{e}_2,$$

according to which for any two vectors $\mathbf{a} = (X, Y, Z)$ $(= X\mathbf{e}_1 + Y\mathbf{e}_2 + Z\mathbf{e}_3)$ and $\mathbf{b} = (X_1, Y_1, Z_1)$ of the space (Lie algebra) $\mathbf{V}_d$ we have (compare with formulas (5.4b) and (5.5)):

$$[\mathbf{a}, \mathbf{b}] = \left(\begin{vmatrix} Y & Z \\ Y_1 & Z_1 \end{vmatrix}, \begin{vmatrix} Z & X \\ Z_1 & X_1 \end{vmatrix}, \begin{vmatrix} X & Y \\ X_1 & Y_1 \end{vmatrix} \right) \left(= \begin{vmatrix} Y & Z \\ Y_1 & Z_1 \end{vmatrix} \mathbf{e}_1 + \begin{vmatrix} Z & X \\ Z_1 & X_1 \end{vmatrix} \mathbf{e}_2 + \begin{vmatrix} X & Y \\ X_1 & Y_1 \end{vmatrix} \mathbf{e}_3 \right), \tag{6.10a}$$

i.e., $[\mathbf{a}, \mathbf{b}]$ is the ordinary vector product! For the group $\mathfrak{J}$ of plane (direct) isometries (1) with basis $\{\mathbf{e}_1, \mathbf{e}_2, \mathbf{e}_3\}$ the corresponding Lie algebra $\mathbf{V}_e$ is generated by the (infinitesimal) translations in the directions Ox and Oy and the rotations about the origin O (see Fig. 23(b)); in this case it is easy to check that the Lie multiplication is given by the relation $[\mathbf{e}_1, \mathbf{e}_2] = 0$ (this relation is a trivial consequence of the commutativity of translations in the directions of Ox and Oy), $[\mathbf{e}_2, \mathbf{e}_3] = \mathbf{e}_1$, $[\mathbf{e}_1, \mathbf{e}_3] = -\mathbf{e}_2$, so that for arbitrary vectors $\mathbf{a} = (X, Y, Z)$ $(= X\mathbf{e}_1 + Y\mathbf{e}_2 + Z\mathbf{e}_3)$ and $\mathbf{b} = (X_1, Y_1, Z_1)$ we have

$$[\mathbf{a}, \mathbf{b}] = \left(\begin{vmatrix} Y & Z \\ Y_1 & Z_1 \end{vmatrix}, \begin{vmatrix} Z & X \\ Z_1 & X_1 \end{vmatrix}, 0 \right) \left(= \begin{vmatrix} Y & Z \\ Y_1 & Z_1 \end{vmatrix} \mathbf{e}_1 + \begin{vmatrix} Z & X \\ Z_1 & X_1 \end{vmatrix} \mathbf{e}_2 \right). \tag{6.10b}$$

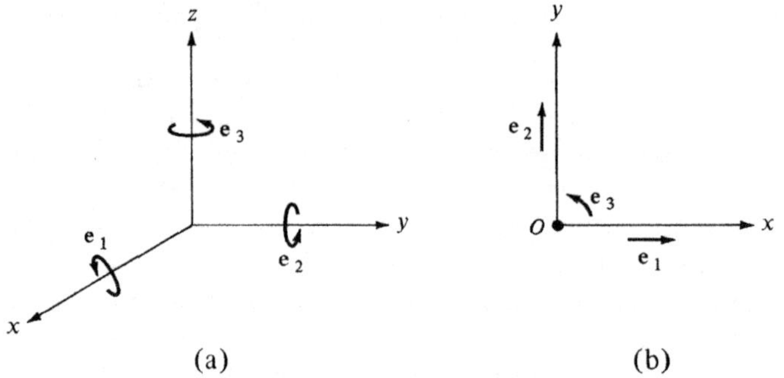

FIGURE 23

The obvious similarity between the formulas (6.10a) and (6.10b) is suggestive. Clearly, if we consider a sphere with center at the fixed point O of the rotations under consideration (in three-dimensional Euclidean space where the transformation group $\mathfrak{B}$ acts), then we can interpret the group $\mathfrak{B}$ as the group of *spherical isometries* or the group of *isometries of Riemann's elliptic plane* (see Chapter 4). On the other hand, if we denote by $\mathbf{e}_1$ and $\mathbf{e}_2$ the vectors corresponding to the infinitesimal translations of the hyperbolic (Lobachevskian) plane in the mutually perpendicular directions of the axes Ox and Oy (these translations no longer commute) and by $\mathbf{e}_3$ the vector corresponding to the small rotation of the hyperbolic plane about the origin of coordinates O, then we come to the following system of relations, which govern the commutativity laws of the non-Euclidean isometries under consideration:

$$[\mathbf{e}_1, \mathbf{e}_2] = -\mathbf{e}_3, \qquad [\mathbf{e}_2, \mathbf{e}_3] = \mathbf{e}_1, \qquad [\mathbf{e}_1, \mathbf{e}_3] = -\mathbf{e}_2.$$

Thus in the Lie algebra $\mathbf{V}_H$ corresponding to the group $\mathfrak{H}$ of hyperbolic isometries, the "vector product" of two vectors $\mathbf{a} = (X, Y, Z)(= X\mathbf{e}_1 + Y\mathbf{e}_2 + Z\mathbf{e}_3)$ and $\mathbf{b} = (X_1, Y_1, Z_1)$ is defined as follows

$$[\mathbf{a}, \mathbf{b}] = \left(\begin{vmatrix} Y & Z \\ Y_1 & Z_1 \end{vmatrix}, \begin{vmatrix} Z & X \\ Z_1 & X_1 \end{vmatrix}, -\begin{vmatrix} X & Y \\ X_1 & Y_1 \end{vmatrix}\right)\left(= \begin{vmatrix} Y & Z \\ Y_1 & Z_1 \end{vmatrix}\mathbf{e}_1 + \begin{vmatrix} Z & X \\ Z_1 & X_1 \end{vmatrix}\mathbf{e}_2 - \begin{vmatrix} X & Y \\ X_1 & Y_1 \end{vmatrix}\mathbf{e}_3\right). \tag{6.10c}$$

The marked similarity among formulas (10a–c) determines the close relationship between the three types of plane geometry of constant curvature (in Riemann's terminology; see Chapter 4), i.e., between Euclidean, elliptic, and hyperbolic geometries.

Lie developed the theory of Lie groups and algebras with rare completeness and thoroughness. He devoted to it a series of books and many articles. Sophus Lie exemplifies that rare figure in the history of science, a scientist with only one love. All his colossal research and its products were devoted to the development of a single topic, the theory of continuous groups. Lie analyzed in detail the relationship between Lie groups and Lie algebras, a connection which enabled him to carry over to the theory of Lie algebras all the specifically group-theoretic notions. Thus to a subgroup $\mathfrak{g}$ of a Lie group $\mathfrak{G}$ there corresponds a subalgebra $\mathbf{v}$ of the algebra $\mathbf{V}$, i.e., a set of vectors in $\mathbf{V}$ closed with respect to the operations of addition of vectors and multiplication of vectors by numbers (i.e., a subspace of the vector space), as well as with respect to vector multiplication, i.e. such that $\mathbf{a}, \mathbf{b} \in \mathbf{v} \Rightarrow [\mathbf{a}, \mathbf{b}] \in \mathbf{v}$. If the subgroup $\mathfrak{g}$ is a *normal* subgroup, i.e., $g^{-1}\mathfrak{g}g = \mathfrak{g}$ for all $g \in \mathfrak{G}$, then the subalgebra $\mathbf{v}$ is an *ideal*, i.e., for $\mathbf{a} \in \mathbf{v}$ and any $\mathbf{b} \in \mathbf{V}$ we have $[\mathbf{a}, \mathbf{b}] \in \mathbf{v}$. If the group $\mathfrak{G}$ is *simple*, i.e., it has no nontrivial normal subgroups, normal subgroups other than the group $\mathfrak{G}$ itself and the identity subgroup $n = \{\varepsilon\}$, where ε is the identity element of the group $\mathfrak{G}$, then the algebra $\mathbf{V}$ is also *simple*, i.e., has no nontrivial ideals, ideals other than the algebra $\mathbf{V}$ itself and the zero ideal $\mathbf{o} = \{\mathbf{0}\}$, etc. Similarly, Lie defines *solvable Lie algebras* corresponding to so-called *solvable Lie groups*,[249] analogous to the discrete solvable substitution groups of Galois (see Note 49), *semisimple Lie algebras and groups*,[249] etc.

In addition to posing the problem of classifying all simple Lie algebras Lie also posed the problem of classifying all simple Lie groups; indeed the two classification problems are equivalent[250]).[251] In Lie's time it was generally acknowledged that this difficult problem was solved by his pupil and follower Wilhelm Killing (1847–1923).[252] Indeed, Killing's list of simple Lie groups was never in doubt. But modern requirements of mathematical rigor—requirements sharpened early in the 20th century in long discussions by such outstanding thinkers as David Hilbert and his pupil Hermann Weyl (whom we shall often mention again), and realized in the work of the Nicolas Bourbaki group—differ notably from the standards of rigor applied to proofs in the time of Lie and Klein. (We shall return to this point in connection with Lie and Klein themselves.) This being so, modern mathematicians tend to feel that the complicated constructions carried out by Killing in 1888–1890[253] contain serious and unfillable gaps and can only be viewed as heuristic justifications of the completeness of the list of simple Lie groups given by Killing or as approaches to the proof of the corresponding theorem. The complete solution of the classification problem for simple Lie groups is now attributed to E. Cartan,[254] who solved this problem in his famous dissertation (Thèse). Other, perhaps simpler and more meaningful proofs of the classification theorem for simple as well as semisimple (see Note 251) Lie groups[255] have been given by such outstanding twentieth-century mathematicians as the Dutchman Bartel Leendert van der Waerden (b. 1903), who lived in the Netherlands, then in Germany, then again in the Netherlands[256] and E.B. Dynkin (Moscow, then USA; born in 1924).[257]

The main result of the theory of simple Lie groups, known to Killing and therefore to his contemporary Lie, is not understood completely even today (see the discussion below of the "singular" Lie groups). It can be stated as follows. The complete list of simple groups turns out to be rather short. It consists of *four* large *series* of groups denoted (following Killing) by the symbols A_n, B_n, C_n ($n = 1, 2, 3, \ldots$) and D_n ($n = 3, 4, 5, \ldots$; the groups A_1, B_1, and C_1 coincide, B_2 coincides with C_2 and A_3 with D_3), plus *five* exceptional or *singular* groups (i.e., groups which do not fit into the series). The dimensions of the Lie algebras corresponding to the singular simple Lie groups are 14, 52, 78, 133, and 248 (i.e., the transformations of these groups depend on 14, 52, 78, 133, and 248 parameters respectively). These five singular groups also have standard notations (indicating specific relationships between them) but the present historical review is hardly the place for analyzing them in detail.

The principal series of simple Lie groups can be uniformly described as follows. The groups of the series B and D are simply the groups of direct isometries (i.e., isometries with a determinant Δ equal to $+1$ rather than -1) Euclidean spaces (i.e., vector spaces with the standard metric of $\|\mathbf{a}\| = \mathbf{a}^2 = x_1^2 + x_2^2 + \cdots + x_N^2$) of odd dimension $N = 2n + 1$ and of even dimension $N = 2n$, respectively. In order to determine the groups of the series A, we must pass to the *complex* almost-Euclidean space, where the expression "almost Euclidean" has the following sufficiently clear meaning. A complex vector space differs from a real vector space of the kind described in Chapter 5 only by the fact that in it we have the possibility of multiplying vectors by complex numbers; in the case of an N-dimensional space, this leads us to identify vectors (or points[258]) of the space with finite sequences $(x_1, x_2, \ldots, x_N)$ of N complex numbers—their coordinates. The introduction of the usual scalar product into the space yields the formula

$$\|\mathbf{a}\| = \mathbf{a}^2 = x_1^2 + x_2^2 + \cdots + x_N^2 \tag{A}$$

for the "norm" or the "squared length" of the vector $\mathbf{a}(x_1, x_2, \ldots, x_N)$ corresponding to the *scalar product*

$$(\mathbf{a}, \mathbf{b}) = x_1 y_1 + x_2 y_2 + \cdots + x_N y_N \tag{A'}$$

of the vectors $\mathbf{a}$ and $\mathbf{b} = \mathbf{b}(y_1, y_2, \ldots, y_N)$. However, since the numbers $x_1, x_2, \ldots, x_N$ are not necessarily real, the expression on the right-hand side of (A) may turn out to be nonpositive (or even imaginary), which would lead to unnecessary complications. Therefore a more usual version of a complex vector space is the one in which the scalar product $(\mathbf{a}, \mathbf{b})$ of the vectors $\mathbf{a}$ and $\mathbf{b}$ is given by the formula

$$(\mathbf{a}, \mathbf{b}) = x_1 \bar{y}_1 + x_2 \bar{y}_2 + \cdots + x_N \bar{y}_N. \tag{B'}$$

In such a space, we again have the (positive!) norm $\|\mathbf{a}\|$ and length $|\mathbf{a}|$ of a vector $a = a(x_1, x_2, \ldots, x_N)$:

$$\|\mathbf{a}\| = |\mathbf{a}|^2 = (\mathbf{a}, \mathbf{a}) = x_1 \bar{x}_1 + x_2 \bar{x}_2 + \cdots + x_N \bar{x}_N \; (= |x_1|^2 + |x_2|^2 + \cdots + |x_N|^2). \tag{B}$$

However, the scalar product (B′) is no longer symmetric, for $(\mathbf{b}, \mathbf{a}) = \overline{(\mathbf{a}, \mathbf{b})}$. (In all cases here the bar denotes passage to the conjugate complex number.) The complex vector spaces with the "metric" (B′) (or (B)) play a much more important role in mathematics than the spaces with the metric (A′) (or (A)); a space with the scalar product (A) is usually called *complex Euclidean*, while the "almost Euclidean" space with the scalar product (B′) is known as *Hermitian* (after the great French analyst Charles Hermite (1822–1901)). Now the simple Lie groups of the series A are just the groups of isometries (or, to use a more scientific term, the groups of automorphisms, i.e., of linear transformations preserving the norm (B) of a vector and the scalar product (B′)) of the complex Hermitian spaces.

It is now easy to characterize the simple groups of the series C. Killing and Cartan described them as the groups of isometries (automorphisms) of the real even-dimensional spaces with skew symmetric (and nondegenerate, in the sense of Note 243) scalar product (i.e., such that $(\mathbf{b}, \mathbf{a}) = -(\mathbf{a}, \mathbf{b})$); such a space is called *symplectic* and plays a notable role in modern mathematics and especially in mechanics (see, for example, the book by V.I. Arnold quoted in Note 208). However, today these simple groups are usually described in the same way as the groups of the series A (see, for example, the book by C. Chevalley, *Theory of Lie Groups*, quoted in Note 241), except that now one considers not complex but *quaternion* vector spaces (whose vectors may be described as finite sequences $(x_1, x_2, \ldots, x_N)$ of quaternions) while formulas (B′) and (B) for the scalar product and the norm (as well as the description of the transformation groups as the groups of isometries of the Hermitian quaternion spaces) remain valid.

Thus the simple Lie groups of the "principal" classes A, B, C, and D can be uniformly described as the groups of isometries of real (Euclidean), complex, and quaternion (Hermitian) spaces. But where do the five singular simple Lie groups come from? According to the Dutchman Hans Freudenthal (born in 1905) and the Belgian Jacques Tits (born in 1932) these groups are related to the isometries of *octave* planes. Here the absence of associativity in the algebra of octaves does not allow us to construct a space of arbitrary dimension, so that infinite series of corresponding groups cannot appear. However, the existence both of a "Euclidean" (or "Hermitian") octave plane, as well as a non-Euclidean plane (elliptic or hyperbolic), determines certain simple singular groups.[259] The details of the geometric construction of singular simple Lie groups still remain somewhat mysterious today. The same is true of the question of their connection with octaves and octave geometries. We note, however, that the interest of contemporary physicists in singular simple Lie groups increases their interest (already mentioned in Chapter 5) in octaves and octave geometries. All these factors have created the "octave boom" so characteristic of the present time—a boom not anticipated by as penetrating an analyst of mathematics as Nicolas Bourbaki.[260]

Sophus Lie not only considered continuous groups *per se*, but went on to assign such groups to differential equations. In his variant of Galois theory for differential equations (see Chapter 1), the role of the Galois group (or symmetry group) of a differential equation is played not by a finite but by a continuous group, whose properties enable one to ascertain the existence of solutions of the equation in quadratures (i.e., to write down the solution in terms of elementary functions and integrals). It turned out that *those and only those equations which correspond to solvable continuous groups have solutions in quadratures* (recall the main results of Galois described in Chapter 1). The beautiful Lie theory of differential equations was very highly valued by its creator. For a time it was extremely popular and its exposition was the high point of many large university courses of mathematical analysis.[261] Our generation of mathematicians, however, was to live through several periods in the development of the mathematical sciences, characterized, in particular, by completely different estimates of the significance of Lie's theory of differential equations. Already in the 1930s the main interest of mathematicians moved far away from Lie's constructions, which seemed old-fashioned to the younger researchers and outside the truly interesting problems of the day. Key notions such as those of a Lie algebra and a Lie group retained their importance, but the study of Lie groups of differential equations no longer elicited much enthusiasm. In the 1940s the first computers appeared and the problem of solving differential equations had to be re-evaluated.[262] In the 1960s the search for such solutions was systematically transferred to computers and, in this connection, the notion of solvability of a differential equation in quadratures lost its previous importance and along with it the question of finding out whether a differential equation is so solvable or not; solvability in quadratures and most of Lie's theory were almost completely forgotten. But in the 1970s physicists and after them, of course, mathematicians, suddenly remembered that "the Lie group of a differential equation" not only characterizes the solvability or unsolvability of the given equation in quadratures, but also describes the *symmetries* of this equation (in this connection recall what was mentioned in Chapter 1 about the Galois theory of algebraic equations) and therefore the degree of symmetry (or, what amounts to the same thing, the "character of invariance"; see Chapter 7) of the solutions of this equation. It therefore also describes the symmetries of the real objects described (modelled) by the differential equation. Finding the symmetries (which are intrinsic to the given object) turned out to be the Ariadne thread which not only enables one to find one's way in the hopelessly complex labyrinth of elementary particles but, more generally, in the labyrinth of natural phenomena dealt with by physicists. And so today we observe a new surge of interest in Lie theory, expressed in the huge number of publications on it and the rather surprising number of relevant dissertations.[263]

One other topic in the area of continuous groups which drew Lie's attention was the theory of so-called *contact transformations*. By a contact transformation, Lie meant a transformation κ (say of the plane) which sends each point (or straight line) into, in general, a curve; a curve γ is sent by the transformation κ into a new curve $\gamma' = \kappa(\gamma)$, but κ preserves the *tangency of curves*: i.e., if γ_1 is a curve which is tangent to γ and $\gamma_1' = \kappa(\gamma_1)$, then the curves γ' and γ_1' must also be tangent (see Fig. 24). Contact transformations of the plane can be described as transformations of the set of tangent elements (points with straight lines passing through them): all curves tangent to each other at a point A and having the same (tangent) direction a determine a tangent element $\Lambda = (A, a)$. The transformation κ sends these curves into curves (also tangent to each other) which determine the tangent element $\Lambda' = (A', a') = \kappa(\Lambda)$ (see the same Fig. 24). However, in order that a transformation π of the set of contact elements be a tangent transformation, it must preserve the so-called "contact condition" of tangent elements (for otherwise π could send the set of

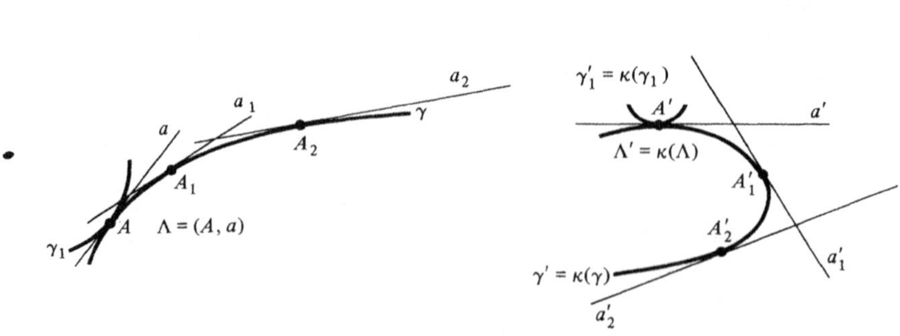

FIGURE 24

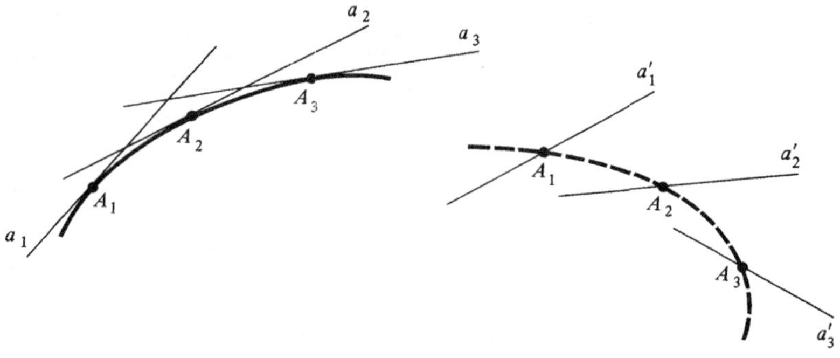

FIGURE 25

tangent elements Λ to a given curve into a set of tangent elements ($\Lambda' = \pi(\Lambda)$) which no longer determine any curve; see Fig. 25).

An important particular case of Lie's contact transformations are the *circular contact transformations* (also discovered by him), which send every circle (including circles of zero radius, i.e. points, and circles of infinite radius, i.e. straight lines) again into a circle (but both points and straight lines transform, in general, into circles). These transformations can be described as follows. Above (see Chapter 3) we have already mentioned "pointwise inversions" (or "Möbius inversions") which send a point into a point and a circle (including a "circle of infinite radius," i.e., a straight line) into a circle (see Chapter 3). Considerations related to the projective duality principle (see Chapter 3) led the French mathematician Edmond Nicolas Laguerre (1834–1886),[264] to the concept of "dual" or "linear" circular transformations, now called *Laguerre transformations*, which send every (directed) line a of the plane into a line but send points (as well as arbitrary circles) into circles. These transformations are generated by the so-called "linear inversions" (or Laguerre inversions) which may be described as follows: an inversion with axis o and degree k sends every line a into the line a' intersecting o at the same point as a and such that $\tan\frac{1}{2}(\angle(a, o)\cdot\tan\frac{1}{2}(\angle(a', o)) = k$. This definition must also be supplemented by a description of the transformation law for those lines which do not intersect o. If we carry out successively a series of Möbius circular transformations and Laguerre transformations we obtain a circular Lie transformation, which no longer preserves either the notion of point or the notion of straight line.[265] Lie's theory of contact transformations turned out to be intimately related to mechanics; it retains its importance today.[266]

In a certain sense, we can say that the entire scientific activity of Sophus Lie and Felix Klein was inspired by the first (and rather special) joint work which they carried out in Berlin before their trip to Paris. This work involved the so-called *W-curves*, whose name they coined.[267]

It is well known that the only homogeneous curves of plane Euclidean geometry, i.e., curves no point of which differs from any other, are either *straight lines* or *circles*. The homogeneity of these curves is related to the

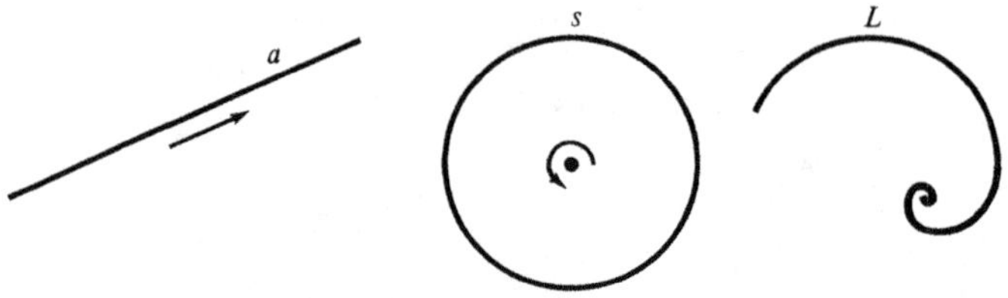

(a) (b) (c)

FIGURE 26

existence of the "self-isometry group" of a curve—the set of isometries which send the curve into itself and each of its points into another of its points. In the case of a straight line a this self-isometry group (the group of "motions along itself" or "symmetries") is the group of translations in the direction of a; for a circle s, it is the group of rotations about the center of s (see Figs. 26(a) and (b)). However, there exists one more plane curve which has almost the same complete degree of homogeneity as the straight line and the circle. This is the *logarithmic spiral* L, whose equation in polar coordinates r, φ is $r = a^{\varphi}$ (Fig. 26(c)). The point here is that L has "similarity motions" along itself, i.e., similarity transformations sending L into itself and every point of L into another one of its points. These transformations can be written in polar coordinates as $r' = a^c r$ and $\varphi' = \varphi + c$; they send the point $M(r, \varphi)$ into the point $M'(r', \varphi')$ and the spiral L into itself.[267] Lie and Klein posed the problem of finding each curve W of the plane that has a "complete group of projective motions along itself" (projective transformations sending W into itself; the adjective "complete" means that there is always a transformation of the group which sends a given point of W into any other one of its points). Klein and Lie called such curves W-curves.[268]

The work on W-curves turned out to be important for Lie's further research because in it he had been studying one-parameter subgroups of a given (continuous) group (the group of projective transformations), which were to play such an important role in the general construction of Lie algebras. It was also important that the authors used the notion of infinitesimal transformation in their search for W-curves. Their paper exhibits key features of Klein's later work, namely the group-theoretic approach (based, in this case, on the study of a group of projective transformations) to an object of projective geometry (such as W-curves) and the definition of W-curves themselves in terms of the group of projective symmetries of the curves. Here the topic itself was the connection between a geometry (in this case projective geometry) and its group of symmetries. This turned out to be so important for Klein, as well as for the progress of geometry in the nineteenth century—even for those phases of the development of geometry which were anterior to Klein and which we discussed in Chapters 3–5—that the entire topic warrants a special chapter.

CHAPTER 7

Felix Klein and His Erlangen Program

The nineteenth century was a period of intensive development of geometric research, and whereas at the beginning of the century the uniqueness of Euclidean geometry was universally taken for granted, so that "geometry" was identified with the notion of Euclidean geometry, by the time of Klein and Lie the situation had changed radically. The 1820s and the 1830s brought the first publications of Lobachevsky and Bolyai on hyperbolic geometry. At the end of the 1860s, Riemann's famous lecture, delivered in 1854, was finally published. It postulated, among other things, the equal validity of the three geometries of constant curvature: Euclidean, hyperbolic, and elliptic geometries. Beginning with Poncelet's treatise, the study of projective geometry became an autonomous topic, whose complete independence from Euclidean geometry was established by von Staudt. We can say that Möbius discovered inversive or circle geometry (or Möbius geometry; see Note 265). Finally in the works of Cayley, and especially Klein, the idea of general projective metrics was stated, covering classical Euclidean geometry as well as the non-Euclidean geometries of Lobachevsky (hyperbolic) and Riemann (elliptic).

This turbulent expansion of geometry, and the growing mathematical territory under its control, made the question of finding a general description of all the geometric systems[269] considered by mathematicians the central question of the day. And no one understood the importance of this problem better than Klein who had actively participated in enlarging the list. The influence of Jordan, who taught Klein the importance of the notion of group and the concept of symmetry, played an important role in the attempt to find a group-theoretic approach to the notion of geometry itself. This is how Klein argued.

The content of any science can be specified by naming the objects, and the properties of those objects, that the science studies. The properties studied by a specific science are always only a few of the many properties possessed by real objects. For example, the physicist is interested in the so-called *physical properties* of bodies, such as their mass, the forces applied to these bodies, and the velocities and accelerations of their motion, and he is not concerned with the inner structure of the body and the elements of which the body is composed; the latter relate rather to the interests of the chemist. Likewise, the natural numbers, say, first arose as characteristics of arbitrary but finite systems of objects. However, mathematics is only interested in one property of such systems—the *number* of objects within the system. Arithmetic arose out of the preoccupation with this characteristic of systems of objects and the refusal to consider all the data concerning such systems that are not related to this characteristic.

This last example is convenient in the sense that it allows us to understand the character of the conditions which single out the family of properties of interest to the given scientific discipline. In order to understand what properties we are interested in, it suffices to indicate what properties do not interest us, what properties we ignore. In the case of natural numbers, such properties are all those which do not relate to the number of objects in the system under consideration and are not related to the character of this number. Therefore, for example, the possibility of splitting the given system into two parts with the same number of elements does interest the mathematician, while the possibility of splitting it into two parts of equal weight is of no concern to him. In other words, from our viewpoint, all sets containing the same number of objects, say the set of eleven soccer players on a team, of eleven cars in a parking lot, of eleven geese in the backyard, or of eleven grades in a student's grade book must be considered identical (indistinguishable or equal). All the properties which interest us in any of these systems are also possessed by any of the others. These systems are identical or equal only from the point of view considered here; in all other respects, they differ essentially from each other (say, a soccer fan will distinguish two teams of eleven soccer players and will be amused if he is asked to replace the eleven soccer players by an equal number of geese). But all this does not concern the mathematician, who studies each system of objects only from the purely arithmetical point of view.

Now it is easy for us to clarify the content of geometry. In order to characterize Euclidean or school geometry, one must indicate the object of study and the family of properties under consideration. The objects of concern here are all possible plane figures and solids; but instead of the terms "plane" or "space figure," we can also speak of point sets in the plane and in space.[270] The properties of figures considered in geometry are entirely specified by an indication of what figures we consider identical (having the same properties), or indistinguishable, or *equal*. It is well known that in school geometry two figures are considered *equal* (or *congruent*) *if there exists an isometry* (a transformation of the plane or of space preserving the distance between points) *sending one figure into the other*. Thus we can say that geometry studies those and only those properties of the figure F which are shared by F and all the figures which are equal to F or (this last formulation will be especially useful for us) that geometry studies the *properties of figures which are preserved by isometries*.

Klein once noted that in most of the problems and theorems of elementary Euclidean geometry we identify not only congruent but also similar figures.

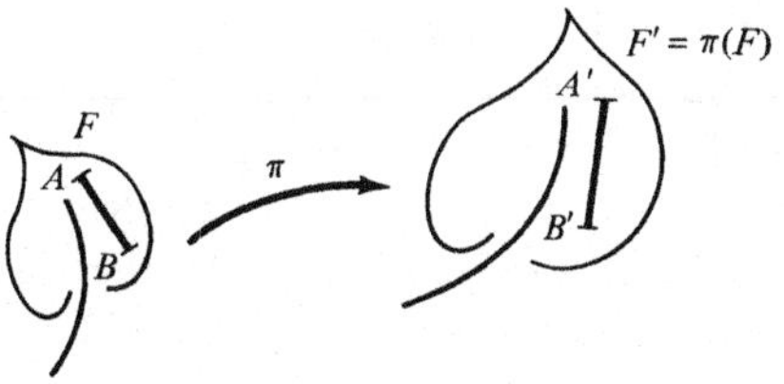

FIGURE 27

Two figures F and F' are said to be *similar* (with ratio k) if they differ only in size: the shape of the figure F' does not differ from the shape of the figure F but the measurements of F' are k times larger (when $k < 1$, k times smaller) than the corresponding measurements of the figure F. It is possible to establish a one-to-one correspondence π between F and F' such that if A and B are points of F and $A'(=\pi(A))$ and $B'(=\pi(B))$ are the corresponding points of F', then $A'B' = kAB$ (Fig. 27). An equivalent formulation is the following: F' is similar to $F \Leftrightarrow F' = \pi(F)$, where π is a similarity transformation, or similitude, i.e., the figure F' is obtained from F by a similarity transformation π, defined as a transformation of the plane or space which changes all the distances by constant factor k or, equivalently, a transformation that preserves the ratio of the lengths of segments (the ratio of distances between points).

The important role of similitudes in geometry is related to the fact that the sizes of figures (determined by comparing the distances between points of the figure with a chosen unit of length, say the meter, centimeter, or inch) cannot be taken into consideration in geometric theorems. Indeed, the choice of the unit of length necessarily appeals to considerations that have nothing to do with mathematics[271] and thus the notion of the length of a segment, i.e., the number indicating how many times a segment of unit length "fits into" a given segment goes beyond geometry. On the other hand, the measure of an angle measured in degrees (or radians or, say, in fractions of a right angle) is relevant to geometry. The same is true of the ratio of segments which, in geometry, has an objective meaning (i.e., is not related to the choice of the unit of length) and therefore can certainly appear in geometric theorems.[272] Clearly, similitudes which change the size of segments but preserve both the ratio of lengths and the size of angles, cannot change any "geometric" properties of figures; *similar* figures (which have the same shape but possibly different sizes) are clearly identical or *equal* for the geometer. It is this impossibility of distinguishing between similar figures which is implicit in the request of a teacher who asks his students to "exactly reproduce" in their notebooks a picture which he has drawn on the blackboard, although this cannot be done without proportionally decreasing its size. However, in some cases in geometry we do have to deal with theorems in which the choice of a unit of length is assumed in advance; such is the case, for example, in theorems on the *measurement of areas*, since the very notion of area presupposes a given unit for measuring area (it is clear that if we do not distinguish between similar figures, then the notion of area loses its meaning), and in *construction problems*, where it is assumed in advance that certain segments are given and the length of these segments must be taken into consideration in the construction of the required figures (and therefore, in this case, of two similar but unequal figures only one can be the solution of the given problem[273]). This circumstance is also underscored by the fact that the school geometry course usually begins with theorems on the *conditions for the equality of triangles* (which, in different expositions of geometry, are either theorems or axioms). These conditions would be meaningless if we did not distinguish between similar figures.

Thus, in certain theorems and problems in geometry, we must begin with the convention that properties of figures are called "geometric" if they are not changed by isometries. In most cases it turns out to be natural to assume that the object of geometry is the study of properties of figures which are preserved under similarity transformations or similitudes. In other words, we can say that in school we do not study one subject, "geometry", but rather a somewhat involved combination of two different scientific disciplines that investigate, respectively, the properties of figures preserved under similitudes and the properties preserved under isometries. (We could call these "geometry of similitude" and "geometry of isometry"[274].) This circumstance is the basis of Klein's general approach. He proposed to fix a certain family $\mathfrak{G}$ of transformations, to study those properties of geometric figures which are preserved by these transformations, and to take this to be a definite branch of geometry, controlled, so to speak, by the family of transformations $\mathfrak{G}$.

For such a general definition of geometry, we must consider identical or equal any two figures which can be sent into one another by a transformation in the family $\mathfrak{G}$. But if this notion of equality of figures is to be meaningful it must satisfy three conditions, which are valid, without exception, for *all* equality relations—the equality of numbers, algebraic expressions, distances, angles, vectors, geometric figures; the equality of forces, velocities, electrical currents or magnetic fields, potentials, heat conduction, valencies, calories; the equality of talents, school achievements, courage, artistic or other qualities, successes, coordination or cleverness, etc.—and which determine the very possibility of using the word "equality". The three conditions are:

1. Each figure F is "equal" to itself (reflexivity);
2. If the figure F is "equal" to the figure F_1 then, conversely, F_1 is "equal" to F (symmetry);
3. If the figure F is "equal" to F_1 and F_1 is "equal" to F_2, then F is "equal" to F_2 (transitivity).

It is clear that in the case of an arbitrary family $\mathfrak{G}$ of transformations the notion of "equality" defined by $\mathfrak{G}$ will not, in general, have these properties. In order to guarantee their validity, it is natural to require that

1a. The family $\mathfrak{G}$ contains the identity transformation ε (which sends each figure F into itself (Fig. 28(a));

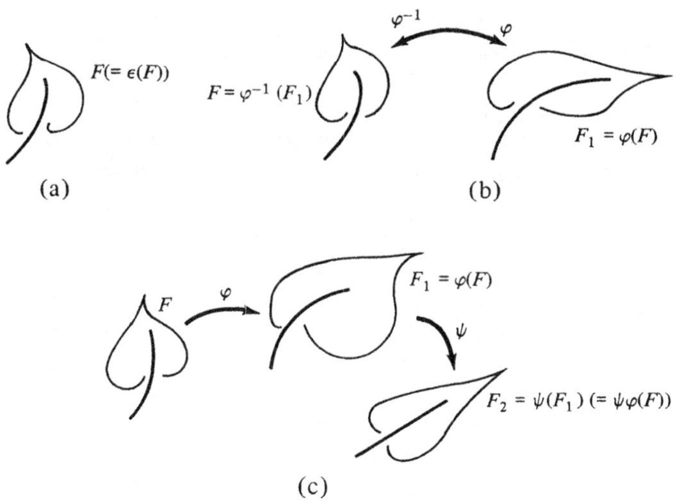

FIGURE 28

2a. Together with each transformation $\mathfrak{G}$ sending the figure F into the figure F_1, the family $\mathfrak{G}$ contains the "inverse" transformation φ^{-1} (which sends F_1 into F_2 (see Fig. 28(b));
3a. Together with any two transformations φ and ψ, of which φ sends the figure F into the figure F_1 and ψ sends the figure F_1 into F_2, the family $\mathfrak{G}$ contains their "product" $\psi\varphi$ (first φ, then ψ) (which sends F directly into F_2 (Fig. 28(c)).

But conditions 1a–3a are obviously none other than the requirements which define a *group* of transformations (see Chapter 1); for the "group operation" in a family of transformations we take the "multiplication" or composition of transformations. Thus we arrive at the following general definition of geometry, due to Felix Klein, which came to be known (we will tell about this below) as the "Erlangen program":

Geometry is the science which studies the properties of figures preserved under the transformations of a certain group of transformations, or, as one also says, *the science which studies the invariants of a group of transformations.*

The notion of group which appears in this definition can also be clarified as follows. The choice of the properties studied by a specific science reduces to the identification or "gluing" of all objects under consideration when all the properties which we consider for these objects coincide. These objects constitute an equivalence class of indistinguishable objects, and these classes are the real objects of study in the given science. Thus in arithmetic the number 4 may be understood as the common property of all sets of four arbitrary elements or, simply, as the set of all systems consisting of four elements.

Likewise the object called "triangle" in elementary geometry is really the class of all congruent triangles; for, with any other understanding of the term "triangle" a statement like "for two given segments a and b and the angle φ we can construct a unique triangle ABC such that $BC = a$, $AC = b$, $\angle C = \varphi$" loses its meaning. But it is well known that the decomposition of any set $\mathfrak{M} = \{\alpha, \beta, \gamma, \ldots\}$ of objects into arbitrary equivalence classes is really the same as the choice of an equivalence relation on $\mathfrak{M}$ having the properties 1a–3a of reflexivity, symmetry, and transitivity: $\alpha \sim \alpha$ for all $\alpha \in \mathfrak{M}$; $\alpha \sim \beta \Rightarrow \beta \sim \alpha$; $\alpha \sim \beta$ and $\beta \sim \gamma \Rightarrow \alpha \sim \gamma$—here the relation $\alpha \sim \alpha_1$ means that the objects α and α_1 belong to the same equivalence class.[275] In arithmetic, as we have already indicated, the relation $\sim$ links finite sets α and α_1 and means that it is possible to establish a one-to-one correspondence between α and α_1 (Fig. 29(a)). In geometry the relation $F \sim F_1$ means that there is a correspondence $\varphi\colon F \to F_1$ or $F_1 = \varphi(F)$, where φ belongs to the given family $\mathfrak{G}$ of transformations (Fig. 29(b)). In order that the relation "$\sim$" which we have introduced between figures be an equivalence relation, the family $\mathfrak{G}$ of transformations should be a group.

Thus, according to Klein, a geometry is determined by a "domain of action" $\mathscr{A}$ (the plane, space, etc.) and a "group of automorphisms" (or a symmetry

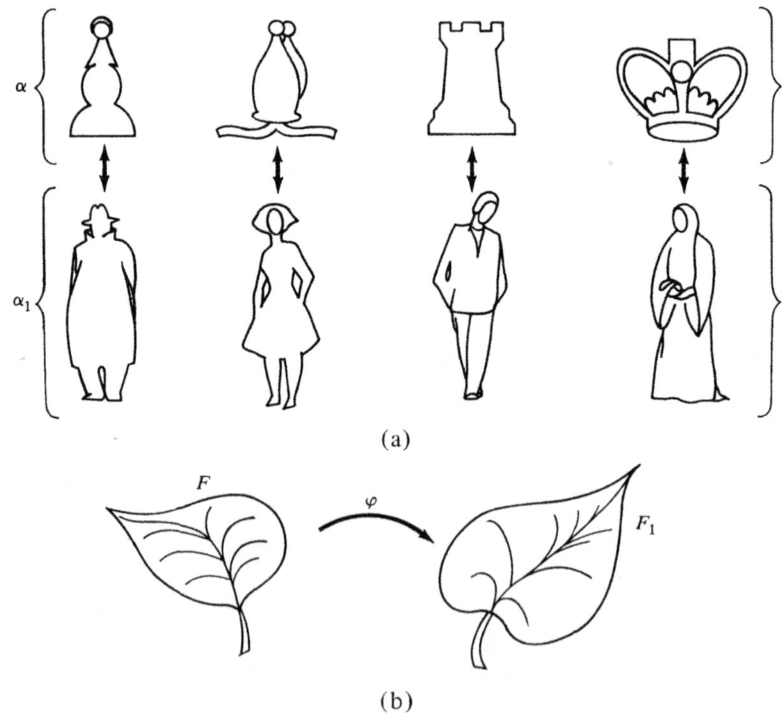

FIGURE 29

group) $\mathfrak{G}$ acting on the domain $\mathscr{A}$. When we change the group $\mathfrak{G}$ we change the geometric scheme under consideration, i.e., we obtain a new "geometry." Thus, for example, plane Euclidean geometry is determined by the group of isometries $\mathfrak{J}$ (or the group $\mathfrak{S}$ of similarity transformations) acting on the plane Π_0. Plane affine geometry is determined by choosing as the principal group of transformations of the plane $\mathscr{A}$ the group of affine transformations $\mathfrak{A}$, generated by all possible "parallel projections of Π_0 onto itself" (i.e., projections on the same plane Π_0, which we view as situated in different positions in space).[276] Plane projective geometry is determined by the group $\mathfrak{P}$ of projective transformations of the (projective) plane Π generated by all possible central (and parallel) projections of Π onto itself.[276] Lobachevsky's non-Euclidean geometry (hyperbolic geometry) is determined by the family $\mathfrak{L}$ of projective transformations of the plane Π which send a circle K (Fig. 13) into itself (more precisely, send the interior $\mathfrak{B}$ of the circle K into itself, so that the disc $\mathfrak{B}$ can be viewed as the "domain of action" of Lobachevskian geometry). Riemann's elliptic geometry in the Cayley–Klein model is determined by another subgroup of the group of projective transformations, and so on. Thus, according to Klein, the main difference between, say, Euclidean and hyperbolic geometry is not the possibility of constructing one or more lines passing through a point A and not intersecting a given line a (this is a secondary difference) but the difference in the structure of the respective groups of symmetries of Euclidean and hyperbolic geometry.

Finally we note that if two geometric systems Γ_1 and Γ_2 with the same "domain of action" $\mathscr{A}$ are determined by two groups of "isometries" $\mathfrak{G}_1$ and $\mathfrak{G}_2$, where $\mathfrak{G}_1 \supset \mathfrak{G}_2$ (i.e., the group $\mathfrak{G}_1$ is larger than the group $\mathfrak{G}_2$, in other words, $\mathfrak{G}_2$ is a subgroup of $\mathfrak{G}_1$), then the geometry Γ_2 is, in a certain sense, larger than the geometry Γ_1. Each notion that is meaningful in Γ_1 is also meaningful in Γ_2—for if this notion is not destroyed by transformations in $\mathfrak{G}_1$, then it will obviously be preserved by all the transformations in the group $\mathfrak{G}_2$, which constitute only a part of the transformations in $\mathfrak{G}_1$. Each theorem of Γ_1 dealing with some property of figures in $\mathscr{A}$ preserved by the transformations of the group $\mathfrak{G}_1$ may be viewed in the framework of Γ_2 (since the transformations from $\mathfrak{G}_2$ also preserve this property), etc. This relation between geometries can be conditionally written as $\Gamma_2 \supset \Gamma_1$. For example, if $\mathfrak{J}$, $\mathfrak{E}$ and $\mathfrak{A}$ are, respectively, the group of isometries, the group of similarities and the group of affine transformations of the plane, then $\mathfrak{A} \supset \mathfrak{E} \supset \mathfrak{J}$ (it is clear, for example, that the group (6.2) of affine transformations contains the group (6.1) of isometries, while the group of similitudes occupies an intermediate position). Therefore $\Gamma_d \supset \Gamma_s \supset \Gamma_a$, where the symbols Γ_d, Γ_s and Γ_a denote ordinary Euclidean geometry, similarity geometry and affine geometry.[277] Thus each notion of affine geometry (that of a parallelogram, a trapezoid etc.) and each "affine" theorem (such as the theorem that the medians of a triangle intersect in a single point which divides them in the ratio 2 : 1) remain valid in ordinary school geometry (as well as in similarity geometry) but, of course, the converse is not true. Each theorem of similarity geometry has an "ordinary" or "school" significance, i.e., it relates to the geometry of isometric transformations; whereas, say, the school theorem that the area of a triangle is equal to half the product of the length of one of its sides by the length of the corresponding altitude is meaningless in similarity geometry since it deals with notions which do not exist there.

Thus Klein's Erlangen program is usually understood to say that a geometry is determined by a certain domain $\mathscr{A}$ (the domain of action of the geometry) and a group of transformations $\mathfrak{G}$ acting on the domain $\mathscr{A}$. The object of the geometry is the study of those properties of the domain $\mathscr{A}$ which are preserved by the transformations in $\mathfrak{G}$. However, such a description of all possible geometries is not quite complete.

In order to clarify this it is sufficient to recall what we said at the end of Chapter 3 about Plücker's geometry of line elements. We indicated there that if the domain $\mathscr{A}$ is projective space (where $\mathfrak{G}$ is the group of projective transformations of the space $\mathscr{A}$) then, if we take the term "geometric point figure" in $\mathscr{A}$ to mean any point set belonging to $\mathscr{A}$, and "plane figure" in $\mathscr{A}$ to mean a set of planes, then we obtain the same geometric system. But this will not be the case if by a figure we mean a set of lines: then we obtain the "geometry of line elements" which considers figures consisting of straight lines in projective space.[278] Similarly, whereas in plane projective geometry (as noted in Chapter 3 in connection with Plücker's geometry of line elements) the geometry of points and the geometry of straight lines coincide apart from names (due to the fact that the role of points in plane line geometry is played by straight lines while the role of straight lines is played by points), in plane Euclidean geometry, where there is no duality principle, the geometry of points and geometry of lines are entirely different subjects. Thus, for example, one of the main properties of a line triangle consisting of three straight lines a, b, and c of the Euclidean plane is the fact that the sum of "the differences from a to b, b to c and c to a," i.e., the sum of the directed angles $\angle(a, b) + \angle(b, c) + \angle(c, a)$ is equal to 2π, i.e., $360°$ (Fig. 30(a)), whereas for a triangle, understood as a triple of points A, B, C, the sum of the distances $d_{AB} + d_{BC} + d_{CA}$ (i.e., the perimeter of the triangle—Fig. 30(b)) is different for different triangles. The set of lines m characterized by the same "deviation" from the line Q, i.e., by the same directed angle $\angle(q, m) = \rho$ is a sheaf of parallel lines (Fig. 31(a)) and does not resemble the set of points M such that $d_{QM} = r$, where Q is a given point and r is a fixed number, i.e, the circle with center Q and radius r (Fig. 31(b)), etc.

All these considerations bring us to the following more precise formulation of Klein's program. In order to determine any geometric system, we must indicate three traits

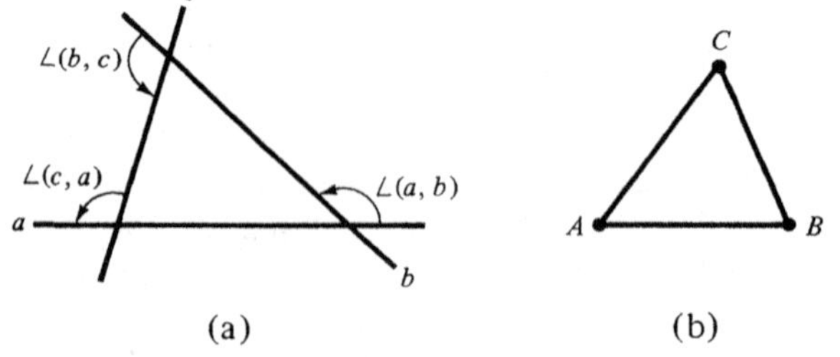

FIGURE 30

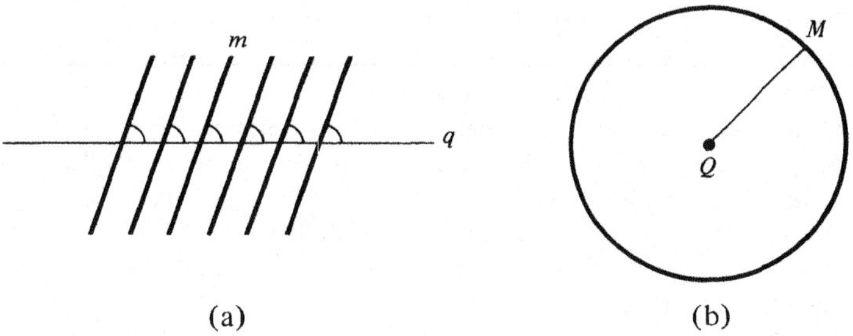

FIGURE 31

which characterize it: the geometric space or the domain of action $\mathscr{A}$ of the geometry; the group $\mathfrak{G}$ of "isometries," i.e., of all the transformations of $\mathscr{A}$ which preserve all the properties of the figures in $\mathscr{A}$ which interest us; and the generating element ξ, i.e., an "atom" or simplest element of the domain $\mathscr{A}$ which is the building block of all the figures F to be considered. Thus in the case of plane geometry the element ξ can be a point, an oriented or unoriented line, an oriented or unoriented circle of fixed or arbitrary radius, or perhaps even some more exotic geometric object, say, a parabola or a line element, i.e., a point with a fixed direction at this point. It is natural to build figures F from identical (i.e., equal) elements ξ. Therefore, it is sufficient to choose a single element ξ which is then "scattered" by means of transformations in the group $\mathfrak{G}$ over the entire domain $\mathscr{A}$. Thus if the group $\mathfrak{G}$ is the group of plane isometries $\mathfrak{J}$, then we can choose as ξ a point, a line, or a circle with fixed radius a, but *not* a circle of arbitrary radius, since two circles with different radii are not equal in the sense of the geometry of the group $\mathfrak{J}$. On the other hand, if $\mathfrak{G}$ is the group $\mathfrak{S}$ of similarity transformations, then we can choose as ξ a point, a line, or a circle of arbitrary radius, but *not* a circle of fixed radius, since if we "scatter" (by means of the group of similitudes) a circle of radius a over the entire plane, then we obtain circles with all possible (finite) radii.

In the language of group theory, the notion of a generating element of a given geometric system may also be described as follows. Consider all the transformations g in the group $\mathfrak{G}$ which preserve the geometric object ξ (i.e., which send it into itself). These transformations also form a certain group $\mathfrak{g}$ of transformations: the identity transformation e sends ξ into itself; if the transformation g sends ξ into itself then the inverse transformation g^{-1} sends ξ into itself; if the transformations g_1 and g_2 send ξ into itself, then their product $g_2 g_1$ sends ξ into itself. This group is only a part of the group $\mathfrak{G}$, i.e., a subgroup, called the *stabilizer subgroup* corresponding to the geometry with the given generating element. (If the subgroup $\mathfrak{g}$ consists of all transformations of the group $\mathfrak{G}$, i.e., if all the isometries of our geometry fix the generating element, then the domain $\mathscr{A}$ will not contain any generating elements different from the element ξ; it is clear that such a geometry will be without content.) Thus, for example, if the group $\mathfrak{G}$ consists of all possible isometries of the plane[279] and the element ξ is a point, then the subgroup $\mathfrak{g}$ consists of all rotations about ξ, while if the element η is a straight line η, then the subgroup $\mathfrak{g}$ consists of the translations in the direction of η and the reflections in the points of the line η.[279]

We note that if the generating elements ξ and η of two geometries with the same group of isometries are such that the corresponding stabilizer subgroups are the same, then the geometries under consideration are also the same. Consider, for example, two geometries whose respective domains of action $\mathscr{A}$ coincide with the plane, whose respective groups $\mathfrak{G}$ coincide with the group (6.1) of isometries of the plane, and whose respective generating elements ξ and η are a point and a circle with fixed radius a. It is clear that the subgroup $\mathfrak{g}$ of the group of isometries $\mathfrak{G}$ which fixes the point ξ coincides with the subgroup of the group $\mathfrak{G}$ which fixes the circle η with center ξ. This implies that the geometry with generating element η coincides with the usual "pointwise" geometry. In the geometry with generating element η we can define all the notions which exist in pointwise geometry. Clearly, in this geometry, the role of a line will be played by the strip between parallel lines filled with circles of radius a; the role of an "angle" with "vertex" η_0 will be played by two such "lines" with common circle η_0, and its measure will be the same as the measure of the "ordinary" angle between the midlines of the two corresponding strips; the distance between two circles η_1 and η_2 or, equivalently, their tangential distance (the length of their common tangent) will be the ordinary distance between their centers, etc. Then all the statements of this "geometry of circles" will coincide with the statements of ordinary Euclidean geometry. For example, here too the distance between the two most distant of three circles on a "line" is equal to the sum of the two other pairs of distance between circles. Again, the circle η_0 of radius a which does not belong to the "line" l belongs to a unique "parallel line" l_0 to l, i.e., a "line" which has no common circles with l, etc. This identity of the geometries with generating elements ξ and η follows from the possibility of mapping one of these geometries onto the other: it suffices to assign to each circle η of radius a its center ξ (or to each point ξ the circle η, of radius a and center ξ) and then each notion in one of these two geometries will be transformed into the corresponding notion in the other and each statement in one geometry into the analogous statement in the other.

The fact that geometries with the same groups of isometries $\mathfrak{G}$ and identical stabilizer subgroups coincide emerges from the following general construction. Consider the set of all transformations of the group $\mathfrak{G}$ as a certain "geometric" object which will eventually play the role of the domain of action $\mathscr{A}$ of the geometry having $\mathfrak{G}$ as its group of isometries. This space $\mathscr{A}$ is called the *group space* corresponding to the group $\mathfrak{G}$ (see Chapter 6). Thus if $\mathfrak{G}$ is the group (6.1) of direct isometries (denoted earlier by $\mathfrak{J}$), then $\mathscr{A}$ is the layer $0 \leqslant \alpha \leqslant 2\pi$ of three-dimensional space with coordinates (a, b, α), where the planes $\alpha = 0$ and $\alpha = 2\pi$ that bound the layer must be identified (glued), since the values $\alpha = 0$ and $\alpha = 2\pi$ in the formulas (6.1) correspond to the same transformation (a translation not followed by any rotation).

We assign to the generating element ξ of our geometry the subset of elements of the group $\mathfrak{G}$ which form the stability subgroup $\mathfrak{g}$. Now consider another generating element ξ_1 of our geometry. If g_1 is one of the transformations of the group $\mathfrak{G}$ sending ξ into ξ_1 (see Fig. 32(a), where the elements ξ and ξ_1 are shown as points), then all the transformations in $\mathfrak{G}$ sending ξ into ξ_1 form the set $g_1\mathfrak{g}$ of transformations; here by $g_1\mathfrak{g}$ we mean the family of all transformations $g_1 g$, where $g \in \mathfrak{g}$. Indeed, any transformations which can be represented in the form $g_1 g$ sends ξ into ξ_1 (since g sends ξ into itself, and g_1 sends ξ into ξ_1). On the other hand, if the transformation g' sends ξ into ξ_1, and g is a transformation in $\mathfrak{G}$ such that $g' = g_1 g$, then g fixes ξ, i.e., g belongs to the subgroup $\mathfrak{g}$. Indeed, $g_1 g\xi = \xi_1$ implies that $g\xi = g_1^{-1}\xi_1 = \xi$.

Thus to every generating element ξ_1 in our geometry (different from ξ) there

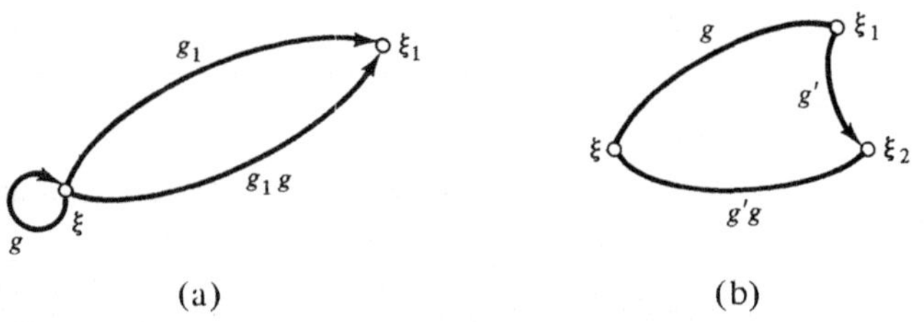

FIGURE 32

corresponds a set $g_1\mathfrak{g}$ of transformations in $\mathfrak{G}$. Each set

$$g_1\mathfrak{g} \equiv \{\text{the set of all } g_1 g, \text{ where } g \in \mathfrak{g}\}$$

of elements in $\mathfrak{G}$, as well as the subgroup $\mathfrak{g}$ itself, which can be written in the form $e\mathfrak{g}$, where e is the identity transformation, is called a *coset* of the group $\mathfrak{G}$ with respect to the subgroup $\mathfrak{g}$ (see Chapter 1, in particular Fig. 3).

We see that in the language of the group $\mathfrak{G}$ the set of generating elements ξ of the geometry under consideration can be described as the set of cosets $g_1\mathfrak{g}$ of the group $\mathfrak{G}$ with respect to the stabilizer subgroup $\mathfrak{g}$. Now suppose g' is an arbitrary transformation of the group $\mathfrak{G}$. Consider the two generating elements ξ_1 and ξ_2 of our geometry such that g' sends ξ_1 into ξ_2 (Fig. 32(b)) and the corresponding cosets $g_1\mathfrak{g}$ and $g_2\mathfrak{g}$. We claim that

$$g_2\mathfrak{g} = g'(g_1\mathfrak{g}) \quad (=(g'g_1)\mathfrak{g}).$$

In other words, if we multiply all the transformations of the coset $g_1\mathfrak{g}$ on the left by the transformation g' (thus forming all possible products $g'g$, where g is in the coset $g_1\mathfrak{g}$), then we obtain transformations belonging to the coset $g_2\mathfrak{g}$. Indeed, since the transformation g in $g_1\mathfrak{g}$ sends ξ into ξ_1, and the transformation g' sends ξ_1 into ξ_2, their product $g'g$ sends the generating elements ξ into the element ξ_2, i.e., $g'g$ belongs to the coset $g_2\mathfrak{g}$ corresponding to ξ_2. (Conversely, if the transformation $g_2' = g'g$ is in the coset $g_2\mathfrak{g}$, then g is in $g_1\mathfrak{g}$.) Clearly, the set $g'(g_1\mathfrak{g})$ of transformations obtained in this way coincides with the coset $g_2\mathfrak{g}$.

Finally, we come to the following geometric scheme, which gives an exhaustive description of the geometry under consideration. As the *domain of action* $\mathscr{A}$ of our geometry we take the set $\mathfrak{G}$ of all transformations. The role of the *generating elements* ξ is played by the cosets $g\mathfrak{g}$ of the group $\mathfrak{G}$ with respect to the subgroup $\mathfrak{g}$ (so that "geometric figures" are sets of cosets). The role of *isometries* in our geometry is played by the transformations in the group $\mathfrak{G}$, and a transformation g in this group sends each coset $g_1\mathfrak{g}$ into the coset $(gg_1)\mathfrak{g}$. Geometric figures (sets of cosets) Φ_1 and Φ_2 are considered *equal* if they are sent one into the other by some transformation g in the group $\mathfrak{G}$ (which acts on the set of cosets in the manner described above).[280] Since the geometry just described (more precisely, one of its models whose distinction is that it accomodates all the notions and statements pertaining to this geometry), depends only on the groups $\mathfrak{G}$ and $\mathfrak{g}$, it is clear that two geometries with identical groups $\mathfrak{G}$ and $\mathfrak{g}$ will be identical.

We suggest that the reader try to find out what this general scheme reduces to in the case when $\mathfrak{G}$ is the group $\mathfrak{J}$ (6.1) of direct isometries of the plane (and then $\mathscr{A}$ is the layer $0 \leqslant \alpha \leqslant 2\pi$ of three-dimensional space (a, b, α)), while ξ is a point and, respectively, an oriented[278] line, i.e., $\mathfrak{g}$ is the group of rotations

$$x' = x\cos\alpha + y\sin\alpha, \qquad y' = -x\sin\alpha + y\cos\alpha$$

about the origin O of the coordinate system, and, respectively, the group $x' = x + a$, $y' = y$ of translations along the x-axis.[281]

We conclude this chapter with the observation that Klein's definition of a geometry, which assigns a key role to the notion of a geometric transformation that preserves the properties of figures of interest to us (i.e., of a transformation that plays the role of an isometry of the corresponding geometry), has an interesting reflection in physics. Recall the so-called *Galilean principle of relativity*, which plays a fundamental role in mechanics and asserts that no physical experiments carried out within a mechanical system are capable of revealing uniform rectilinear motion of the system. According to this principle, if we carry out any experiments, say, on a ship which moves in a fixed direction with constant velocity, then we shall not be able to discover any effects which are due to the motion of the ship. It follows from Galileo's principle of relativity that all physical properties are preserved under transformations of the physical system which impart to the system a constant velocity (these transformations are called *Galilean transformations*). In other words, the *physical properties* of bodies can be described as those properties which remain unchanged under Galilean transformations—just as the geometric properties of figures in Euclidean geometry are those that are unchanged by isometries.

Galileo's principle of relativity may be stated in a geometric form which is clearly related to Klein's definition of a geometry. Let us suppose, for the sake of simplicity, that we limit our study to physical processes which take place in a plane—for example, to the motions of physical bodies in a limited region of the earth's surface (which may be viewed as flat). Let us introduce Cartesian coordinates (x, y) in the plane under consideration. Suppose the mechanical motion of a mass point is given by the formulas

$$\begin{cases} x = f(t), \\ y = g(t), \end{cases}$$

which show how the coordinates of the point in question change with time t. It is clear that transition to another coordinate system cannot influence physical laws, which must, therefore, have the same form in the coordinate system (x, y) and in the coordinate system (x', y') obtained from the (x, y)—system by an arbitrary rotation of the coordinate axes and a translation of the origin (Fig. 33). But the passage from the coordinates (x, y) to the coordinates (x', y') is given by the formulas (cf. (6.1))

$$\begin{cases} x' = x\cos\alpha - y\sin\alpha + a, \\ y' = x\sin\alpha + y\cos\alpha + b, \end{cases} \tag{7.1}$$

in which α denotes the angle formed by the x-axis with the x'-axis, while a, b

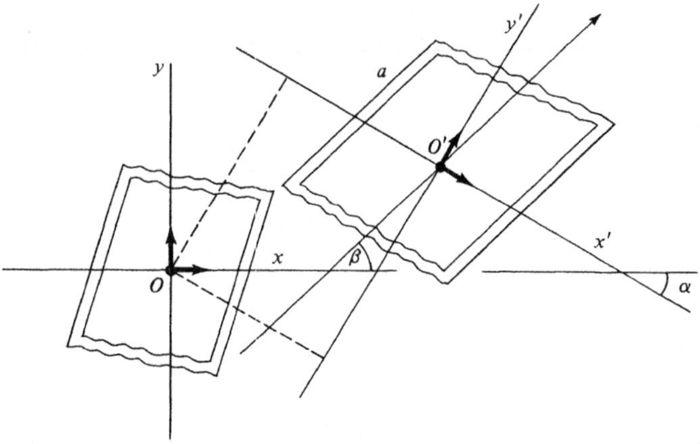

FIGURE 33

are the coordinates of the origin O of the coordinate system (x, y) in the system (x', y'). Hence any statement which has physical meaning must preserve its form under the transformations (7.1). Moreover, Galileo's principle of relativity asserts that even if the origin and the axes of the coordinate system (x', y') move uniformly and rectilinearly with respect to the coordinate system (x, y), the descriptions of all the physical processes will have the same form in both coordinate system. Now if the origin O' of the coordinate system (x', y') moves with velocity v along the line forming an angle β with the x-axis (see Fig. 33), then the connection between the coordinates (x', y') and (x, y) is given by the equations

$$\begin{cases} x' = x\cos\alpha - y\sin\alpha + \cos\beta\cdot vt + a, \\ y' = x\sin\alpha + y\cos\alpha + \sin\beta\cdot vt + b; \end{cases} \tag{7.2}$$

so that the descriptions of all the phenomena having physical meaning must preserve their form under the transformations (7.2). Also, since our formulas involve time t and the choice of the "time origin" must not influence the physical essence of any process, it follows that we can rewrite the formulas (7.2) in the following somewhat more complete form:

$$\begin{aligned} x' &= x\cos\alpha - y\sin\alpha + \cos\beta\cdot vt + a, \\ y' &= x\sin\alpha + y\cos\alpha + \sin\beta\cdot vt + b, \\ t' &= t + d, \end{aligned} \tag{7.3}$$

here d is the time of the old time origin in the new reference system. Formulas (7.3) give a mathematical description of the Galilean transformations. Galileo's principle of relativity states that physics (more exactly, mechanics) of plane motions can be defined as the science of the properties of three-dimensional spacetime (x, y, t) which are preserved under the transformations (7.3). Since the Galilean transformations (7.3) can be easily shown to form a group, this description identifies the mechanics of plane motions with a certain geometry of a three-dimensional space determined by the choice of the group (7.3) of isometries.

We should also point out that modern physics has replaced Galileo's principle of relativity by the so-called *Einstein principle of relativity* which is the basis of the special theory of relativity. It is therefore necessary to replace the Galilean transformations (7.3) by more complicated transformations, known as the *Lorentz transformations* (Hendrik Anton Lorentz (1853–1928) was an outstanding Dutch physicist). Lorentz transformations depend on a certain parameter c (whose physical meaning is clarified by relativity theory: c is the velocity of light in a vacuum). When $c \to \infty$, these formulas reduce to Galilean transformations. The Lorentz transformations also form a group. Thus when we pass from the classical mechanics of Galileo and Newton to the relativity theory of Einstein and Poincaré, we are actually changing our view of the geometry of the surrounding world and this geometry, in full agreement with Klein's point of view, is determined by prescribing the group of transformations which preserve the form of physical laws.[282]

Chapter 8

Biographical Sketches

We have previously described how Lie and Klein happened to meet in Berlin and travelled to Paris; that chance meeting was to play a crucial part in both their lives. Let us continue the life stories of these two outstanding scientists. We have little left to say about Sophus Lie: his biography was not rich in outwardly striking events. The research he carried out at the turn of the 1870s (in particular, the results obtained in the Fontainebleau prison) brought him wide fame, mostly through Klein's efforts. Klein had the highest regard for Lie and had extensive contacts in the mathematical world. He used them, and, as a result, Lie was offered a professorship at Norway's only university, in Christiania (now Oslo). The following year he taught at Lund in Sweder, but the passionate Norwegian patriot Lie felt out of place there and returned to Oslo.

In 1874, Lie, then 32 years old, married Anna Sophie Birch; the marriage was a happy one. (The younger Felix Klein had married earlier, also happily.) Lie worked in Norway for fourteen years. The university appealed to him by its proximity to the fiords he loved so much, providing opportunities for the long outdoor hikes he favored. But in scientific terms it was then an extremely provincial place, and Lie suffered from a lack of meaningful scientific exchanges and a shortage of competent pupils. Pupils were indeed needed to develop his ideas and solve the problems he made up—Lie could never complain of a shortage of ideas and problems! And so he eargerly accepted Klein's suggestion, made in 1886, that he replace Klein as professor of geometry at Leipzig University. In addition to a higher level of instruction and better students, Leipzig offered Lie the opportunity to supervise the printing of his books. (All of Lie's books were published by the prominent Leipzig mathematics publisher, B.C. Teubner.)

Lie worked in Leipzig for twelve years. These years were very productive in terms of scientific research, but they failed to bring him complete satisfaction. Tall and physically very strong, with an open face and loud laugh (people who knew Lie often said that he was their idea of a Viking), distinguished by rare candor and directness, always convivial with anyone who approached him, Lie produced an impression that did not correspond to his inner nature:

Sophus Lie

actually he was very refined and easily hurt. He was always in sore need of friends to support him, especially in the last Leipzig years, when his best pupils, such as Friedrich Engel (1861–1941), Georg Scheffers (1866–1945), Friedrich Schur (1856–1932), Eduard Study (1862–1930) and Felix Hausdorff (1868–1949), who did not study under Lie for long but whom Lie esteemed highly, matured and left their teacher for different German universities. In Norway, the nature he loved so much was a source of strength for Lie; in Germany he felt himself in large measure an alien. It is also possible that the depression Lie suffered at the end of his stay in Leipzig, and for which he had to take a cure at a psychiatric clinic in Hannover, was due to extreme exhaustion—there will be more about Lie's rare productivity as a mathematician below.

During this period of Lie's nervous and physical malaise there occurred the one unfortunate incident that marred his otherwise closeknit and friendly relationship with Klein: in Volume 3 of his joint book with Engel, *Theorie der*

Transformationsgruppen (1893), Lie pointed out with uncharacteristic bluntness that many people thought he was Klein's pupil, whereas in fact the opposite was true. That rather tactless remark, quite out of place in a purely scientific work, hurt Klein very much, perhaps just because Lie was not very far off the mark. Nevertheless the remark was uncalled for—there can be no doubt that the scientific influence of the friends was mutual. Klein, however, chose not to respond. In a short time Lie, apparently also suffering from his own tactless act—which had been committed in a state of depression—appeared at Klein's house again, and was of course welcomed as warmly as ever. Lie and Klein never returned to this episode, and fortunately their friendship did not suffer from it at all.[283]

In 1892–93 the Kazan physico-mathematical society solemnly celebrated N.I. Lobachevsky's centenary. The celebration included the creation of the International Lobachevsky prize and medal. The prize was awarded for the first time in 1898 and its first recipient was Sophus Lie. A detailed review of his work, requested by the Kazan physico-mathematical society, was written by Felix Klein. (The Lobachevsky prize immediately became quite prestigious. The second, third, and fourth recipients were W. Killing, D. Hilbert, and F. Klein. Later recipients include H. Poincaré, H. Weyl, E. Cartan, and, more recently, G. de Rham, H. Hopf, and H. Buseman.)

In 1898 Lie left Leipzig and returned to his alma mater at Christiania—but not, alas, for long. He still had time to enjoy the welcoming ceremonies held in honor of the man who had made little Norway famous in the scientific world, to breathe the North Sea air he loved so dearly and to enjoy the sounds of the Norwegian language spoken in the streets. But he had very little time left to live and work. He died in Oslo on February 18, 1899.

Although devoid of spectacular events, Sophus Lie's life was filled with intense creative work to the very end. All the work of this outstanding mathematician centered around one subject—the theory of continuous transformation groups—but what passion and capacity for work Lie displayed in developing the mathematical vein he had discovered!

Lie devoted many papers and a number of books to the theory of continuous groups. Most of his papers (and all of his books) were very long. His style was leisurely and polished. He carefully set down details and provided many examples. Today Lie's books and articles may seem archaic in some ways, and not always up to the standards of rigor achieved in the same questions by modern mathematicians.[284] His constructions may occasionally seem rather involved. Yet on the whole Lie—though a sportsman—did not have a competitive attitude in science. He greatly disliked overcoming difficulties for their own sake. He believed, quite sensibly, that any "natural" mathematical theory should be transparent, and he felt that difficulties in mathematics usually arise not from the essence of the problem but from badly conceived definitions at the base.

A committee for the publication of Lie's collected mathematical works was created in 1900, but it was hampered from the outset by the size of the project.

It was not until 1912, when the "Leipzig Scientific Society" and the Teubner publishing house agreed to participate, that work began in earnest. But then post-World War I inflation in Germany rendered the collected funds worthless. Fortunately, the active support of mathematicians from various countries, who stressed the international significance of this endeavor, led to a new fund-raising drive which made possible the successful completion of the project. The overall editing was done by one of Lie's closest pupils and associates, Friedrich Engel, and by the leading Norwegian mathematician of the time, Poul Heegard (1871–1948). B.C. Teubner Publishers in Leipzig, and one of the largest Norwegian mathematical publishing houses in Christiania (Oslo), were responsible for the printing. Publication of the collected works *Gesammelte Abhandlungen* (Samlede Avhandlinger), Bd. 1–10, Leipzig, B.C. Teubner; Kristiania, H. Aschehoug, 1920–1934; second impression: 1934–1960) took fifteen years; the works comprised fifteen large books (five of the ten works consisted of two volumes each) and many thousands of pages. Lie's collected works did not include the books: *Theorie der Transformationsgruppen*, Bd. 1–3, Leipzig, Teubner, 1888, 1890, and 1893 (second edition Leipzig, Teubner, 1930), written jointly with Engel (about 2,000 pages), and three more special books, with Lie's pupil Scheffers as coauthor: *Vorlesungen über Differentialgleichungen mit bekannten infinitesimalen Transformationen*, Leipzig, Teubner, 1891, *Vorlesungen über continuierliche Gruppen mit geometrischen und anderen Anwendungen*, Leipzig, Teubner, 1893, and *Geometrie der Berührungstransformationen*, Leipzig, Teubner, 1896. The striking similarities of language, and even of style, of all six books and of Lie's papers suggest that in all cases he was the chief writer, or that his influence was so great that it even determined the style of the writing.

It can be said that Lie was one of the last great mathematicians of the nineteenth century. There was something of Gauss or Riemann in his scientific profile (although in human terms Gauss, Riemann, and Lie were quite different personalities). Like his great precursors, Lie hardly needed a milieu: of course he valued pupils, but he took nothing from them and gave his ideas generously to the young mathematicians he met on his way.

The nineteenth century gave birth to the legend of the lonely genius—composer, philosopher, mathematician, or writer—creating values far away from people, by the force of spirit alone.[285] Of course there were really no such great hermits even in the nineteenth century and even the likes of Gauss and Balzac, say, were greatly influenced by their times. On the other hand, it was no accident that the image of the ivory-tower philosopher was particularly dear to the people of the nineteenth century.

As for Klein, he had nothing in common with this nineteenth-century image.

We have already mentioned that immediately after the Franco–Prussian War Klein went to live in Göttingen, to which he was attracted, above all, by his friendship with A. Clebsch and W. Weber; but he did not long remain there. In 1872 there was an opening for a professor at the newly organized mathematics department at Erlangen University, and the influential Clebsch, who held Klein in high esteem, recommended him for the post.

Felix Klein

In Germany at that time a prospective professor was required to deliver a public lecture to the Academic Board of the university on a subject chosen by the candidate himself. The decision whether to offer the post to the candidate was made after the lecture was discussed. The twenty-three-year-old Klein chose as his subject a *Comparative review of recent research in geometry*,[286] (just as, in a similar situation, eighteen years before, Riemann had spoken *On the hypotheses that lie at the foundations of geometry*).[287] The principal ideas of Klein's lecture were described in Chapter 7 above. The lecture soon became known as **The Erlangen Program**, a title which underscores both the broad vistas opened by Klein for further progress in geometry and his clear standpoint. It greatly enhanced the author's prestige.

The starting point for the Erlangen program and, at the same time, the application of its ideas, were provided by Klein's and Lie's previous concrete geometric works, beginning with the paper on *W*-curves and ending with Klein's broad vision of non-Euclidean geometries (these could be spoken of in the plural after the paper "Über die sogennante nicht-Euklidische Geometrie"; see Chapter 4). At the present time all works in this area are considered from the viewpoint of the Erlangen Program. At one time, geometric research in this area was very popular and was dealt with in great detail in university geometry textbooks, particularly German ones.[288]

Klein's Erlangen years (1872–1875) were remarkably productive in the

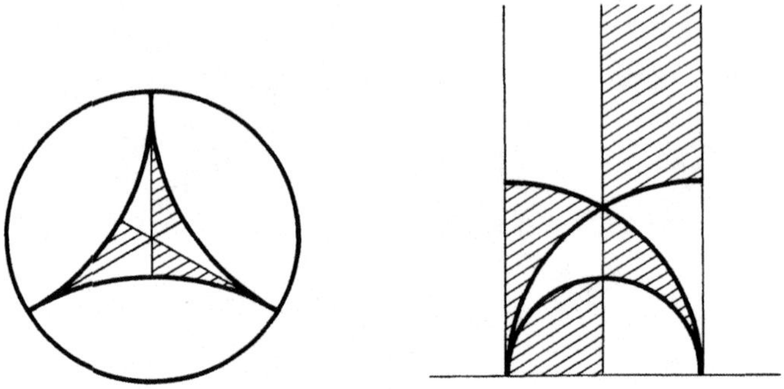

(a) (b)

FIGURE 34

scientific sense. As a result he received a very flattering invitation to the Technische Hochschule in Munich, which enjoyed a high reputation in Germany and where he worked for five more years. In 1880 Klein joined the geometry department of Leipzig University; in 1886 he yielded his post there to Lie and moved to Göttingen, where he would remain till the end of his life.

The period of Klein's greatest scientific productivity was his time in Munich and his first years in Leipzig. His works dealt with geometry, mechanics, and the theory of functions of a complex variable (theory of automorphic functions). He worked with particular intensity in 1880–1882, when he developed the (geometric) theory of automorphic functions, following ideas "combining Galois and Riemann," as he explained later, i.e., attempting to imbue Riemann's geometric approaches with group-theoretic ideas derived from Galois. A significant role in Klein's research was played by pictures of the type shown in Figs. 34(a) or (b) and by (discrete) transformation groups (linear fractional transformations of a complex variable) related to such pictures (transforming Figs. 34(a) and (b) into themselves). These groups proved to be closely linked to certain rectilinear polyhedra and the solution of algebraic equations in radicals (one of Klein's first books was devoted to that range of questions).[289] However, Klein failed to notice that the groups he considered could be interpreted as (discrete) subgroups of the groups of isometries of the Lobachevskian plane $\mathscr{L}$ modeled by the interior $\mathscr{K}$ of a circle (or as the half-plane $\mathscr{H}$). In this model the role of isometries is played by Möbius's circle transformations sending $\mathscr{K}$ (or $\mathscr{H}$) into itself; the "straight lines" of $\mathscr{L}$ are parts of circles and straight lines in $\mathscr{K}$ perpendicular to its boundary (or perpendicular to the boundary of the half-plane $\mathscr{H}$); "angles" are ordinary angles, etc. (see Fig. 35, which shows "straight lines" of the Lobachevskian plane passing through the point A and not intersecting line a).[290]

At the most intense period of his scientific work Klein came across a cycle of articles published in French journals and dealing with much the same subjects. They were written by the young French mathematician Henri Poin-

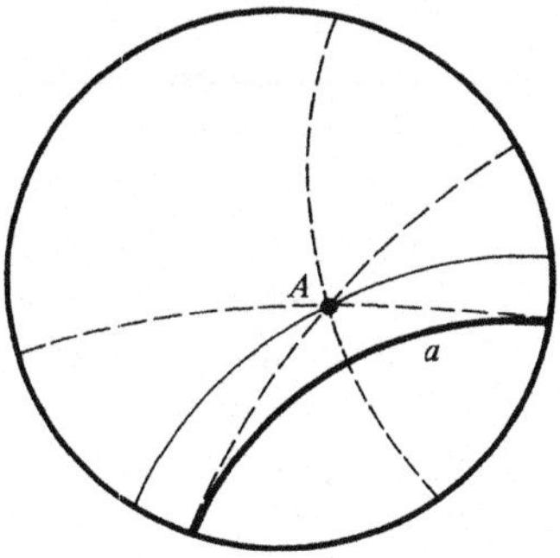

FIGURE 35

caré (1854–1912), who was hardly known at the time.[291] These articles made a profound impression on Klein. In his very first letter to Poincaré Klein communicated to his younger and less famous colleague everything he knew in the field both were interested in—in particular the results of papers he had not yet published. This behavior was in sharp contrast to that of many leading scientists confronted by rivals working in the same field (no need to list them here). But Klein's later activities were marked by acute rivalry with the remarkable French mathematician who was developing the same range of questions at the same time. In particular it was Poincaré who first noticed the relationship between circular transformations of the plane and Lobachevsky's non-Euclidean geometry. It is after him that the "models" of hyperbolic geometry[292] in the circle and the half-plane $\mathscr{H}$, whose "isometries" are circular transformations sending $\mathscr{H}$ into itself, are called *Poincaré models of Lobachevskian geometry*.[293] The discovery of a connection between the theory of automorphic functions and non-Euclidean geometry impressed Poincaré greatly and provided him with a "geometric key" to the entire theory,[294] which he of course immediately put to good use. This circumstance gave Poincaré a certain advantage over Klein. Besides, the difference in age was to have a telling effect. Pioncaré was five years younger than Klein, and mathematics is the domain of the young. In any case, as a result of the stress of this acute scientific rivalry—which was certainly not conducive to a calm creative effort—Klein suffered a serious nervous breakdown, brought on by exhaustion. This illness enabled Poincaré to celebrate a victory that is reflected in his memoirs of that period, and, in particular, in his famous report on mathematical discovery delivered at the Paris psychological society in 1908.[295] Especially popular is Poincaré's story of how the idea of the connection between non-Euclidean isometries and linear fractional transformations $z' = (az + b)/(cz + d)$ of the complex variable z (i.e., Möbius's circle transformations[296] sending a fixed circle into itself) came to him at the moment he put his foot on the footboard of an omnibus, seemingly without any link to his previous thoughts. On the other hand, Poincaré's victory echoes in the disappointment exuding from Klein's recollections about this rivalry in *Vorlesungen über die Entwicklung der Mathematik.* Today, however, conditioned by a long period when the attitude towards joint research is quite different from the attitude typical for the period between the seventeenth and nineteenth centuries, we are perfectly aware of the fact that there are no winners or losers in scientific contests. (Was there a winner in the competition between Newton and Leibniz for priority in the founding of the calculus? It is incredible how much ardor was expended on a rivalry which to us seems senseless.) The view, advanced some twenty years ago by the Moscow mathematician Israel Moisseievich Gelfand (b. 1913) in his seminar, to the effect that Klein achieved greater success than Poincaré since the two mathematicians' most important ideas and theorems were derived from Klein, seemed paradoxical at the time; but that viewpoint (as debatable from our vantage point as the opposite view) now finds support in the ubiquity of the term "Klein groups" in modern mathematics and in the frequent references in the latest research[297] to Klein's old works on the theory of automorphic functions.

Klein's illness, the result of his exhaustion, unfortunately had a marked effect on his scientific activity, which never again reached the level of the early 1880s. Klein became a mature scientist at an early age—not a rare case in our discipline. At seventeen he was Plücker's assistant at the university; at eighteen he was faced with the formidable task of publishing his teacher's unfinished book, a task assigned to him by the also quite young but already very influential Clebsch. Klein also published his own first independent research at that time. In this respect Klein was different from Lie, who developed relatively late as a professional mathematician and hesitated for a long time before choosing his future occupation. But Klein's creative period was much shorter than Lie's (compare their collected works). In this sense Klein is probably closer to Poncelet (of whom he wrote with such regret), and differs sharply from, say, Möbius, whose productivity as a mathematician hardly seemed to decrease over the years.

But being very active by nature, Klein immediately made up for his reduced creative potential by truly extensive teaching, and by literary, organizational, and administrative activities. In 1872 Clebsch, then thirty-nine years old, suddenly died of diphtheria, and Klein immediately took over *Mathematische Annalen*, the journal founded and headed by Clebsch. Klein became the journal's *de facto* editor, and in 1876 its formal editor. Under Klein's leadership *Mathematische Annalen* soon gained the reputation of being the world's leading mathematics journal. Klein's books began to appear, beginning in 1882: *Vorlesungen über Riemann's Theorie der algebraischen Funktionen und ihrer Integrale* (1882), *Vorlesungen über das Ikosaeder und die Auflösung der Gleichungen vom fünften Grade* (Leipzig, Teubner, 1884; see Note 289), four volumes, written jointly with Karl Immanuel Robert Fricke (1861–1930), Klein's pupil and collaborator, of *Vorlesungen über die Theorie der elliptischen Modulfunktionen* (Bd. 1, 2, Leipzig, Teubner, 1890 and 1892) and *Vorlesungen über die Theorie der automorphen Funktionen* (Bd. 1, 2, Teubner, 1897 and

1901, 1911, 1912, the second volume of the book was published in three separate issues); four volumes, written jointly with Arnold Sommerfeld, of *Über die Theorie des Kreisels* (Leipzig, Teubner, 1897–1910), and others. And throughout his time in Göttingen, mimeographed versions of his lecture courses were regularly published, and many of these were subsequently—often posthumously—printed as books and translated into many languages. Thus, for example, the courses mentioned many times in the present book, *Lectures on Non-Euclidean Geometry, Lectures on Higher Geometry*, or *Lectures on the Progress of Mathematics in the Nineteenth Century* were derived from Klein's lecture courses.

All of Klein's books—not only those which were revisions of mimeographed lecture notes—were conversational in tone, retaining echoes of their author's speech. As was usually the custom at German universities,[298] Klein often assigned to his pupils and associates the task of preparing his books for print; in some cases their names (Fricke,[299] Sommerfeld) were placed on the book's title page, but more often they were mentioned in the form *bearbeitet von ...*[300] or simply omitted. Nevertheless, Klein's authorship was never in doubt: he produced his books in the course of long discussions with his associates, and this form of work was apparently more suitable and common for him than writing down the texts himself.[301] It is interesting that Klein found it easier to assimilate new material through informal conversations than through reading scientific works.[301] It was for this reason that the memorable trip to Paris with Lie, about which so much was said above, proved so fruitful for Klein. We mentioned already that Klein first talked about non-Euclidean geometry in Weierstrass's seminar when he knew next to nothing about the subject and, in any case, without having read anything about it. When it became necessary for him to take a closer look at non-Euclidean geometry, he turned to his friend Otto Stolz rather than to Lobachevsky or Bolyai, whose texts were available at the time. In much the same way, von Staudt's fame was largely due to the enthusiasm with which Klein popularized Staudt's ideas—yet Klein apparently never read a single line written by Staudt, his predecessor at the university in Erlangen. In this case too he preferred to turn to Stoltz, who, by the nature of his interests and talent, was better able to read Staudt's ponderous works and more likely to study them (since Staudt was his relative).

The conversational nature of Klein's books (and also of most of his papers), which secured their popularity, had to do, in some sense, with their lack of mathematical rigor. If we adhere to the classification of all mathematicians into "physicists" and "logicians" (see Chapter 2), then Klein undoubtedly belongs to the physicists. The modern mathematician Laurence Young, in the often quoted *Mathematicians and Their Times*, states that Klein did not understand what a proof was, at least not in the modern sense of this important term. The same assertion was made less bluntly by Hermann Weyl in the brilliant article devoted to Klein ("Felix Klein's Stellung in der mathematischen Gegenwart." *Die Naturwissenschaften*, Bd. 18, 1930, S. 4–11, reprinted in Weyl's *Gesammelte Abhandlungen*), where he writes about Klein's almost total lack of interest in questions of mathematical rigor and theory of proof. Of course this in itself does not diminish Klein's achievements nor make his place in the history of mathematics any less considerable; after all, "intuitive" mathematicians such as Riemann and Klein have contributed to mathematical knowledge no less than "logicians" like Weierstrass and Weyl.

In 1898 Klein headed the tremendous undertaking of publishing Gauss's collected works (the edition was finished only in 1918). He also directed work on the *Enzyklopädie der Mathematischen Wissenschaften*, whose editors intended to include all the results and methods obtained up to the early twentieth century in pure and applied mathematics. Today this encyclopedia occupies lots of shelf space in many of the world's largest libraries; but it was never completed (because gradually people realized that the amount of material not included did not decrease with time but, on the contrary, grew!),[302] and unfortunately it has become hopelessly outdated.

It seems to us that Klein's attempt to create a comprehensive encyclopedia of the mathematical sciences deserves particular attention. Klein enlisted the most outstanding scientists of his time to work on this project, which aimed to offer its readers the key achievements of mathematics and of mathematical natural science. Although the attempt failed, it is revealing in many respects. Klein realized better than anyone else that the rapid expansion of knowledge (which delighted many people) represented a real threat to mathematics: it led to excessive specialization and to disregard of adjacent fields.[303] This was the unavoidable result of the increasing difficulty of gaining a comprehensive view of everything achieved even in the restricted field within which a scientist was specializing. It was only natural then that individual creative effort was replaced by collective work, a development typical of the twentieth century. It is well known that it is difficult, as a rule, to attribute outstanding achievements in physics to a particular person: trans-Uranium elements are discovered in our day by groups in Berkeley, at CERN or in Dubna, and not by individual researchers such as Seaborg or Flerov. It is less well known that in mathematics today collective undertakings are absolutely necessary not only in applied research, where it is usual to have large institutes, computer centers with numerous staff—technicians, engineers, mathematicians, and electronics experts—and expensive hardware, but also in a number of purely theoretical fields (see Note 260). The situation is similar for most of the works of art typical of our time—films, TV productions, architectural ensembles, or pop music melodies—which are created by large groups where a key figure can be singled out only tentatively if at all. It was to this universal twentieth-century trend that Klein responded by his encyclopedia, which was meant to present all of mathematics to the reader from a single vantage point, with due attention paid to the variety of existing interrelations. Klein's attempt was unsuccessful, but it had a tremendous influence on, say, the more advanced experiment of the Nicolas Bourbaki group.[304]

Klein also approached his activities at Göttingen from this characteristically twentieth-century standpoint! Over many years, even decades, he did everything he could to transform little Göttingen into a world center of physics and mathematics. His outstanding eloquence attracted to Göttingen talented students from all over the world. He sought to bring together many outstanding scientists there; their joint work and mutual consultations created ideal conditions for research.[305] The leading role was played by several outstanding mathematicians and physicists from Königsberg whom Klein enticed to Göttingen. Among them we must mention, above all, David Hilbert (1862–1943), whom many regard as the greatest mathematician of the twentieth century.[306] Following Hilbert, his friend Hermann Minkowski came to Göttingen. From the late 1890s until the destruction of this outstanding scientific center in the 1930s, Göttingen was firmly associated in the minds of all the world's mathematicians with the names of Felix Klein and David Hilbert. Klein and Hilbert complemented each other very well. Modest in dealing with other people and always engulfed in purely scientific problems, Hilbert directed the scientific activities of the Göttingen mathematicians. Administrative functions were profoundly alien to him. (It was not by accident that, following Klein's death, it was not Hilbert, undoubtedly the leading scientist in Göttingen, but his pupil Richard Courant (1888–1972)[307] who was appointed director of the mathematical institute there.) On the other hand Klein, an outstanding organizer, ruled the mathematical institute—which had been created largely by his efforts—rather autocratically. Reminiscences of Klein often compare him to Jupiter, whose marble bust decorated the stairway leading to Klein's office!

Klein's search for fresh forces in Königsberg was quite natural, for he greatly valued the Königsberg mathematics school (which had been founded largely by Carl Jacobi). That school had a profound influence on Klein; in particular his friend and teacher Adolf Clebsch was a graduate. Subsequently, Göttingen's reputation for excellence, due to Klein, Hilbert, and Minkowski, attracted scientists from all the other German universities. In addition to the Königsberg "input," the arrival at Göttingen of several students and teachers from Breslau (now Wroclaw in Poland) University was particularly important. An exceptional role, comparable to the role of the Breslau mathematicians Richard Courant and Otto Töplitz (1881–1940), was played by the outstanding physicist Max Born (1882–1970), the future Nobel Prize winner and director of the Göttingen physics institute.[308] Born headed the Göttingen school of theoretical physics, which produced such outstanding scientists as the Nobel Prize winner and founder of quantum mechanics Werner Heisenberg (1901–1976) and the American physicist J. Robert Oppenheimer (1904–1967), who later directed work on the atomic bomb.[309]

Klein's method of operation may be epitomized by one typical episode. At the International mathematics congress in Heidelberg (Germany) in 1904, Klein listened to a communication on hydrodynamics by the then relatively obscure German engineer Ludwig Prandtl (1875–1953). Profoundly impressed, he immediately invited Prandtl to Göttingen appointing the twenty-nine-year-old engineer to direct an applied mathematics institute especially founded for him. Subsequently the world-famous Göttingen school of mechanics arose from this institute.

Finally, mention should be made of another aspect of Klein's activities, which presaged many features of mathematical life in our day. We have already pointed out that Klein detested the classical Prussian gymnasium from which he had graduated. He was the first scientist who fully realized the need for a fundamental reform of the whole system of school mathematical education. He succeeded in rallying many prominent mathematicians and teachers to the struggle for such reform. In 1898 he organized an International Commission on Mathematics Education which he headed for a number of years. The first international congresses on mathematics education were held under Klein's direct leadership. He spent much time and effort in popularizing his pedagogical views.[310] The basic principles expounded by Klein and by the commission he headed were: more of the graphic element in teaching, greater attention to the functional viewpoint in algebra and analysis, and application of geometric transformations in the teaching of geometry. He called for the elimination of the "China wall" separating different mathematical subjects, for taking into account the needs of related fields in mathematics courses, and for decreasing the gap between mathematical education and contemporary science. All this was to play an enormous role in furthering progress in that sphere.

Klein gained very high administrative titles in science. In 1913 he was elected corresponding member of the German academy of sciences in Berlin. He was *Geheimrat*, a title received by very few German scientists, and a representative of Göttingen University in the Chamber of Lords, the upper chamber of the Prussian parliament. Klein's political views were moderately conservative. However, in one respect he was very far from the German obscurantism which began its offensive while Klein was still alive: he was never a chauvinist or racist. Thus the destruction of Göttingen University (after his death), immediately following the coming to power of the Nazis, when some of the professors were expelled and others left voluntarily, may well be explained by the fact that Klein's humanistic traditions were still alive then at the university.

Lie's and Klein's memorable trip to Paris and London, which we have dealt with previously, was very rewarding for the two friends. Before the trip, Klein's father, who was influential in government circles, attempted to procure a letter of recommendation at the Prussian ministry which would help his son establish relations with French and English mathematicians. But Felix received in reply an official letter expressing the Prussian officials' firm conviction that a German mathematician had nothing to learn from the French or the English. Klein did not like to recall the incident later, but whenever he did he spoke with extreme indignation about the chauvinism thus revealed.

At the beginning of World War I, a group of leading German scientists, headed by the physicist Wilhelm Ostwald (1853–1932),[311] issued a manifesto notorious for its anti-French and anti-British sentiments. Klein, a member of the Chamber of Lords, refused to sign the manifesto. In the book so frequently referred to above, *Vorlesungen über die Entwicklung der Mathematik*, Klein points out with satisfaction the major contribution to German mathematics made by scientists of French origin (such as Pierre Lejeune Dirichlet) or Jewish origin (e.g., Gauss's constant rival Carl Jacobi). Concerning the previously mentioned *Treatise on Substitutions* by Jordan, Klein says that this outstanding book was written in a dull [*langweilig*] style—in the German manner rather than the French.[312]

Klein's collected works in three volumes were published by his pupils in 1918–1924 and were intended to mark the 70th birthday of *Herr Geheimrat*, as stated in the address opening the first volume, but were completed only in the year he turned 75. Many papers devoted to Klein were published in scientific journals on the appearance of these volumes and on the occasion of his seventy-fifth birthday. They were almost immediately followed by obituaries—Felix Klein died on July 22, 1925.

Notes

Chapter 1

[1] Galois was a conspicuous figure among the Parisian republicans dissatisfied with the results of the July 1830 Revolution, which had merely replaced the Bourbon dynasty (Charles X) by the Orléans one (Louis-Philippe). His first arrest was the result of a toast he proposed to Louis-Philippe at a republican banquet: Galois raised his glass and struck it with an open knife. This was justly perceived as a call to assassinate the king. One of the frightened guests was Alexandre Dumas père, who escaped through an open window of the restaurant so as not to be forced to declare for or against the toast or its proposer. Galois's second arrest (leading to imprisonment) was prompted by his participation in a republican demonstration. He marched, fully armed, at the head of a mob of demonstrators, wearing the uniform of the disbanded Garde Nationale.

On Galois see, for example, the book by Einstein's staff member, the Polish physicist Leopold Infeld (1898–1969), *Whom the Gods Love: the Story of Evariste Galois* (McGraw Hill, 1948); a smaller book by A. Dalmas, *Evariste Galois révolutionnaire et géomètre* (Paris, Fasquelle, 1956); or the detailed and definitive essay by the French historian P. Dupuy, "La vie d'Evariste Galois" (*Ann. de l'Ecole Norm. Sup.*, **13** (2), 1896, pp. 197–266). A chapter is devoted to Galois in the well-known book by the American historian and popularizer of mathematics Eric Temple Bell (1883–1960) *Men of Mathematics*, N.Y., Simon and Schuster, 1937; this also contains essays on some of the other personalities in the present book. Also see the following more recent articles, which present different viewpoints on Galois's life: T. Rothman, "Genius and Biographers: The Fictionalization of Evariste Galois," *Amer. Math. Monthly*, **89** (2), 1982, pp. 84–106; T.R. Rothman, "The short life of Evariste Galois," *Scient. American*, **246** (4), 1982, pp. 136–149; R. Taton, "Evariste Galois and his contemporaries," *Bull. London Math. Society*, **15**, 1983, pp. 107–118. The most complete edition of everything written by Galois is *E. Galois, Ecrits et mémoires mathématique* (editors: R. Bourgne, T.P. Azra), Paris, Gauthier-Villars, 1962. This book contains a brilliant article by J. Dieudonné on Galois's work.

[2] It is said that when the examiner asked Galois what logarithms were, he simply wrote a geometric progression

$$1, q, q^2, q^3, \ldots$$

and an arithmetic progression under it

$$0, 1, 2, 3, \ldots$$

and pointed to the lower row. This answer was correct if rather laconic: the lower numbers are proportional to the logarithms of the upper numbers to any base. The examiner failed to understand what Galois had written and declared the answer unsatisfactory; the angry Galois responded by throwing a blackboard rag at him.

[3] For these and other details see the book by Infeld mentioned in Note 1 or Dupuy's essay. It should be pointed out that the École Normale, which subsequently raised its reputation considerably, now takes pride in Galois's short stay within its walls. This was one of the reasons why a long essay about Galois by Sophus Lie was included in a volume devoted to the 100th anniversary of the École Normale.

[4] The first, and for many years the only, document which referred to Galois as a *mathematician* was the official certificate attesting to the death on June 1, 1832, at noon, of "Evariste Galois, mathematician, unmarried, born in Bourg-la-Reine."

[5] In that age of political instability the range of political sympathies and affinities of French mathematicians was rather wide. Thus Lazare Carnot was a republican and Gaspard Monge a loyal Bonapartist (these names will be encountered below). For these scientists the restoration of the Bourbons was a personal calamity. Carnot was exiled while Monge was dismissed from all his posts and soon died. As noted above, Cauchy's political views were quite different—he was consistently conservative. In contrast to the three men mentioned above, the outstanding scientist Pierre Simon Laplace (1749–1827) was completely unprincipled and prospered under all regimes.

[6] Cauchy's collected mathematical works only began to appear in 1882; the edition, finally completed in 1958, includes 26 volumes.

[7] A major part was played here by Cauchy's works of 1844–1846, which focussed on the notion of a group, the most basic concept in Galois's work (see below; Cauchy did not use the term "group"). Cauchy's prestige helped kindle interest in that range of questions; in particular, it brought about the publication in 1846 of most of Galois's works by Joseph Liouville (1809–1882) in the mathematical journal which he edited.

[8] This refers, above all, to problems famous since the ancient Greeks: the

trisection of an angle (i.e. its division into three equal parts), and the doubling of a cube (i.e. the construction of the side of a cube whose volume is equal to double the volume of the cube with the given side). In the 19th century it was shown that these problems can not be solved by ruler and compass alone, because the cubic equations to which these problems reduce cannot be solved in quadratic radicals. The 16th century's leading algebraist, François Viète (or Vieta in the Latin transcription; 1540–1603) found that every cubic equation can be interpreted as corresponding either to the trisection problem for some angle or to the problem of constructing two geometric means between two given segments (i.e., of constructing two segments x and y such that $x^2 = ay$, $y^2 = bx$, where a and b are given; when $b = 2a$, this reduces to the problem of doubling the cube).

[9] In particular, the classification of cubic equations and the discussion of the number of, and bounds on, their roots were the subjects of a treatise by the outstanding Muslim mathematician (and great Persian poet) Omar Khayyam (1048–1131).

[10] Division of the terms of the general cubic $ax^3 + bx^2 + cx + d = 0$ by a yields an equation of the form $x^3 + \alpha x^2 + \beta x + \gamma = 0$; further, the substitution $x = x_1 - \alpha/3$ transforms the latter into $x_1^3 + px_1 + q = 0$, where $p = \alpha^3/3 + \beta$, $q = -\alpha^3/27 - \alpha\beta/3 + \gamma$.

[11] At different times, this university (whose astronomy department alone had 16 professors by the end of the 15th century) was attended by Albrecht Dürer and Nicolas Copernicus, and, before them, by Luca Pacioli (ab.1445–ab.1515), one of the founders of Renaissance mathematics. Pacioli was the tutor and friend of Leonardo da Vinci.

[12] This mathematician's real name seems to have been Fontana; the nickname Tartaglia (Italian for stammerer; it is also the name of one of the main heroes of Italian folk theatre—the plump, sanguine hare) was given to him because of a speech defect, the result of his having been wounded in the face in childhood during the occupation of his native town (Brescia) by the French. Poor, and brought up by strangers (his mother died in the same attack that disfigured him), Tartaglia revealed outstanding abilities at an early age and became a first-rate scholar. It was he who introduced the (restricted) binomial theorem, sometimes attributed to Newton: $(a + x)^n = a^n + \binom{n}{1}a^{n-1}x + \cdots + x^n$ for any natural number n. Unknown to Tartaglia, the formula had been discovered earlier by Arab mathematicians. Newton extended it to arbitrary fractional exponents.

[13] An outstanding mathematician, naturalist, and philosopher, Cardano was typical of his stormy age. Famous for his scientific, medical, and literary achievements (see his absorbing autobiography, *The Book of My Life*, Dover

reprint, 1962), he was also an astrologist, a common adventurer, and perhaps a murderer; he introduced complex numbers and the first ideas of probability theory (see O. Ore, *Cardano: the Gambling Scholar*, Princeton University Press, 1953; G. Cardano, *The Book on Games of Chance*, Holt, Rinehart and Winston, 1961). His life was filled with adventure and he had an explosive temper. It is said that having drawn up his own horoscope, Cardano committed suicide in order to ensure that the day of his death was predicted correctly.

[14] Cardano's *Artis magnae, sive de regulis algebraicis liber unus* appeared in 1545 and was a major event in the history of algebra. (In those years, algebra was called *ars magna* as opposed to *ars minor*—arithmetic.)

[15] Tartaglia's idea was to represent the unknown root of equation (1.1) as the sum of two auxiliary quantities u and v; substitution of the expression $x = u + v$ into (1.1) immediately leads to the equality $(u + v)^3 + (3uv + p)(u + v) + q = 0$; therefore if u and v are taken such that $uv = -p/3$, then u^3 and v^3 may be found from the system $u^3 + v^3 = -9$, $u^3v^3 = -p^3/27$. This system is obviously equivalent to the quadratic equation $z^2 + qz - p^3/27 = 0$. In this way the solution of the initial cubic equation (1) reduces to the solution of a quadratic equation. The latter might well be called *Tartaglia's resolvent* of the cubic equation (1.1).

[16] The values of the cube roots $R_{1,2} = [-q/2 \pm \sqrt{(q/2)^2 + (p/3)^3}\,]^{1/3}$ on the right side of (1.2) must satisfy $R_1R_2 = -p/3$ (see Note 15).

[17] For the dramatic story of the discovery of the formulas for solving cubic and quartic equations see, for example, H.-G. Zeuthen, *Geschichte der Mathematik im XVI und XVII Jahrhundert*, Leipzig, Teubner, 1903.

[18] Thus, for example, if $p = -1$, $q = 0$, i.e., if equation (1.1) is of the form $x^3 - x = 0$ (obviously the latter has the roots $x_1 = 0$, $x_{2,3} = \pm 1$), then formula (1.2) unexpectedly yields $x = \sqrt[3]{\sqrt{-1/27}} + \sqrt[3]{-\sqrt{-1/27}}$.

[19] In order to reduce the general quartic $ax^4 + bx^3 + cx^2 + dx + e = 0$ or $x^4 + \alpha x^3 + \beta x^2 + \gamma x + \delta = 0$ to the form (1.3), it suffices to put $x = x_1 - \alpha/4$ (cf. Note 10).

[20] "A rosy young man with a gentle voice, merry visage, tremendous abilities and the temper of a devil," according to Tartaglia's description in *Questi et inventioni diversi* (1546), which contains a detailed (though evidently biased) account of the discovery of formula (1.2). The pro-Cardano side of the story is told in the book *Cartelli* by L. Ferrari (1547–1548). The latter had first been Cardano's servant, but Cardano noticed his outstanding abilities and soon began to study with him.

[21] There was a time in the memory of people of the older generation when the Middle Ages, also known as the Dark Ages, were regarded as a thousand-year gap in European culture. By now it is widely held that this viewpoint is untenable. What is clear to most of us is that medieval culture was based on principles other than those of ancient culture, and that the Renaissance built on the heritage of antiquity. In particular, mathematics was a serious concern of the leading thinkers of antiquity, Plato and Aristotle, and played a most important part in the investigation of the universe by ancient Greek philosophers, whose basic assumption, due to Thales of Miletus and Pythagoras of Samos (6th century B.C.), was that the laws of nature were knowable and that the universe was harmonious. Byzantine and medieval European culture were based on Christianity and differed in many respects from the pagan culture of Greece and Rome. Mathematics played no important part in it, and so progress in mathematics was modest over the centuries. The Renaissance marked a spiritual revolution in the history of European culture which again brought mathematics and the natural sciences to the forefront of European thought.

[22] In the 18th and early 19th centuries, the centers for mathematics and the natural sciences in Britain, France, Italy and in other countries were not the universities, which were preoccupied with philosophy, theology, and the humanities, but rather the military, engineering, and naval institutions of higher learning, (not quite appropriately) referred to here as military academies. Examples of the latter are, in addition to the Turin artillery school, the French military school in Mezières (see Chapter 3) and the British royal military school in Woolwich, London.

[23] Leonhard Euler, the leading mathematician of the 18th century, was born in Basel (Switzerland) into the family of pastor Paul Euler. The conservative views and deep religious faith characteristic of his family assumed a rather naive form in Leonhard (as when he attempted to prove the existence of God mathematically). The mathematician Euler was to retain these views all his life.

The Euler family was on close terms with the "mathematical" Bernoulli family. The elder of the Bernoulli brothers, Jacob (1654–1705) had, rather reluctantly, taught Paul Euler mathematics. Euler Sr. wanted his son to become a clergyman like himself and, conforming to his father's desire, Leonhard diligently studied theology. However, unfortunately for himself and fortunately for the rest of humanity, Paul Euler also gave mathematics lessons to Leonhard; the latter turned to Johann Bernoulli (1667–1748), Jacob's younger brother, for help. Johann was amazed by Leonhard's rapid progress and aptitude. He agreed to teach him free of charge once a week. Leonhard devoted most of his spare time to mathematics, discussing his most recent lesson with Johann's sons, the future outstanding mathematicians Daniel Bernoulli (1700–1782) and Nikolaus Bernoulli (1695–1726), and preparing for the next lesson. With great difficulty, Johann Bernoulli persuaded Paul Euler that his son would make a great mathematician and that it would be a sin to hold back the development of his remarkable talent in order to make him into an ordinary pastor.

At that time a mathematician could not count on a regular income in Switzerland. For example, Johann Bernoulli earned a living by medical practice until the death of his brother Jacob, who held the only mathematics chair at Basel University. (Incidentally, Johann made the first attempts to apply the newly created differential calculus to the medical problems of muscle contraction.) As for Johann's sons, they went to Russia, to the newly founded St. Petersburg Academy of Sciences. They suggested that their friend Euler follow their example and informed him of a vacancy in physiology at the academy. Euler industriously studied biology and medicine. However, when he arrived in St. Petersburg, he immediately began working in the mathematical sciences and never again returned to physiology, which was generally alien to him. (At one time, following the example of his tutor Johann, he intended to apply mathematical methods to physiology.)

Euler spent most of his long life in St. Petersburg, except for the period between 1741 and 1766, when he decided that Russia, then ruled by Biren, a favorite of the reigning empress Anna Ioannovna, was no longer a safe place. At the invitation of Frederick II he moved to Berlin, where he headed the physics and mathematics department of the Prussian Academy of Sciences. But even during his stay in Berlin Euler did not break with Russia. He continued to publish papers in editions of the St. Petersburg Academy of Sciences and even received a small stipend from the Russian government.

Euler was exceptionally prolific; it seems that no other scientist made a comparable contribution to mathematics (in volume *and* in content). We encounter *Euler formulas*, *Euler theorems*, and *Euler relations* in the differential and integral calculus, in differential equations, in the theory of functions of a complex variable, in the theory of series, in the theory of numbers, in geometry and, of course, in the calculus of variations (a branch of mathematics largely created by Euler and Lagrange). Euler published many long books and innumerable papers. He himself estimated that the articles unpublished in his lifetime (but ready for the press) would last the St. Petersburg Academy of Sciences for 20 years after his death. Actually, he underestimated his own legacy; publication continued until 1862, four times longer than he had expected. Even the blindness which afflicted him at the end of his life did not interrupt the stream of his works and apparently did not reduce his creative powers: he dictated his last works to his sons, pupils, and colleagues.

Leonhard Euler passed away instantly and easily, in his 77th year. Blind, he wrote out on a blackboard his calculation of the orbit of the newly discovered planet Uranus; he then had some tea, and was playing with his grandson when the stroke came—the old mathematician "ceased to calculate and live," in the words of de Condorcet, who wrote his obituary in a publication of the Paris Academy of Sciences (Euler was a member of the Paris Academy of Sciences and the London Royal Society, as well as the academies in St. Petersburg and Berlin).

In Switzerland, just before World War I, it was decided to publish Euler's complete works. They were issued in a small number of copies, just enough for each country taking part in the undertaking to receive a copy. The number of (hefty!) volumes grew from an estimated 40 to 70. Then new notes and letters by Euler were discovered in Leningrad—and the number of volumes became greater still. The enormous and extremely valuable *Collected Works* have not been completed to this day, so the total number of volumes is still unknown.

[24] In his invitation to Lagrange the king of Prussia, Frederick II, who insisted on being called "the Great", wrote, overbearingly, that he would like the greatest of geometers to work near the greatest of kings (the word "geometer" meant mathematician then).

Frederick II, who valued science highly and was well-disposed towards Lagrange, died in 1786. This immediately made Lagrange's position in Berlin intolerable. Matters were exacerbated by the fact that Lagrange had to some extent exhausted the subject he had been working on during his stay in Berlin, and this led to a certain disenchantment with mathematics. All this resulted in a depression, which might have proved quite serious. Under the circumstances, the invitation to Paris was a blessing for Lagrange. The new flourishing of his creative activity in the Paris period was linked to his public duties (in particular, his work with the commission which introduced the metric system), to his scientific and literary work (described below) and, especially, to his professorial post at the École Polytechnique; this last he accepted at the suggestion of Napoleon, who held him in high esteem. At the École Polytechnique Lagrange found a circle of friends and people with the same interests (for example, he won the admiration of Jean Baptiste Joseph Fourier, 1768–1830, one of the pillars of the analytic trend); he also found new topics for research. His lectures on the calculus, delivered at the École Polytechnique, were a major event in the history of mathematics and served as a springboard for the revision of its foundations by another École Polytechnique professor—Augustin Cauchy. Another major event was the publication of Lagrange's textbooks on theoretical mechanics (*Mécanique Analytique* in two volumes, 1788) and on the calculus (*Théorie des fonctions analytiques*, 1797; and *Leçons sur le calcul des fonctions*, 1801). The principal ideas of the *Mécanique Analytique* date back to the Turin period in Lagrange's life, but the book's publication was mainly due to the efforts of an outstanding mathematician and admirer of Lagrange named Adrien Marie Legendre (1752–1833). The book appeared before Lagrange became a staff member at the École Polytechnique, in the period when he was still disillusioned with mathematics; Lagrange took so little interest in it that it remained unopened on his desk for two years. On the other hand, his two-volume calculus textbook (part one, *Theory of analytic functions*; part two, *Lectures on the calculus of functions*) exactly followed Lagrange's lectures at the École Polytechnique. In the very last years of his life Lagrange paid much attention to the publication of the *Mécanique Analytique*, which had previously left him indifferent.

[25] In modern terms (originating with Galois) Lagrange's theorem asserts that the order (i.e., number of elements) of any subgroup of a finite group is a divisor of the order of the group itself. Lagrange proved the theorem for what is known today as the symmetric group S_n (of order $n!$)—the group of all permutations of n elements; but his proof holds for any finite group.

[26] The mathematicians who read the book (e.g., Poisson) found it unclear, perhaps because Ruffini's ideas were clearly ahead of their time.

[27] It is not for a man to know the fate awaiting his works! In his lifetime Malfatti tended to regard himself as Italy's foremost mathematician and looked down on Ruffini. Today, Ruffini is the pride of Italian science, while Malfatti's name is mostly mentioned only in connection with one minor problem in elementary geometry, which Malfatti stated incorrectly and solved in an extremely involved algebraic manner. Nowadays, the following is called the "Malfatti Problem": in a given triangle ABC inscribe three nonintersecting circles S_A, S_B and S_C, each of which touches two sides of the triangle and the two other circles (circle S_A touches the sides AB and AC of the triangle and the circles S_B and S_C, and so on). By now there are many brilliant, purely geometric, solutions of this problem. One of the most beautiful is due to J. Steiner (of whom more will be said below). Actually, Malfatti's note *Concerning a Space Geometry Problem* stated a different question, which is easily reduced to the problem of finding in a given triangle ABC three nonintersecting circles with the largest possible total area. Malfatti took it for granted that the required circles are the circles S_A, S_B, and S_C described above. However, it was proved not long ago that there is no triangle ABC in which it is impossible to inscribe three nonintersecting circles whose overall area is greater than the total area of Malfatti's circles S_A, S_B, and S_C!

[28] On Abel see, for example, O. Ore, *Niels Henrik Abel: Mathematician Extraordinary*, University of Minnesota Press, 1957, and the chapter devoted to him in the book mentioned in connection with Galois (E.T. Bell, *Men of Mathematics*).

[29] It is interesting that the very same mistake was made later by Galois, who also believed for a time that he had found a formula for solving fifth-degree equations.

[30] The special notation $\sqrt[k]{a}$ for the kth root of a, and the fact that the extraction of the kth root is an operation inverse to raising to the kth power (i.e., iterating multiplication), hide what is actually meant by finding an

algebraic formula for the equation's roots. It means reducing the equation $P(x) = 0$ to a chain of auxiliary equations of the type $y^k - a = 0$, where k is a positive integer (why must we reduce the initial equation to these particular equations?). It is not *a priori* clear that this can be done for an arbitrary polynomial $P(x)$. Lagrange, Ruffini, Abel, and Galois were well aware of this circumstance which went virtually unnoticed before them. (Perhaps only the great Carl Friedrich Gauss, 1777–1855, realized it clearly. His research on the solution of algebraic equations in quadratic radicals was a major contribution to this trend in scientific thought.)

[31] Abel himself understood that shortcoming very well, pondering criteria for determining whether a given equation can be solved in radicals or not.

[32] The following quotation is a persuasive endorsement of this view: "I believe indeed that the concepts of *number*, *set*, *function* and *group* are the four cornerstones on which the entire edifice of modern mathematics rests and to which any other mathematical concept reduces" (P.S. Alexandroff, *Introduction to the Theory of Groups*, Blackie/Hafner, 1959). Pavel Alexandroff (1896–1982), an outstanding Soviet mathematician, contributed greatly to the early development of modern (general) topology; for more on this subject see below.

For an elementary introduction to the theory of groups see F.J. Budden, *The Fascination of Groups*, Cambridge University Press, 1972; this book includes a detailed bibliography.

[33] See, for example, the following books and essays, generally intended for a reader more advanced than the one to whom the present book is addressed: J.E. Burns, "The Foundation Period in the History of Group Theory," *American Math. Monthly*, **20**, 1913, pp. 141–148; H.L. Wussing, *The Genesis of the Abstract Group Concept*, MIT Press, 1984; H. Burkhardt, "Die Anfänge der Gruppentheorie und Paolo Ruffini," *Abhandlungen zur Geschichte der Mathematik*, Heft 6, 1892, S. 119–159; G.A. Miller, *History of the Theory of Groups to 1900*, Collected Works, Vol. 1, pp. 427–467, University of Illinois Press, 1935.

[34] See Note 41 below. Ruffini's works in effect established a connection between the algebraic properties of an equation related to its solvability and certain properties of the permutation group of the equation's roots, later called *intransitivity* (i.e., nontransitivity; a transformation group $\mathscr{G}$ of some set M of elements is said to be *transitive* if for any two elements of M it contains a transformation which sends the first element into the second) and *imprimitivity* (i.e., nonprimitivity: a transformation group $\mathscr{G}$ of a set M is *primitive* on that set if there is no partition of M into "blocks" $M_1, M_2, M_3, \ldots$, different from M and from the partition into "single-element" blocks $\{\alpha\}, \{\beta\}, \{\gamma\}, \ldots$, which are preserved by the transformations in $\mathscr{G}$, i.e., are such that for any block M_i

and any $g \in \mathcal{G}$ we always have $gM_i = M_j$, where M_j is also a "block". On this subject see the article by Burkhardt cited in Note 33.

[35] It is clear that if $x = a + br$ and $y = c + dr$, where $a, b, c, d \in Q$ and $r = \sqrt{2}$ or $r = i(= \sqrt{-1})$, then $x \pm y = (a \pm c) + (b \pm d)r$ and $xy = (ac + bdr^2) + (ad + bc)r$, where $r^2 = 2$ or $r^2 = -1$; on the other hand, multiplying the numerator and denominator of the fraction x/y by the number $y^* = c - dr$, we obtain $x/y = [(ac - bdr^2)/(c^2 - d^2r^2)] + [(ad + bc)/(c^2 - d^2r^2)]r$. Thus when $y \neq 0(= 0 + 0r)$ the ratio x/y also has the form $A + Br$, where $A, B \in Q$ (note that when $c, d \in Q$, the relation $c^2 - d^2r^2 = 0$ holds only if $c = d = 0$).

[36] The first description of Galois theory in monograph form was Jordan's book mentioned earlier. One of the best popular accounts in Russian was written by M. Postnikov (see the English translation: M. Postnikov, *Foundations of Galois Theory*, Oxford University Press, 1962); for a more advanced presentation, see, for example, E. Artin, *Galois Theory*, Notre Dame, 1948.

[37] In Moscow University's mathematics department, the Galois Theory course was taught for a number of years by the outstanding algebraist (who also specialized in the theory of numbers and discrete geometry) Boris Delonay (1890–1980). The course followed exactly Galois's letter to his friend Auguste Chevalier, written just before Galois's death. The letter is only about ten printed pages long, but the course based on it, in which the lecturer gave rigorous definitions of all the concepts and presented all the proofs Galois left out for lack of time, required a full academic year. Galois's complete works constitute a very slim volume (*Œuvres mathématiques*, Paris, Gauthier-Villars, 1897. See also Note 1); it was issued on the initiative of the Société Mathématique de France and edited by the highly regarded Emile Picard (1856–1941).

[38] Two elements α, β in a group $\mathcal{G}$ (for example, two permutations) *commute* if $\alpha\beta = \beta\alpha$ or $\alpha^{-1}\beta^{-1}\alpha\beta = \varepsilon$, where ε is the identity element of $\mathcal{G}$ (the identity permutation). The element $[\alpha\beta] = \alpha^{-1}\beta^{-1}\alpha\beta$ is called the *commutator* of α and β. The subgroup $\mathcal{Z} = \mathcal{Z}(\mathcal{G})$ of the group $\mathcal{G}$ generated by all commutators $[\alpha\beta]$, where $\alpha, \beta \in \mathcal{G}$, is called the *commutator subgroup* of $\mathcal{G}$; its "size" serves as a "measure of commutativity" of $\mathcal{G}$ (if $\mathcal{G}$ is commutative, then the commutator subgroup $\mathcal{Z}$ consists of the single element ε). For the commutator subgroup $\mathcal{Z} = \mathcal{Z}_1$ of $\mathcal{G}$ we can form its commutator subgroup $\mathcal{Z}_2 = \mathcal{Z}(\mathcal{Z}_1) = \mathcal{Z}(\mathcal{Z}(\mathcal{G}))$ (the second commutator, or the commutator of the commutator of $\mathcal{G}$); the third commutator $\mathcal{Z}_3 = \mathcal{Z}(\mathcal{Z}_2)$ is formed in the same way, and so on. The group $\mathcal{G}$ is called *solvable* if its sequence of commutator subgroups $\mathcal{G} \supset \mathcal{Z}_1 \supset \mathcal{Z}_2 \supset \mathcal{Z}_3 \supset \cdots$ ends with the identity element ε of the group: $\{\varepsilon\} = \mathcal{Z}_n$ (for a commutative group this condition holds trivially, and $n = 1$). Other definitions of a solvable group use the concept of a group's normal series, introduced by Jordan; see Note 49.

[39] The fullest review which Galois could make of what he had done in the theory of equations was in the letter to his friend Auguste Chevalier, mentioned in Note 37, written on the eve of the duel: Galois spent that night writing the letter. Time, wrote Galois, was running short. He had kept what he was putting down on paper in his head for about a year, but lacked the opportunity to set it down in greater detail. He was aware and afraid that he would be accused of stating theorems for which he did not have full proofs. The letter ended with the request that it be published in the *Revue encyclopédique* (Chevalier carried out this last wish, but, obviously, the item was not understood by anyone at the time and was ignored) and to ask Jacobi or Gauss to appear in public, not to state whether the theorems are correct, but to appraise their importance; after that, Galois hoped, people would be found who would clear up *ce gachis*. However, there is nothing to show that Jacobi or Gauss ever learned about Galois's ideas, and the first person to seriously undertake to clear up *ce gachis* was Camille Jordan, who is often considered (unjustly, I think) to be the real author of Galois theory, since Galois's own exposition was undoubtedly defective.

[40] Of course, Cayley's tables "begin" with such a row and column only if the group's elements begin with the identity element ε. Since this need not be the case, it would be more correct to say that these tables include such a row and column.

[41] The considerable progress achieved by Galois (as compared with Ruffini and Abel) was largely due to the fact that he did not consider commutative groups only (we note that the equations solvable in radicals considered by Abel can nowadays be characterized by the commutativity of their Galois groups (and by certain additional conditions); according to the definition in Note 38, any commutative group is certainly solvable).

[42] It is easy to see that, apart from notation, there is only one group of order two and only one group of order three (both groups are commutative). But there are two (very different) groups of order four, namely, the groups (1.6) (or (1.6′)) and (1.7) (or 1.7′)); the first is called the "cyclic" group (of order four) and the second the Klein group. The number of groups of order N grows quite rapidly with N: thus, for example, the number of different groups of order 2, 4, 8, 16, 32 and 64 is equal to 1, 2, 5, 14, 51, and 267 respectively. The interest in "finite" objects, typical of present-day mathematics, is illustrated by a monograph by the well-known American algebraist Marshall Hall, Jr. and his colleague K. James, Sr., published in 1964 and containing a detailed description of all these 340 ($= 1 + 2 + 5 + 14 + 51 + 267$) groups (see M. Hall, Jr., K. James, Sr. *The Groups of Order* $2^n (n \leqslant 6)$, N.Y., MacMillan, 1964).

[43] The permutation $\binom{1\ 2\ 3\ \cdots\ n}{i_1\ i_2\ i_3\ \cdots\ i_n}$ is called even (odd) if the finite sequence i_1, $i_2, i_3, \ldots, i_n$ of natural numbers obtained by rewriting the initial numbers 1,

2, 3, ... , n in a different order contains an even (odd) number of transpositions, i.e., pairs (i_a, i_b) of numbers appearing in the "wrong order" (as compared with their natural order): pairs (i_a, i_b) such that $a < b$ but $i_a > i_b$.

[44] On the structure of groups of isometries and similitudes see, for instance, the very elementary books I.M. Yaglom, *Geometric Transformations* I, II, N.Y., Random House, 1962, 1968, volumes 8 and 21 in the New Mathematical Library series, the slightly more advanced, but accessible, book G.E. Martin, *Tranformation Geometry*, N.Y. Springer, 1982, and a book, also meant for beginners, which makes rather wide use of the terminology and notations of group theory, M. Jeger, *Transformation Geometry*, London, Allen and Unwin, 1966.

[45] Recent decades have been marked by a veritable assault on the topic of finite groups, involving a wide use of computers by a large number of researchers. This peculiar boom (see Note 42) is a very good illustration of the switch, characteristic of the second half of the 20th century, from "continuous" mathematics (the differential and integral calculus of Newton and Leibniz) to "discrete" mathematics; this switch is closely linked to the appearance of computers, devices which are discrete in principle. No topic in the theory of finite groups is as important as that of the mysterious subject of finite simple groups (see Note 260).

[46] Often the definition of a field does not include the requirement that multiplication be commutative, and then the multiplicative group of the field, consisting of all the nonzero elements, can be nonabelian (non-commutative); if the multiplicative group is commutative, then the field is also called commutative. If the field is not commutative, then two distributivity conditions are required:

$$a(b + c) = ab + ac \quad \text{and} \quad (a + b)c = ac + bc.$$

[47] The elements of a two-element field can be represented as "even" (0) and "odd" (1) numbers; then the addition and multiplication tables of the field reflect rules of operation with integers such as: even × even = even, and so on. More generally, a field with p elements, where p is a prime, can be represented as the set $0, 1, 2, \ldots, p - 1$ of residues modulo p, i.e., the set of classes of integers characterized by the remainder in the division of its elements by p. (If $p = 2$, then we arrive at the classes of even and odd numbers.) The sum $a + b$ and the product ab of elements in the field coincide with the respective remainders in the division by p of the arithmetical sum $a + b$ and product ab.

[48] Its continuing scientific importance was reaffirmed by the reprinting of that classic in 1957 by the Paris publishing house Gauthier–Villars, which first issued it in 1870.

[49] The Jordan–Hölder theorem (Otto Hölder, 1859–1937, was a German mathematician) asserts that the number of terms in a normal series (of which it is natural to require maximal "compression" in the sense that (a) any two adjacent terms of the series are distinct; and (b) no intermediate group $\mathcal{G}$ can be inserted between two consecutive groups $\mathcal{G}_i$ and $\mathcal{G}_{i+1}$ in the series (a chain $\mathcal{G}_i \supset \mathcal{G} \supset \mathcal{G}_{i+1}$ would increase the number of subgroups in the normal series by one)) is completely determined by the initial group and not by the choice of the normal series, which can in fact be chosen in different ways even if our additional conditions are fulfilled. If the factors (already "compressed to the utmost") of the normal series are abelian (commutative), then the group is solvable.

Chapter 2

[50] Of course, the objects considered in differential geometry (curves, surfaces) must be "smooth," i.e., must be determined analytically by smooth (differentiable) functions, otherwise the methods of the calculus cannot be applied to them.

[51] His appearance being so close to the type admired by the German Nazis, Lie might perhaps have been admired by Hitler but the sympathy would hardly have been mutual.

[52] This episode is described in greater detail in the last chapter of the book.

[53] It is curious that, unlike Klein, the Italian mathematician Federigo Enriques (1871–1946), a well-known expert on the foundation of geometry who was slightly younger than Klein and also graduated from a classical gymnasium, regarded the study of ancient languages, and, in particular, of ancient Greek grammar, as a very important element of education, instrumental in the development of logical thinking. Apparently Enriques did not think that the geometry taught in Italian schools could develop logical thinking!

[54] A monument to Gauss and Weber was unveiled in Göttingen in 1899; it shows them working jointly on the invention of the telegraph. A special memorial volume was published on the occasion. It contained two long papers: D. Hilbert, *The Foundations of Geometry* (regarded as continuing Gauss's research) and E. Wiechert, *The Foundations of Electrodynamics* (a continuation of Weber's work). Today, the latter has been almost completely forgotten, while Hilbert's work was reissued almost immediately as a separate book, which has now gone through more than ten continuously improved editions in German, has been translated into practically all European languages and seems destined to live a long and glorious life. From the standpoint of the psycho-physiological differences between Weierstrass and Klein (as well

as between Weierstrass and Riemann; see below) it is only natural that Weber was on friendly terms with Klein and Riemann, both very "physics-minded" scientists.

[55] Things are slightly different in the case of hieroglyphic writing, which is more like a picture; this, however, is not the place to deal with the relevant details.

[56] The lecture in question is "Topologie und abstrakte Algebra als zwei Wege mathematischen Verständnisses." It was read by Weyl in 1931 (he was then a Swiss gymnasium teacher) and published in *Unterrichtsblätter für Mathematik und Naturwissenschaften*, Bd. 38, 1933, S. 177–188; reprinted in H. Weyl, *Gesammelte Abhandlungen*, Bd. 3, Heidelberg, Springer, 1968.

[57] See, for example: I.M. Yaglom, *Hermann Weyl*, Moscow, Znaniye Publishers, 1967 (in Russian).

When Weyl's paper was written, topology was definitely considered a part of geometry; opposing abstract algebra to topology in the title was the same as contrasting algebra with geometry. But, in the second half of the century, topology largely changed its character (this is reflected in the term "algebraic topology," designating one of the most important parts of the science); this undermined some of the most geometrically-minded researchers in topology and, in some cases, even resulted in serious psychic disorders.

[58] Further grounds for that antipathy were provided by the fact that Leibniz, who valued Newton highly as a mathematician, underestimated his achievements in physics, of which Newton was particularly proud.

[59] And equally successful: accepted to the physics department by Plücker to prepare for research in physics, Klein became an outstanding mathematician. Sommerfeld, who was asked to be an assistant in the mathematics department, subsequently became famous as a physicist. Of course, this did not prevent Klein and Sommerfeld from being grateful to their respective teachers throughout their lives, and from acknowledging how much they owed them.

[60] The issue of the journal in which the article was published was entirely devoted to Felix Klein. Let us quote a few more sentences from the same article, in which Sommerfeld very vividly describes his tutor's style of thinking and teaching: "... his conclusions made his thoughts stand out in bold relief. His thoughts but not his calculations. The latter played a very modest part in Klein's lectures. It was one of the points in which his approach came close to Riemann's way of thinking. He defined functions by their properties, regardless of their formal representation; the formula was not the basis, but only the source of mathematical knowledge!"

Chapter 3

[61] Paradoxically, only recent research in psychology concerning the different functions of the left and right hemispheres of the brain sheds light on the difference between algebraic and geometric thinking, their complementary nature and their equal importance in the cognition of reality. And, of course, it would be completely incorrect to imagine ancient Greek mathematics as pure geometry and the mathematics of the Renaissance as pure algebra: it is only a question of one or another trend prevailing, and tentatively at that, because there can be a purely algebraic approach to geometric problems (more will be said about this in connection with projective geometry) as well as a geometric approach to algebraic subjects. The importance attached to the concept of (natural) number in the school of Pythagoras of Samos (6th century B.C.), the number mysticism characteristic of that school and its odd reflections in the later works of the neo-Pythagoreans—*The Arithmetic* (or *Introduction to Arithmetic*) by Nicomachus of Gerasa (circa 100 A.D.), *On the Arithmetic of Nicomachus*, a work by a Syrian Christian, Iamblichus of Chalcis (about 250–about 330 A.D.) commenting and continuing Nicomachus—all this contradicts the view of Greek mathematics being pure geometry. Even Euclid's *Elements*, with which our idea of ancient mathematics is primarily associated, includes much brilliant material in number theory (like the proof that the sequence of prime numbers is infinite), mostly derived from the Pythagoreans. Above all, the most brilliant figure in the Greek mathematics of the Roman period was Diophantus of Alexandria (most likely 3rd century A.D.), whose interests focussed exclusively on arithmetic and algebra.

[62] One of the consequences of Pascal's geometric works being ahead of their time and quickly forgotten was the irremediable loss of Pascal's treatise on the theory of conic sections, a treatise admired by Leibniz, who had urged Pascal's heirs to publish that work (unfortunately, this was not done). Unlike Leibniz, René Descartes (1596–1650) underestimated Pascal's geometric works. This was one of the reasons for the tense relations between Descartes and the group of scientists around Blaise's father, Etienne Pascal (1588–1651) and Gilles Personne (who called himself de Roberval, 1602–1675), the group from which the French academy of sciences subsequently arose (and, eventually, l'Institut; see Note 65). The intellectual conflict between Descartes and Pascal is sometimes explained by the intrinsic incompatibility of Descartes's formal algebraic thinking with Pascal's "physical" thinking. Yet the great logician Leibniz welcomed Pascal's works enthusiastically (in this connection see the main text on Weierstrass's attitude to Riemann).

[63] Thus, for example, Euler introduced the study of a figure's affine properties, and so initiated affine geometry. (The term "affine" is also due to Euler and derives from the Latin *affinitas*—relation by marriage—Euler sought to emphasize that although a figure and its affine image are not, strictly speaking, similar, they are, nevertheless, related.)

[64] The decline of interest in geometry characteristic of the 20th century, which has relegated it to the position of a second-rate science, led to a paradoxical situation: for a time, the five-year course at the Mathematics department of Moscow University did not include a single compulsory geometric subject (except for a preliminary course in analytic geometry in the first semester). This also is to blame for the worldwide confusion concerning the teaching of geometry in secondary school, which has led to numerous proposals to abolish geometry in secondary school altogether (this viewpoint was vigorously supported in the introduction to Jean Dieudonné's *Algèbre linéaire et géomètrie élémentaire* (Paris, Hermann, 1968)).

[65] The hard times in which Carnot lived and worked were clearly reflected in the abrupt changes in his position within the French scientific hierarchy. In the years of the revolution, as a member of the Committee for Public Salvation (in effect the revolutionary government of France), Carnot took an active part in organizing the new scientific center, L'Institut, which replaced the Royal Academy of Sciences. When the Institut was finally set up in 1796, he naturally became one of its members. But by then his position was very precarious (this applies to the period after the Thermidor coup d'état, when only the honorary title of "organizer of victory" saved him from trial and death). In the very next year, 1797, Carnot was expelled from the Institut which he had founded, and was replaced by a young general—Napoléon Bonaparte. In 1800 Napoleon, seeking reconciliation with Carnot, reinstated him at the Institut, and although Carnot's sharp criticism of Napoleon's adopting the title of Emperor spoiled his relations with the new ruler, he still remained a member of the Institut during Napoleon's entire rule. After Napoleon abdicated, the fiery republican Carnot headed a Directoire which assumed full power for a short period of time, attempting to organize resistance against the interventionist forces in France. Naturally the Bourbons never forgave him these actions. Once again, he was expelled from the Institut (as was Monge, in 1816) and sent into exile; he died abroad (in Magdeburg).

[66] Typically, the officers who proposed the problem refused at first to consider Monge's solution, being certain that his mathematical training was insufficient for solving it.

[67] Monge's role in the founding of the École Polytechnique was such that he could later tell one of his pupils, with good reason: "I created the school as I deemed necessary." Monge also took part in the setting up of the Paris École Normale for the training of future teachers (see above) but here his role was minor; this is probably why the École Normale was initially far behind the École Polytechnique.

[68] It is only natural that Klein should have paid a great deal of attention to the École Polytechnique in his *Vorlesungen über die Entwicklung der Mathematik im XIX Jahrhundert* (Teil 1, Heidelberg, Springer, 1979; we shall re-

peatedly return to his remarkable book below). Chapter 2 of Klein's book is entitled "France and L'École Polytechnique in the First Decades of the 19th Century," and begins with a section on "The Rise and Organization of the School" which describes the principles underlying the institution and its curricula. Chapter 3 of Klein's book, "The Flourishing of Mathematics in Germany," begins with an introduction called "An Attempt to Create a Polytechnical School in Berlin;" here Klein sadly describes the unsuccessful endeavor to carry over the French experience to German soil: Alexander von Humboldt's (1769–1859) attempt to set up a German counterpart of the École Polytechnique to be headed by Gauss. The attempt failed through Gauss's refusal to take part. After lengthy discussions, efforts to create an alternative school for the training of mathematics teachers for the higher forms of the gymnasium (similar to the Paris École Normale) were also abandoned. This was due to the death of Abel, who was to have been the leading mathematician there.

[69] The original name of the descriptive geometry course was *Texte des leçons de géomètrie descriptive données à l'École Normale*, the original title of the course in differential geometry was *Feuilles d'Analyse appliquée à la géomètrie à l'usage de l'École Polytechnique*; the first edition consisted of 28 feuilles, the second of 34 feuilles (in the 3rd and 4th editions the feuilles were stitched together and therefore entitled *Analyse appliquée à la géomètrie*).

[70] It was at this time that Monge succeeded in persuading Napoleon to visit the École Polytechnique, his pride and joy, for the first (and last) time.

[71] Poncelet's only substantial contribution to science and pedagogy after the appearance of the *Traité* was the publication of his *Cours de Mécanique* (1826) based on his lectures at the École Polytechnique; this outstanding book played a major role in the further progress of mechanics.

[72] See, for example, Chapter IV in R. Courant and H. Robbins, *What is Mathematics*? London, Oxford University Press, 1948; or Chapter 6 in H.S.M. Coxeter and S.L. Greitzer, *Geometry Revisited*, N.Y., Random House, 1967; or H.S.M. Coxeter, *Projective Geometry*, Waltham (Mass.), Blaisdel, 1964; J.W. Young, *Projective Geometry*, 1938; H.S.M. Coxeter, *Introduction to Geometry*, N.Y., Wiley (Chap. 14); G. Ewald, *Geometry: an Introduction*, Belmont (Calif.), Wadsworth, 1971 (Chap. 5); D. Pedoe, *A Course of Geometry for Colleges and Universities*, Cambridge, Cambridge University Press, 1970 (Chap. 7); D.J., Struik, *Lectures on Analytic and Projective Geometry*, Cambridge (Mass.), Addison-Wesley, 1953; H.S.M. Coxeter, *The Real Projective Plane*, Cambridge, Cambridge University Press, 1955 (the books are listed in order of increasing difficulty). The following articles may also serve as elementary introductions to projective geometry: G. Pickert, R. Stender, and M. Hellwich, "From projective to Euclidean Geometry" in *Fundamentals of Mathematics*,

Vol. II (Geometry), The MIT Press, 1974, pp. 385–436 and I.M. Jaglom and L.S. Atanasjan, *Geometrische Transformations* in: *Enzyklopädie der Elementarmathematik*, Bd. IV (Geometrie), Berlin (DDR), Deutscher Verlag der Wissenschaften, 1980, S. 43–151 (cf. the book mentioned in Note 73). The following (brilliant) books are quite different: F. Klein, *Vorlesungen über höhere Geometrie*, Heidelberg, Springer, 1968 and W. Blaschke, *Projektive Geometrie*, Basel, Birkhäuser, 1954.

[73] The study of the invariant properties of figures under parallel projection is a necessary element of Monge's descriptive geometry. Such an approach to affine geometry, and a similar approach to projective geometry, based on central projection, are consistently applied in I.M. Yaglom, *Geometric Transformations* III, N.Y., Random House, 1973, the beginning of which can serve as an introduction to affine geometry. See also Chapter 13 of H.S.M. Coxeter's *Introduction to Geometry* mentioned in Note 72 and books by I.M. Yaglom and V.G. Ashkinuze (see Note 268).

[74] Apparently Gergonne and Poncelet discovered the duality principle independently and proved it in different ways. Gergonne believed that the duality principle holds because of the complete symmetry between the basic properties of points and lines (which he stressed by writing statements dual to each other in two columns, graphically demonstrating the parallelism). Today we can refer to the "self-duality" of the axioms of projective geometry, which automatically implies the duality principle. But the logical structure of geometry was not so clearly understood in Gergonne's time, and a complete list of axioms had not yet been drawn up. As for Poncelet, he proved the duality principle by showing the existence of a "dual" or "polar" transformation, briefly, a *polarity*, sending each point of the plane (proper or at infinity) into a line, and any set of collinear points into a sheaf of (concurrent, i.e., intersecting in one point, or parallel) lines. A polarity maps the configuration associated with each theorem to the configuration associated with its dual. If we proceed from the model of the projective plane as the sheaf of lines and planes passing through a fixed point O of space, then the correspondence $l \leftrightarrow \pi$, where $l \perp \pi$ (the line l and the plane π pass through O), is a polarity. The properties of polarities of the (projective) plane are analyzed in detail, for example, in the article by Yaglom and Atanasjan and the book by Yaglom mentioned in Notes 72 and 73. For another proof of the duality principle, due to Möbius, see Note 81 below.

[75] Characteristically, in the context of the declining interest in geometry mentioned previously (see also Note 64), projective geometry largely retained its position because of its close alliance with algebra; as a result it is possible to regard modern projective geometry as a branch of algebra rather than geometry. On this see the classical works of R. Baer, *Linear Algebra and Projective Geometry*, N.Y. Academic Press, 1952 and E. Artin, *Geometric*

Algebra, N.Y. Interscience, 1957, as well as the textbooks of R. Hartshorne, *Foundations of Projective Geometry*, N.Y., Benjamin, 1967; R. Artzy, *Linear Geometry*, Reading (Mass.), Addison-Wesley, 1965; E.W. Stevenson, *Projective Planes*, San Francisco, Freeman, 1972; K.W. Gruenberg and A.J. Weir, *Linear Geometry*, Princeton (N.J.), van Nostrand, 1967; G. Pickert, *Projektive Ebenen*, Berlin, Springer, 1955 (Hartshorne's textbook is the easiest of these).

[76] It is undoubtably to Chasles's credit that he was the first fully to appreciate the works of ~~Descartes~~ and Pascal, which had been forgotten by that time. It is an indication of the abnormal relations between the founders of projective geometry that Poncelet considered Chasles's book as hostile to him: he regarded the appraisal of works by 17th-century scholars as a means of playing down his own achievements.

M. Chasles, the head of the geometry department of the famous *École Polytechnique*, was the first French mathematician whom Lie and Klein visited in Paris. The elderly Chasles (then 77) was quite friendly to the two young foreigners and appreciative of their work: as an *Institut* member, he presented their joint paper on *W*-curves (to which we will return below), as well as Lie's important article "On Geometric Transformations," to the *Comptes Rendus* of the French Academy.

[77] See M. Chasles, *Traité de la géometrie supérieure*, Paris, 1852.

[78] In his remarkable lectures on the development of mathematics in the 19th century (see Note 68), Klein compares Chasles with Steiner, the head of the synthetic school in German geometry. These two are indeed similar in some respects: both were exceptionally hard working, prolific and rather ill-tempered; both became professors relatively late, but immediately became heads of mathematical schools, which they supervised rather stringently, and so on.

[79] Lacking any interest in people and any desire to support young talent, Gauss—who offended both Bolyai and Abel by saying about their respective works that praise for them would be immodest, in that he had come to the same thoughts himself a long time before—lacked the ability to perceive the potential of young scholars. He sought to direct Möbius to astronomy and to computational mathematics; just as unreasonably, he attempted to direct another outstanding geometer—von Staudt—to algebra and to the theory of numbers.

[80] Möbius's paper on one-sided surfaces was called *On the Volume of Polyhedra*. Discussing the volume problem, Möbius concluded that some polyhedra, e.g., the one now known as the Möbius heptahedron (obtained from a regular octahedron *EABCDF* with opposite vertices *E* and *F* by removing the two "upper" faces *EAB* and *ECD* and two "lower" faces *FAD*

and FBC, but then adding the three diagonal faces $EAFC$, $EBFD$ and $ABCD$) cannot be assigned any volume at all, and carefully analyzed the reasons for this phenomenon. We note that the Möbius strip was independently discovered in Göttingen by Johann Benedict Listing (1808–1882) in the same year (1858). This clearly points to the objective nature of the development of mathematics, in which discoveries are made "when the time has come" and very often independently by several scholars at the same time.

81 The new approach to projective geometry shed new light on the duality principle. Möbius interpreted that principle in the following way. In the equation $a_0x_0 + a_1x_1 + a_2x_2 = 0$ of the line a it is natural to consider the numbers a_0, a_1, a_2 as (fixed!) coefficients of the equation and the numbers x_0, x_1, x_2 as (variable) coordinates of the point $X = (x_0 : x_1 : x_2)$ on the line a. Since the line a is completely determined by the coefficients a_0, a_1, a_2 of its equation (more precisely, the ratio of these coefficients, $a_0 : a_1 : a_2$), the numbers a_0, a_1, a_2 can be viewed as coordinates of the line $a = (a_0 : a_1 : a_2)$ (Möbius called them tangential coordinates). But, conversely, we can regard the numbers x_0, x_1, x_2 in our equation as fixed and a_0, a_1, a_2 as (variable) tangential coordinates of lines. Then the set of all lines $a = (a_0 : a_1 : a_2)$ whose coordinates satisfy the linear equation $a_0x_0 + a_1x_1 + a_2x_2 = 0$ with the coefficients x_0, x_1, x_2 is the set of lines in the projective plane passing through the fixed point $(x_0 : x_1 : x_2)$. Thus the same equality is now regarded as the equation of a sheaf of lines with center at $(x_0 : x_1 : x_2)$, i.e., as an equation of a point of the projective plane in tangential coordinates. Finally, if the numbers x_0, x_1, x_2 and a_0, a_1, a_2 are viewed as fixed, then the same relation between these numbers is a condition for the incidence of the point X and the line a: $X \in a \Leftrightarrow a_0x_0 + a_1x_1 + a_2x_2 = 0$. It is this symmetry of the analytic (coordinate) interpretation of the concepts "point of the projective plane" and "line of the projective plane" which proves the equivalence of the properties of points and of lines.

82 On circle geometry (Möbius geometry) see, for example, Chapter 5 of the book by Coxeter and Greitzer in Note 72; Chapter 5 of Ewald's book; Chapter 6 of Coxeter's *Introduction to Geometry*; Chapter 6 of Pedoe's book; and part A of the article by I.M. Jaglom, *Geometrie der Kreise* in the *Enzyklopädie der Elementarmathematik*, Bd. IV (S. 459–488).

83 Thus, for instance, in connection with projective transformations Steiner (and virtually hundreds of authors after him) speak of *Gebilde erster*, *zweiter und dritter Stufe* where it would be natural to speak about one-dimensional, two-dimensional, and three-dimensional manifolds.

84 Even Klein, who did more than anyone else to popularize and propagate Staudt's ideas, admitted that these books were difficult for him to understand. Only his friendship with Otto Stolz (1842–1905), a mathematician close to

Staudt in spirit, who explained Staudt's ideas to Klein, enabled him to appreciate these outstanding works (which Klein himself, however, never read!).

[85] The discovery of cathode rays was completed (under Plücker's guidance) by his pupil Johann Wilhelm Hittorf (1824–1914). It is also to be noted that Plücker was the founder of spectral analysis (in physics).

[86] A systematic account of that theory is contained in Steiner's long book (which unfortunately was published only posthumously) *Die Theorie der Kegelschnitte gestützt auf projektive Eigenschaften*, Leipzig, Teubner, 1867; 3rd edition, 1898).

[87] Here is Steiner's definition: consider two sheaves of lines (with different centers Q and Q') such that these sheaves can be obtained from one another by a chain of central projections; then the set of points at which corresponding lines of the sheaves intersect constitutes a (possibly degenerate) conic section.

[88] A short but essentially complete and understandable account of von Staudt's ideas is given by Klein in Chapter V of his *Vorlesungen über nicht-euklidische Geometrie* (Heidelberg, Springer, 1968) published posthumously. A more detailed description of these ideas, reflecting more fully their inner perfection, is given in Chapter VIII of Young's little book *Projective Geometry*, mentioned in Note 72.

[89] Let line a be the intersection of two planes

$$A_1x + A_2y + A_3z + A_0 = 0,$$
$$B_1x + B_2y + B_3z + B_0 = 0;$$

in this case, the second-order minors of the matrix $\begin{pmatrix} A_1 & A_2 & A_3 & A_0 \\ B_1 & B_2 & B_3 & B_0 \end{pmatrix}$, i.e., the numbers

$$p_{12} = \begin{vmatrix} A_1 & A_2 \\ B_1 & B_2 \end{vmatrix}, \quad p_{13} = \begin{vmatrix} A_1 & A_3 \\ B_1 & B_3 \end{vmatrix}, \quad p_{10} = \begin{vmatrix} A_1 & A_0 \\ B_1 & B_0 \end{vmatrix},$$
$$p_{23} = \begin{vmatrix} A_2 & A_3 \\ B_2 & B_3 \end{vmatrix}, \quad p_{20} = \begin{vmatrix} A_2 & A_0 \\ B_2 & B_0 \end{vmatrix}, \quad p_{30} = \begin{vmatrix} A_3 & A_0 \\ B_3 & B_0 \end{vmatrix},$$

are called the *Plücker coordinates* of a. It is not difficult to see that the numbers $p_{12}, \ldots, p_{30}$ are determined by the line a up to a common factor, so that the point $A(p_{12} : p_{13} : p_{10} : p_{23} : p_{20} : p_{30})$ in five-dimensional projective space P^5 corresponds to a; the Plücker coordinates of any line satisfy the Plücker relation:

$$p_{12}p_{30} - p_{13}p_{20} + p_{10}p_{23} = 0, \qquad (*)$$

which determines the so-called Plücker quadric Σ in P^5; the points of this (four-dimensional!) surface correspond bijectively to the lines in three-dimensional projective space.

Chapter 4

[90] Among Greek geometers, Claudius Ptolemy (c. 100–c. 170 A.D.) paid particular attention to spherical geometry; his principal work *The Large Mathematical Construction in Thirteen Books*, better known under the Arabic title *Almagest* (from the Greek *μεγιστη*, "the greatest," as the Arabs judged Ptolemy's work), begins with a detailed account of the plane and spherical geometry needed for subsequent studies, including the elements (largely worked out by Ptolemy himself) of plane and spherical trigonometry.

[91] It was this fact, involving the history of mathematics and not at all mathematics itself, which explains the exceptional popularity of hyperbolic geometry. Innumerable books and articles have been devoted to it (see Note 116; there are even several Russian books on so exotic a subject as *the theory of geometric ruler and compass constructions in the Lobachevskian plane*!). Other equally important and, in many cases, simpler geometric non-Euclidean systems (some of which will be dealt with below) have been accorded much less attention in the scientific and popular literature and in textbooks.

[92] G. Nöbeling's recent book on non-Euclidean geometry (see Note 116) includes F.K. Schweikart among the discoverers of hyperbolic geometry; this is an unusual point of view, but it cannot be rejected as completely groundless.

[93] See, for example, E.T. Bell's book mentioned in Note 1 (the chapters on Gauss, Lobachevsky, and Bolyai) and other books on the history of mathematics. Of the latter we mention only some of the most substantial, such as, Morris Kline, *Mathematical Thought from Ancient to Modern Times*, N.Y., Oxford University Press, 1972 (Chapter 36); the classical work of F. Engel and P. Stäckel, *Die Theorie der Parallellinien von Euklid bis auf Gauss*, 2 Vols., Leipzig, Teubner, 1895; books on Lobachevsky's non-Euclidean geometry, some of which contain a detailed history of the issue (see, for example, M.J. Greenberg's book mentioned in Note 116 and subtitled *Development and History*); Russian books by V.F. Kagan (an English translation of one of them is currently available: V.F. Kagan, *N. Lobachevski and his Contribution to Science*, Moscow, FLPH, 1957) and B.A. Rosenfeld's *History of Non-Euclidean Geometry*, now being translated into English.

[94] These three postulates should be supplemented by the requirement that the intersection points of two given lines (given, in each case by, say, two points), of a given line and a given circle, and of two given circles are assumed to be given. For an axiomatic approach to the theory of geometric constructions see, for example, L. Bieberbach *Theorie der geometrischen Konstruktionen*, Basel, Birkhäuser, 1952.)

[95] It is possible that Euclid's unexpected Postulate IV refers to the plane's

"degree of mobility," in the sense that it claims that the plane can be moved so as to superpose any right angle (with a given order of sides) on any other right angle. Seeking to avoid direct reference to motion (because of the metaphysical principles derived from Aristotle and, earlier, from Zeno of Elea), Euclid may have laid down that basic principle in the form of a postulate on the congruence of all right angles.

[96] The best known of the simpler forms of Euclid's postulate V reads: through a given point not lying on a given line there can be drawn only one line parallel to the given line. Apparently this form of the parallel axiom first appeared in an old school version of the *Elements* (which was used in English secondary schools as a standard geometry textbook) issued in 1795 by the English teacher John Playfair; hence it is sometimes called the *Playfair axiom.* It is the latter which appears (instead of Euclid's much more complicated fifth postulate) in the modern version of the axioms proposed in 1899 by the great David Hilbert (1862–1943; see his *Foundations of Geometry*, La Salle, Ill., Open Court, 1971). In British schools, at least until the 1920s, geometry was mostly taught according to Euclid; many schoolchildren must have thought that Euclid was an English schoolmaster who, unfortunately for them, had written a mathematics textbook. It is possible that Euclid deliberately chose such a clumsy and intuitively unclear form for his fifth postulate in order to underscore its special place in the system of facts on which geometry is founded.

[97] Most of these attempts led to the replacement of Euclid's fifth postulate by another proposition which would turn out to be equivalent, provided all the other axioms explicitly stated or tacitly assumed by Euclid were accepted. Many of these propositions equivalent to the fifth postulate were thought to be more obvious by some mathematicians, i.e., their validity seemed intuitively clearer; one such is the assumption that *through every point in the interior of an angle it is possible to draw a line intersecting both sides of the angle.* The outstanding 18th-century French mathematician Alexis Claude Clairaut (1713–1765) published a school textbook (*Eléments de géomètrie.* Paris, 1741) that replaced the fifth postulate by the assumption that *rectangles exist*; this he justified by referring to "the shape of houses, rooms, walls and so on." However, from the standpoint of logic, such assumptions are, of course, in just as much need of proof as Euclid's postulate. (When geometry is approached as a mathematical science founded on the scheme: undefined notions → subsequent definitions; axioms → theorems, the very idea of "more obvious" and "less obvious" propositions no longer makes sense.)

[98] Lambert's most famous achievements were probably his proofs that the base e of the system of natural logarithms and the ratio π of the circumference of the circle to the length of its diameter are irrational.

[99] Saccheri's quadrilateral had been considered earlier by a number of leading Arab (or, to be more precise, Arabic-speaking) mathematicians. These included Omar Khayyam (c. 1048–c. 1131), Persian (or Tajik) mathematician, born in Nishapur, Khorasan, who worked in Samarkand, Bukhara, Isfakhan, and Merva, astronomer, philosopher, and famous poet, author of the celebrated *Rubaiyat* widely known in Edward Fitzgerald's translation; and Nasir-Eddin (1201–1274, born in the city of Tusa in Khorasan), who worked in Quhistan, the capital of the "country of assassins" (founded by the "old man of the mountains" Khasan-ibn-Sabakh and his terrorist sect of drug addicts and assassins). Following the conquest of that dreadful state by the Mongols in 1256 he became court astrologer and adviser to the Mongol Khan Khalug, organizing an excellent observatory and science school in the capital Marage of Khulag's state (Southern Azerbaijan). Lambert's quadrilateral had been considered earlier by another Arabic-speaking mathematician, the Egyptian al-Hasan ibn al-Haitham (965–1039), better known in Europe as Alhazen. All these mathematicians considered the same three hypotheses (of the acute angle, the right angle, and the obtuse angle) and showed that the fifth postulate is equivalent to the hypothesis of the right angle; but they failed to progress as far as Saccheri and Lambert in analyzing the two other hypotheses.

[100] Saccheri's basic work was *Euclides ab omni naevo vindicatus: Sive conatus geometricus quo stabiliuntur prima ipsa universae geometriae principia*, Milan, 1733; Lambert's treatise was *Theorie der Parallellinien*, Leipzig, 1786.

[101] It is not difficult to see that in Euclidean, hyperbolic and spherical (or elliptic; see page 63) plane geometries the "angular defect" $\delta = (n - 2)\pi - \angle A_1 - \angle A_2 - \cdots - \angle A_n$ of the polygon $A_1 A_2 \cdots A_n$ has the *invariance* property (congruent polygons have equal angular defects) and the *additivity property* (if a polygon M is split into two nonintersecting polygons M_1, M_2, then the angular defect of M is equal to the sum of the angular defects of M_1 and M_2). This already implies that *the angular defect $\delta(M)$ of the polygon M must be proportional to the area $S(M)$ of the polygon*: $\delta(M) = kS(M)$; here it is assumed that area satisfies the conditions of invariance, additivity and nonnegativity (as well as a normalizing condition, fixing the unit of area). The factor k is positive in hyperbolic geometry, negative in elliptic geometry and zero in Euclidean geometry.

[102] Here too, a comparison of modern and ancient mathematics shows that the ancient Greek thinkers possessed tremendous insight. Discussing the place of basic assumptions in deductive systems, Aristotle considers two possibilities: either the sum of the angles of a triangle equals π, and then the diagonal of a square is incommensurable with its side; or one could assume that the sum of the angles of a triangle is not equal to π, and then the diagonal of the square may turn out to be commensurable with its side. (Indeed, in spherical

geometry the length of the diagonal of a "square"—a quadrangle with congruent sides and congruent angles—can be twice that of the side.) However, says Aristotle, the corollaries of the assumption that the sum of the angles in a triangle is not equal to π are so ugly that they are not worth considering. To this, the famous Jewish philosopher Moses Maimonides (1135–1204), who was born in Spain and lived and worked in Spain and Egypt, objected: "We do not claim that God is impotent because he is unable to ... create a square whose side is equal to its diagonal," and considered Aristotle's references to the "ugliness" of one or another logical system illegitimate on the ground that God is the only judge of such matters. Perhaps in some other worlds God has actually created a different geometry, in which the sum of the angles of a triangle differs from π (compare the "astral" geometry of Schweikart discussed in the main text). Maimonides's argument was later repeated by the outstanding English clergyman and statesman Thomas Becket (1118–1170) who was later canonized as a Catholic saint. (It is quite likely that Becket, who studied in Moslem Spain, borrowed these ideas from Maimonides.)

[103] Schweikart began his mathematical research in 1808 by publishing a "proof" (of course, incorrect) of the fifth postulate. It is possible that the change in his viewpoint was partly due to the influence of Timofei Osipovsky (1765–1832), rector of Kharkov University (where Schweikart taught), whose "rector's address" of 1807 on space–time was directed against Kant's idea that the notions of space and time are given *a priori*, implying (according to Kant) that only one geometric system is possible.

[104] See, for example, the booklet V.G. Shervatov, *Hyperbolic Functions*, Boston, Heath, 1963, written in very simple language.

[105] See F. Klein, *Nicht-Euklidische Geometrie*, I. *Vorlesungen, gehalten während des Wintersemesters 1889–1890*, Göttingen, 1893.

[106] P. Stäckel and F. Engel, *Die Theorie der Parallellinien von Euklid bis Gauss. Eine Urkundensammlung zur Vorgeschichte der nicht-euklidischen Geometrie*, Leipzig, Teubner, 1895; F. Engel and P. Stäckel, *Urkunden zur Geschichte der nicht-euklidischen Geometrie*, Bd. 1, 2, Leipzig, Teubner, 1898, 1913. The latter work consists of two independent parts: F. Engel, *Nikolay Ivanovitsch Lobatschefsky, Zwei geometrische Abhandlungen aus dem russischen übersetzt mit Anmerkungen und mit einer Biographie des Verfassers*, Leipzig, 1898 (in writing Lobachevsky's biography and commenting on his works. Engel was helped by a mathematician from Kazan, Alexander Vasilyev, 1853–1929); and P. Stäckel, *Wolfgang und Johann Bolyai. Geometrische Untersuchungen. I. Leben und Schriften der beiden Bolyai. II. Stücke aus den Schriften der beiden Bolyai*, Leipzig und Berlin, 1913 (when writing this book Stäckel was partly assisted by the Hungarian mathematician J. Kurschák). In order to write *Die Theorie der Parallelinien von Euklid bis Gauss*, Engel learned

Russian and Stäckel learned Hungarian. Publication of Lobachevsky's and Bolyai's works (with commentaries by Stäckel and Engel) showed their complete independence of Gauss. If we bear in mind that Farkas Bolyai was unable to appreciate his son's great work, and Bartels had a sharply negative attitude toward Lobachevsky's research, then we see that the claim that Gauss's ideas were communicated to Lobachevsky and J. Bolyai by Bartels and F. Bolyai is without substance.

Stäckel also studied Gauss's geometric legacy in great detail; see his *Gauss als Geometer*, Gesellschaft der Wissenschaften, Göttingen und B.G. Teubner, Leipzig, 1923, which was published as a supplement to the second part of the 10th volume of Gauss's 12-volume *Werke* (which appeared in 1863–1933 under Klein's editorship) and was also issued as a separate book. In this work Stäckel carefully analyzes Gauss's achievements in non-Euclidean (hyperbolic) geometry and, very favorably inclined to Gauss, divides his activities into four stages: thoughts in his youth (1792–1795); early steps in the foundations of geometry (1795–1799); vacillations and doubts (1799–1805); creation of non-Euclidean geometry (1805–1817).

[107] See F. Klein, *Vorlesungen über nicht-euklidische Geometrie*, Berlin, Springer, 1968 (1st edition 1928).

[108] Various authors give different dates for Gauss's discovery of non-Euclidean geometry, but it is hardly possible to indicate a definite date before which Gauss doubted the existence of a geometry differing from Euclid's and after which he no longer doubted it. Gradually, over many years, Gauss arrived at the conclusion that there exist two equally valid systems of geometry. Sometimes it is assumed that Gauss still had his doubts as to the existence of non-Euclidean geometry as late as 1816. Authors often refer to the remarkable letter he wrote in that year to Gerling, in which he said that "as can be easily shown, if Euclidean geometry is not the true geometry, then no similar figures exist at all; in an equilateral triangle the angle changes with the length of the side, and I find nothing absurd in this. In this case the angle is a function of the side, and the side a function of the angle, however, in the latter function a certain constant length appears. It seems paradoxical that there exists a line segment, a length, which seems to be given *a priori*; but I find nothing contradictory in this. It would even be desirable that Euclidean geometry be untrue, because we would then have at our disposal a general *a priori* measure. For example, for the unit of length we could choose the side of the equilateral triangle whose angle equals $59°59'59.99999$." We could argue that since Gauss says that it would be desirable that Euclidean geometry be untrue, then he apparently thinks that it is true. However, we feel that this argument is not very convincing: Gauss thinks of the truth or untruth of geometry not in the purely logical sense but from the viewpoint of its relevance to the properties of physical space. Since he has no convincing experimental data, Gauss hesitates between two possibilities, equally valid from the logical

standpoint, but contradicting each other from the point of view of physics: (1) Euclidean geometry is true (i.e., it actually describes the physical world); (2) non-Euclidean geometry is true (see the next note).

[109] Of course, it would be more accurate to say that a positive result of Gauss's experiment (as well as Lobachevsky's later experiment), i.e., the discovery of a triangle with angle sum differing from 180°, would mean that Lobachevsky's non-Euclidean (hyperbolic) geometry gives a better model of the properties of the physical world than Euclidean geometry when points are interpreted as tiny parts of the Universe and straight lines as the trajectories of light rays. On the connection between mathematics and the natural sciences, see, for example, P.J. Davis and R. Hersh, *The Mathematical Experience*, Boston, Birkhäuser, 1981 or I.M. Yaglom, *Mathematical Structures and Mathematical Models*, N.Y., Gordon and Breach, 1986.

[110] Non-Euclidean geometry was at the heart of all of Lobachevsky's scientific research—he gave it most of his strength and abilities. However the results obtained by Lobachevsky in, say, pure algebra, also had indubitable scientific value. Also, having learned from the first publications of Bernhard Bolzano (1781–1848) the very general definition of a function as a mapping of a set X on a set Y such that to every $x \in X$ there corresponds no more than one $y \in Y$, Lobachevsky immediately (before Dirichlet, with whom the definition is usually associated) began to use this definition in his lectures.

[111] The high level of instruction at Kazan University was largely due to the efforts of the first superintendent of the Kazan public education district, the astronomer S.Y. Rumovsky, Euler's pupil and staff member.

[112] Magnitsky's name is associated with that of Rumovsky and an unfortunate period in the history of Kazan University. Positions largely lost under Magnitsky were regained under Lobachevsky, who for many years was the rector of the university and, for some time, also stood in for the district superintendent of public education. Lobachevsky's *Geometry* was one of the few textbooks written in answer to Magnitsky's suggestion in a letter sent in autumn 1822 to the university rector G.B. Nikolsky: "It would be fitting if in the future... each professor would send at least one good item to the Publishing Committee every year."

[113] That book was published in Kazan only in 1909, when Lobachevsky's name was known widely enough. It was reprinted in his Russian collected works.

[114] In the Encyclopedia entry mentioned in the main text d'Alembert sharply criticized the "chimeric" rigor of Euclid's school as harmful for textbooks. Lobachevsky, for whom this point of view was congenial, never writes

down the axioms on which his constructions are based; a truly complete list of axioms of Euclidean and hyperbolic geometry was first proposed by D. Hilbert in his *Foundations of Geometry*, as late as 1899.

[115] Fuss's harsh verdict on Lobachevsky's *Geometry* (the name of the author was not known to Fuss) was partly due to a basic mistake he made in assessing the purpose of the manuscript, which unfortunately had no author's introduction. Fuss regarded the *Geometry* as a textbook for beginners, while Lobachevsky had obviously intended it as a refresher course in school mathematics.

[116] At present, elements of hyperbolic geometry of Lobachevsky are sometimes included in school geometry textbooks (see, for example, H.R. Jacobs, *Geometry*, San Francisco, Freeman, 1974, pp. 635–664). The general literature on hyperbolic geometry is vast. In addition to such old but still well-known books as R. Bonola, *Non-Euclidean Geometry*, N.Y., Dover, 1955 (this edition also contains an English translation of Lobachevsky's and Bolyai's memoirs); D.M.Y. Sommerville, *The Elements of Non-Euclidean Geometry*, N.Y., Dover 1958; H.S. Carslaw, *The Elements of Non-Euclidean Plane Geometry and Trigonometry*, London, Longmans, 1916; H. Liebmann, *Nichteuklidische Geometrie*, Berlin, 1923; R. Baldus, F. Löbell, *Nichteuklidische Geometrie*, Sammlung Göschen, Berlin, 1964; A.P. Norden, *Elementare Einführung in die Lobatschewskische Geometrie*, Berlin (DDR), Deutscher Verlag der Wissenschaften, 1958; H. Meschkowski, *Non-Euclidean Geometry*, N.Y., Academic Press, 1964, we note some more advanced books: H. Busemann and P.J. Kelly, *Projective Geometry and Projective Metrics*, N.Y., Academic Press, 1953, and, in particular, H.S.M. Coxeter, *Non-Euclidean Geometry*, Toronto, University of Toronto Press, 1965 (and Chapter 16 in Coxeter's *Introduction to Geometry* mentioned in Note 72). One may also consult such articles as H. Karzel and E. Ellers, "The Classical Euclidean and the Classical Hyperbolic Geometry," in *Fundamentals of Mathematics*, Vol. II (Geometry), The MIT Press, 1974, pp. 174–197; B.A. Rosenfeld and I.M. Jaglom, *Nichteuklidische Geometrie*, *Enzyklopadie der Elementarmathematik*, Bd. V (Geometrie), Berlin (DDR), Deutscher Verlag der Wissenschaften, 1971, S. 385–469; a few newer books: Nöbeling, *Einführung in die nichteuklidischen Geometrien der Ebene*, Berlin–N.Y., Walter de Gruyter, 1976; M.J. Greenberg, *Euclidean and Non-Euclidean Geometries*, San Francisco, Freeman, 1974; P. Kelly and G. Matthews, *The Non-Euclidean Hyperbolic Plane*, N.Y., Springer, 1981; G.E. Martin, *The Foundations of Geometry and the Non-Euclidean Plane*, N.Y., Springer, 1982; B. Klotzek and E. Quaisser, *Nichteuklidische Geometrie*, Berlin (DDR), Deutscher Verlag der Wissenschaften, 1978; and a more general book, W. Prenowitz and M. Jordan, *Basic Concepts of Geometry*, N.Y., Blaisdell, 1965.

[117] Bolyai's absolute geometry is apparently the first axiomatically treated example of an *incomplete* set of axioms in the history of mathematics, i.e., a

set which admits several nonisomorphic interpretations. In his book *Eléments d'histoire des mathématiques* (Paris, Hermann, 1974), the famous "Nicolas Bourbaki" points out that the only fundamental difference between ancient and modern mathematics lies in the extensive use in modern mathematics of *incomplete* axiom systems (for example, nearly all the algebraic structures analyzed by mathematicians—groups, rings, fields, lattices, and others), while the ancient mathematicians (in effect) recognized only *complete* axiom systems (such as the axioms of plane Euclidean geometry or those of real numbers).

[118] Note that formula (1) remains valid in spherical geometry.

[119] Actually we are dealing here with the "relative consistency" of hyperbolic geometry: Beltrami's model, based on Euclidean concepts, proves that *hyperbolic geometry is consistent (free of contradictions) provided Euclidean geometry is consistent.* [However, what is known as the arithmetical model of Euclidean geometry—points $\equiv$ pairs (x, y) of real numbers; lines $\equiv$ sets of points (i.e., pairs (x, y)) satisfying a linear equation $ax + bx + c = 0$—establishes that the axioms of Euclid's geometry are consistent provided the axioms underlying the number system are consistent (say, the natural numbers, because all the other types of numbers can be developed from the natural numbers).] We note that all three discoverers of hyperbolic geometry had proofs of the fact that *the consistency of Euclidean (plane) geometry follows from that of hyperbolic (space) geometry.* This proof also follows from the fact (known to Lobachevsky, Bolyai, and Gauss) that *Euclidean plane geometry is realized on the horosphere in hyperbolic space,* i.e., on a surface which can be described as the limit of a sphere passing through a point A and tangent at this point to a plane α when its radius tends to infinity. (In Euclidean space this limit is, of course, the plane α itself, but in hyperbolic space we obtain a surface β which differs from α.)

[120] "The Prince (or King) of mathematicians" was the inscription on the medal issued by the Göttingen scientific society when Gauss died, but that was what he was called during his lifetime as well. (There was an occasion when the famous French mathematician Pierre Simon Laplace (1749–1827) named Johann Friedrich Pfaff (1765–1825) as Germany's top mathematician. When his interlocutor expressed surprise that he had not chosen Gauss, Laplace replied: "Gauss is the first mathematician in the world.")

[121] The books and articles devoted to Lobachevsky usually state that his father was a land surveyor. However, it seems that he held that job only for a short time and that it was the high point of his career. (Apparently Lobachevsky's father drank a great deal—whence the family's extreme poverty).

[122] Only the civilian ranks of *privy councillor* and *acting privy councillor* bestowed on ministerial officials were higher than that of *acting state coun-*

cillor (in the table of ranks the highest rank was actually that of *chancellor*, but this was very rarely used in practice).

[123] The revolutionary-minded Chernyshevsky, who trusted the reviews in journals and the opinions of the Academy of Sciences, regarded the fact that a scientist ridiculed in the journals was a high-ranking figure in science and education as proof that Russian society had to be radically changed. Of course, the nonmathematician Chernyshevsky had no way of knowing that Lobachevsky was a great scientist and the future pride of Russian science!

[124] Gauss had outstanding linguistic abilities and the learning of a foreign language often served as a pastime for him.

[125] Here we are being deliberately inexact. Actually Farkas Bolyai's *Tentamen* came out in 1832, not in 1831. However, a few copies of the *Appendix* to the *Tentamen* appeared slightly earlier, in 1831. The latter year is always mentioned in references to the second publication (after Lobachevsky's *On the Elements of Geometry*) devoted to the non-Euclidean geometry of Lobachevsky and Bolyai. János Bolyai's *Appendix* was sent to Gauss in 1831, but never reached him; Gauss only received the long book by the older Bolyai, with János's *Appendix*, when it was sent through a common acquaintance in 1832.

It is difficult to understand Gauss's attitude to J. Bolyai in connection with the one problem the elder man set before the younger one: *to find the volume of a non-Euclidean tetrahedron given its six dihedral angles.* Gauss himself had failed to solve this problem (subsequently solved by Lobachevsky). That he proposed it to J. Bolyai shows how much he respected the young man. On the other hand, Gauss seems to have given no thought to how discouraged J. Bolyai would be if he failed to solve the problem.

Incidentally, Gauss's interest in the problem stated above is yet another indication that his intuition guided him unfailingly to significant problems. Indeed, the problem is one of current interest; see, for example, H.S.M. Coxeter, "The functions of Schläfli and Lobatschewsky," *Quart. J.* of *Math.*, **6**, 1935, pp. 13–29, and J. Milnor, "Hyperbolic Geometry: the first 150 years," *Bull. Amer. Math. Soc.*, **6**, (1), 1982, pp. 9–24.

[126] This was apparently related to the fact that Gauss, who had by then learned Russian, had become familiar with the voluminous and often revealing Lobachevsky papers, undoubtedly superior in their depth of understanding of the subject to Bolyai's brief publication. It is possible that this was also due to the fact that Lobachevsky was closer to Gauss in his methodological principles and in his psychological stance than was János Bolyai, whose style of writing and results were far ahead of their time. Gauss probably appreciated the insistence with which Lobachevsky included a discussion of "the geometry of physical space" in all his publications. Lobachevsky suggested experiments

to verify whether "imaginary" (hyperbolic) geometry, rather than Euclidean geometry, is the geometry of our space. He first thought that it was the former but in his last years leaned towards the latter. As we have pointed out, such an approach would have been completely alien to Bolyai—a fact which, in our opinion, speaks in his favor rather than against him.

[127] A full treatment of that subject (beginning, of course, with the theorem in Farkas Bolyai's *Tentamen* on reassembling polygons of equal area) can be found in V. Boltyanskii, *Hilbert's Third Problem*, N.Y., Wiley, 1978.

[128] The name of János Bolyai is now part of the name of the Hungarian Mathematical Society. Hungary, which in the days of Farkas and János Bolyai was, in the scientific sense, very provincial, in the 20th century gave the world a whole constellation of first class mathematicians and physicists.

[129] The reference to the first third of the nineteenth century (Lobachevsky's *On the Elements of Geometry* was published in 1829–1830 and Bolyai's *Appendix* in 1831) is not really justified here, since initially Lobachevsky's and Bolyai's remarkable works were hardly noticed by anyone. The turning point came only after the posthumous publication (in the 1860s) of Gauss correspondence. What attracted immediate attention were letters which made it clear that Gauss had a high opinion of Lobachevsky's work (but at the same time insisted that he, Gauss, had all the relevant ideas in 1792—the year when he just began to seriously think about the theory of parallel lines and about the nature of space! It was only later that Gauss became convinced of the existence of a second geometry differing from Euclid's. These letters drew attention to Lobachevsky's works, and (in the late 1860s) their first translations into foreign languages (and a translation into Russian of *Geometrische Untersuchungen*) appeared, together with the first papers explaining and commenting on Lobachevsky's works. Lobachevsky's name was first mentioned in the scientific literature by the English algebraist and geometer Arthur Cayley in his *Note on Lobachevsky's Imaginary Geometry* (1865). That note compared the trigonometry of a triangle in Lobachevsky's geometry and in spherical geometry. Although it is obvious from the note that Cayley had failed to understand the essence of Lobachevsky's discovery (compare this with what is said about Cayley's attitude towards another work on non-Euclidean geometry, with which he was much more directly involved), that note undoubtedly contributed to the growth of interest in the new geometry.

[130] Gauss was not a good teacher and rarely appreciated the merits of his pupils (as we have already mentioned in connection with Möbius and von Staudt). Riemann took Gauss's course on the method of least squares, but apparently they had no personal contacts. However, Gauss's works influenced Riemann greatly (see main text). So it can be assumed that the very fact of

Riemann's attending Gauss's lectures, regardless of their quality and content, stimulated Riemann's work.

[131] In dealing with Riemann's career in mathematics, one cannot pass up the two years (1847–1848) of his stay in Berlin, where he went after his first year in Göttingen to get to know new people and their views. Temporary stays at different universities were very popular among German students in those years. The trip to Berlin proved very fruitful: Riemann made friends with a talented young mathematician (who unfortunately died young), Ferdinand Gotthold Max Eisenstein (1823–1852), whom Gauss held in high esteem. (Gauss is alleged to have said that only three mathematicians were really epoch-making—Archimedes, Newton and Eisenstein.) Riemann found his talks with Eisenstein very inspiring. But what impressed Riemann most in Berlin was a lecture by Pierre Gustave Lejeune Dirichlet (1805–1859) (whom we will mention again). The two outstanding mathematicians were apparently of similar psychological type, and immediately established scientific and personal contacts. Riemann gladly acknowledged Dirichlet's influence; its relation to Weierstrass's sharp criticism of Riemann's work will be discussed below (see Note 147).

[132] Of French extraction, Lejeune Dirichlet came from an émigré family. He was a link between German and French mathematics. After a secondary education in Germany he spent several years in Paris, where he earned a living as a tutor in a wealthy family. He received a kind of informal education from the analysts of the École Polytechnique. Above all, he was influenced by the great Jean Baptiste Joseph Fourier (1768–1830). Dirichlet came to Prussia on the recommendation of the famous naturalist and public figure Alexander von Humboldt (1769–1859), who was very influential in government circles. His brother, the well-known geographer and traveller Wilhelm von Humboldt (1767–1835), founded Berlin University in 1810, when the city was occupied by Napoleon's troops. (In that idyllic age it was possible to found a university in an occupied city.) Alexander von Humboldt lived for a long time in Paris and was in close contact with French scientists, who held Dirichlet in high esteem on the strength of his very first works. Dirichlet was offered the post of assistant professor first in Breslau (now Wroclaw in Poland) and then in Berlin, where he subsequently became a professor. Dirichlet's houses, in Berlin and later in Göttingen, were gathering places for the local scientific and artistic intelligentsia, largely because of Dirichlet's wife Rebecca, who came from a wealthy, distinguished German-Jewish family. Her grandfather was the famous philosopher, author, and public figure Moses Mendelssohn (1729–1826), her brother was Felix Mendelssohn-Bartholdy (1809–1847), the popular composer and conductor, while her cousin was the economist Ernest von Mendelssohn, who was later to acquire the reputation of being one of Bismarck's most influential advisers. In Berlin Rebecca Dirichlet-Mendelssohn's

salon was visited by Carl Jacobi (of whom more below) who, like the hostess, came from a wealthy and famous German-Jewish family and was interested in art, history, and culture. However, Dirichlet himself was usually very restrained and modest at the receptions in his house. "The continuous small ripples of the intellectual world surrounding him apparently did not conform to the profound oceanic depths of his spirit" as Klein wrote in the book on the history of mathematics mentioned in Note 68.

[133] In 1837, Göttingen, one of the most prestigious of German universities suffered a blow which can only be compared with the destruction of that university one hundred years later by the Nazis (more about this below). In that year the new king of Hannover, Ernst-Augustus II, abolished the democratic constitution of 1833 and introduced a new one, doing away with almost all the civil rights granted by the previous constitution. All government officials, including university professors, had to pledge allegiance to the new constitution, and while Gauss, who took no interest in public affairs, readily consented, seven outstanding scientists (known in the history of German social thought as the Göttingen Seven) refused outright and were forced to leave the university, which immediately lost its reputation as Germany's best institution of higher learning. The seven—who included the physicist Wilhelm Weber and the famous German philologists and folklorists, the brothers Jacob and Wilhelm Grimm—were immediately offered positions at other German universities. In 1848, frightened by the revolutionary wave which swept through all of Europe, the King of Hannover agreed to restore the former constitution. Then many of the exiled scientists, including Weber, returned to Göttingen.

[134] It would probably be more accurate to call the post that of a *Privatdozent*, since the holder was not paid by the university. He had the right to lecture for a fee agreed upon with the students.

[135] We have already pointed out Gauss's inadmissible attitude towards the scientific achievements of Taurinus, János Bolyai, and Lobachevsky. Gauss's treatment of Abel's work was just as reprehensible—timely support might have lengthened Abel's tragic life. When he received Abel's outstanding work on the theory of elliptic functions from Crelle in 1828, Gauss's sole response was a letter in which he claimed to have known these ideas since 1798 (i.e., for thirty years). When he was sent Abel's paper with the revolutionary discovery of the unsolvability in radicals of the general nth-degree equation for $n \geq 5$, Gauss failed to respond at all. This delayed the paper's publication for a long time. The contrasting behavior of Jacobi bears recounting. When he received Abel's memoir on the theory of elliptic functions Jacobi, who had studied the subject deeply, wrote to Crelle: "Abel's work is above my praise, as it is above my works," and immediately began to use the terms *Abelian functions* and *Abelian integrals* in his subsequent publications. As Jacobi was

then reputed to be Germany's second-best mathematician (after Gauss), this was very flattering for the poor, unknown Norwegian student.

[136] See B. Riemann, *Über die Hypothesen, welche der Geometrie zu Grunde liegen*, Berlin, Springer, 1919; second edition 1923. Weyl's comments were included in most foreign-language versions of Riemann's lecture.

[137] The applicant for a position at a German university was required to submit a competition paper (Habilitätsschrift), which could include some computations, and to deliver a competition lecture (Habilitätsvortrag) with few (if any) computations. Riemann's *Habilitätsschrift* was the remarkable memoir *Über die Darstellbarkeit einer Funktion durch eine trigonometrische Reihe*. His *Habilitätsvortrag* was the famous *Über die Hypothesen*, which was very general in form and contained practically no computations. What is striking about this lecture is the depth of its concepts, its prophetic connections with physics (deciphered by Einstein 60 years later), as well as the purely verbal specific results, which were obviously based on an impressive analytical apparatus. Today we know from the paper *Über eine Frage der Wärmeleitung*, submitted by Riemann in 1861 for the competition sponsored by the Paris Academy of Sciences (but unappreciated at the time; the paper was neither published nor awarded a prize) that the founder of Riemannian geometry had complete command of the analytical apparatus and had proofs for all the facts stated in the *Habilitätsvortrag*.

[138] The first publication (*Annalen der Physik, Vierte Folge*, Bn. 49, S. 769–822) was reprinted many times (in particular, it went through several separate editions) and was translated into many languages.

[139] Einstein gladly recalled his many discussions with Weyl of ideas from the general theory of relativity, discussions which went back to their joint work in the Zürich *Technische Hochschule*. But when he spoke about his familiarization with Riemann's geometric ideas it was not Weyl he always gratefully remembered but a far lesser mathematician, his friend from student days Marcel Grossman (1878–1936), who also worked in the Zürich *Hochschule* at the time (it was to Grossman that Einstein owed his job in that school), and who coauthored with Einstein one of the papers on the general theory of relativity preceding the long memoir *Die Grundlagen der allgemeinen Relativitätstheorie*.

[140] Both a journal publication of this fundamental work and a separate edition (Göttingen, Kön. Gesellschaft der Wissenschaft) appeared in 1828. It was later translated from Latin into practically all the modern European languages. In particular, the German translation was published twice in the famous series *Oswald's Klassiker der exakten Wissenschaften* (C.F. Gauss, *Allgemeine Flächentheorie*, Leipzig, Teubner, 1889 and 1890). The work was included in Gauss's *Werke* (see Note 106), Bd. IV, Göttingen, 1873, S. 217–258.

[141] See also E.A. Abbott, "Flatland," in J.R. Newman (ed.), *The World of Mathematics*, Vol. 4, N.Y., Simon and Schuster, 1956, pp. 2383–2396; D. Burger, *Bolland*, 's-Gravenhage, 1957.

[142] The *curvature* k of a smooth *curve* φ can be defined as the velocity of rotation of its tangent: indeed, it is clear that if the tangent to a curve does not change its direction, then the curve must be a straight (entirely uncurved) line; but the faster the tangent changes its direction, the more the curve "curves". The curvature can be defined rigorously as follows: consider a small arc δ of length s of a curve φ. Through a fixed point P draw lines parallel to all the tangents at all the points of the arc δ; the ends of the unit segments (unit vectors) of these lines with common beginning P describe an arc δ_1 of a (unit) circle of a length σ. The ratio $k_m = \sigma/s$ is the "mean curvature" of the curve φ over the arc δ; the limit k of this ratio as $s \to 0$ (and δ shrinks to a point of φ) is the curvature of φ at that point. In a similar way one defines the *curvature* K of a smooth *surface* Φ. Here one also takes a small part Δ, of area S, of a surface Φ; all the tangent planes to Φ at the points of Δ are characterized by the perpendiculars to these planes, the so-called normals to the surface; they are all drawn from one point P and the ends of the unit segments (unit vectors) of these normals with common origin P describe a part Δ_1 of a unit sphere of area Σ. The ratio $K_m = \Sigma/S$ is the mean curvature of Φ over the part Δ of the surface, while the limit K of this ratio as $S \to 0$ (and Δ shrinks to a point) is the curvature of Φ at that point. It follows from these definitions that the curvature of a circle of radius r is $k = 1/r$, while the curvature of a sphere of radius r is $K = 1/r^2$; the greater the radius of the circle or sphere, the smaller the curvature. (The sign of the curvature of a surface is determined by comparing the "orientations" of the corresponding parts Δ, Δ_1 of the surface Φ and the unit sphere; we do not treat this question here.) This definition, however, does not suggest that the curvature K of the surface Φ is a fact of its intrinsic geometry i.e., can be defined by using only those notions which have an intrinsic meaning on Φ. In his *Disquisitiones generales* Gauss called this deep fact the *Theorema egregium*, and this remains its name in most modern textbooks.

[143] At this point it is appropriate to recall Lambert's prophetic statement about "some imaginary sphere" on which hyperbolic geometry holds (see page 49). According to Note 142, the (positive) curvature of a sphere of radius r is $K = 1/r^2$; therefore, a (fictional) "imaginary sphere" of radius $r = ui$ should have curvature $K = -1/u^2$. It turns out that the intrinsic geometry of this "sphere of constant negative curvature" is hyperbolic geometry (cf. pages 67–68).

[144] This construction brings us to the interesting question of the possible global forms of various (say, two-dimensional) geometric systems (Euclidean,

hyperbolic, elliptic) first stated (in connection with Euclidean geometry) by the outstanding English geometer William Kingdon Clifford (1845–1879), professor at London University (who died of tuberculosis on the island of Madeira at the age of 34) and, after Clifford, by F. Klein. Today this question is known as the Clifford–Klein problem and the possible global forms of geometries are called *Clifford–Klein forms.* The issue here is that whereas, say, the sphere and the elliptic plane described above have the same structure locally (in the neighbourhood of any of their points, "in the small"), they may be quite different "in the large," i.e., over regions of greater size. The general Clifford-Klein problem was investigated by Klein, by the German Wilhelm Karl Joseph Killing (1847–1923), one of the most brilliant followers of Sophus Lie, and by the outstanding Swiss topologist Heinz Hopf (1884–1971). It was discovered that there are only two spatial forms of two-dimensional elliptic geometry (the two mentioned above: the *sphere* and the *elliptic plane*), but there are as many as five forms of two-dimensional Euclidean geometry (the ordinary *Euclidean plane*, the infinite *Möbius band*, the infinite *cylinder*, the *torus* i.e., the surface of a doughnut, and the so-called *Klein bottle*). It is indeed clear that a piece of paper may be rolled into a cylinder which, in the neighbourhood of any of its points, does not differ from the plane, but certainly does not resemble it overall. Finally, there are infinitely many forms of two-dimensional hyperbolic geometry (in this connection, see Chapter 9 in Klein's book on non-Euclidean geometry referred to in Note 107).

We conclude with the observation that the hemisphere with identified antipodal points on its boundary equator (see Fig. 11) can also be used as a model of the *projective plane.* Indeed, we saw above (see page 37 and Note 74) that we can consider the *sheaf of straight lines* passing through the point O in space as a realization of the projective plane. But if we choose for the point O the center of the hemisphere shown in Fig. 11, then each line of the sheaf, except the horizontal ones, intersect the hemisphere in precisely one point, while the horizontal lines intersect it in two diametrically opposed boundary points, viewed here as single points of the elliptic plane. Thus we have established a one-to-one correspondence (or *bijection*, as mathematicians say) between the points of the elliptic and projective planes, which allows us to consider the elliptic plane (without its metric, i.e., without the possibility of measuring distances and angles) as being the projective plane and, conversely, to treat the elliptic plane as the projective plane supplied with a metric, i.e., with formulas allowing one to measure the distance between any two points of this plane and the angles between any two of its lines (the Klein approach to plane elliptic geometry; it is discussed on pages 65–67).

[145] Riemann was elected a member of the Göttingen Scientific Society and a corresponding member of the Berlin (Prussian) Academy of Sciences. In 1866, the year in which he was to die, he received notices of his election as a member of the Berlin Academy (early in the year), the Paris (French)

Academy of Sciences (in March), and finally (on June 14, a month before his death) the London Royal Society which had once had Newton as its president. Riemann died in full command of his faculties on July 20, 1866.

146 Riemann was a subject of the Hannover kingdom, i.e., a foreigner in Prussia (Hannover and Prussia united in the year Riemann died).

147 Weierstrass's attitude to Riemann was not unequivocal; it was marked by tremendous respect, because Riemann's results really astonished Weierstrass. For example, Weierstrass withdrew a profound paper on the theory of abelian functions which he had submitted to the Berlin Academy only because Riemann's article on the same subject appeared in Crelle's journal; Riemann's results coincided only partly with the results Weierstrass had obtained using a different approach. But at the same time Weierstrass completely rejected the "physical demonstration" style basic in Riemann's work. In this connection, see Weierstrass's letter to his favorite pupil, the Berlin professor Carl Hermann Amandus Schwarz (1843–1921), quoted in H. Weyl's article mentioned in Note 56, in which he compares his own approach to the theory of functions (based exclusively on algebra—in his view the only correct approach) and the "physical mathematics" of Riemann (to use the complimentary expression of Klein's pupil A. Sommerfeld). Weierstrass regarded Riemann's approach as insufficiently rigorous, and therefore as unacceptable.

Weierstrass was especially vehement in his criticism of the "Dirichlet principle" on which Riemann based many of his constructions. This principle involves the use of the solution of a certain optimization (or variational) problem in the theory of functions of a complex variable. Weierstrass did not doubt that the optimum appearing in the Dirichlet principle exists. His criticism concerned the fact that this statement was proved neither by Riemann nor by Dirichlet. Therefore the results obtained by using the principle could not be considered proved. Weierstrass suggested to G.A. Schwartz that he should work out a proof of the Dirichlet principle for certain particular cases which interested him, and this Schwartz was able to do. A complete proof of the validity of the Dirichlet principle was given by D. Hilbert in 1899. Since Weierstrass did not accept Riemann's approach to the theory of functions of a complex variable, i.e., to that branch of science where the two mathematicians competed, it is all the more to Weierstrass's credit that he helped publicize Riemann's work in the scientific community; in particular, he twice took the initiative of getting Riemann elected to the Berlin Academy of Sciences, where he was one of the most influential members. In this connection we should also note the constant support which Weierstrass gave to Georg Cantor (1845–1918), the creator of set theory, despite the fact that Cantor's scientific, methodological, and even philosophical and religious attitudes—the latter involving both mysticism and a certain fanaticism—were alien to Weierstrass. The latter was a Catholic, very formal in the practice of his religion, while the former, who was of Jewish descent, was a passionate Lutheran.

[148] The article was published in the *Göttinger Nachrichten*, Bd. 14, 1868, S. 193–221; it was reprinted in the second volume of Helmholtz's Works (*Wissenschaftliche Abhandlungen*), Leipzig, 1887, Bd. II, S. 618–639. In this connection we note that Helmholtz was one of the first scientists to speak out publicly in favor of the as yet little-known Lobachevskian geometry. We refer to his lecture "On the origin and meaning of geometric axioms", which he delivered at the University of Heidelberg in 1870 and which appears in the second volume of his *Lectures and Speeches* (H.v. Helmholtz, "Über den Ursprung und Bedeutung der geometrichen Axiome," in the book *Vorträge und Reden*, Bd. II, Braunschweig, 1884). Here we are not only interested in the fact that the physicist and physiologist Helmholtz actually defended the hyperbolic geometry of Lobachevsky–Bolyai and Gauss, but also in his arguments Helmholtz begins with an expressive characterization of the views on geometry held in antiquity: "Among the fields of human knowledge there is none which, like geometry, has appeared before us in a completely perfect, entirely conclusive form and in such complete scientific armor, like Athena Pallada from the head of Zeus ... The long and tedious compilation of new experimental facts, as practiced in all the natural sciences, is alien to geometry. The only method of its scientific development is the method of deduction: one logical conclusion follows from the other." But the main point of his rhetoric is to disprove these views on geometry, which are due to Plato, Aristotle, and Kant. If both systems (the one due to Euclid and the one due to Lobachevsky) are logically consistent, then the truth of one or the other must be established not by deduction but by physical experiment. Thus Helmholtz (as we should expect!) occupies a position akin to that of Gauss and Lobachevsky, who viewed geometry as one of the natural sciences—the study of the specific properties of real space. The logical ("mathematical", as we would say today) approach to geometry so natural for the brilliant János Bolyai was alien to Helmholtz.

[149] It is curious to note that Riemann's very last research paper, on which he worked until his very death, and which was published only posthumously, was inspired by Helmholtz's work in the physiology of sound. Riemann was also interested (again under Helmholtz's influence) in the physiology of vision.

[150] The second important reaction to Riemann's lecture—like Helmholtz's article, a reaction to Riemann's ideas on the natural sciences—was the speech on the spatial theory of matter delivered to the Cambridge Philosophical Society at the beginning of 1870 by W.K. Clifford, whom we mentioned in Note 144. (This speech is published in W.K. Clifford, *Lectures and Essays*, Vol. 1, 2, London, Macmillan, 1901.) Here Clifford actually developed a somewhat more general point of view on the geometry of space than Riemann's, assuming, as did the latter, that the new "Clifford geometry" (which from the mathematical viewpoint was not clearly described in the lecture) allows one to construct a theory of physical space by including the actual structure of

space among the properties which are not purely geometric but physical as well. Clifford's speech was highly regarded by Einstein; but it did not influence him very much, since his theory of relativity (the geometric theory of gravitation) is entirely based on the theory of *Riemannian* spaces. However, some of the later ideas on space developed by the American physicist John Archibald Wheeler and, especially, by the English physicist Stephen Hawking can be viewed as a certain return to the ideas of Clifford.

151 It is curious that Klein's books on non-Euclidean geometry, mentioned in Notes 105 and 107 are especially concerned with hyperbolic, Euclidean, and elliptic spaces of constant curvature, as the only spaces likely to model our real physical space—since the latter is homogeneous and isotropic. Although this is in keeping with the physical picture of the world in 1893 (the year the book mentioned in Note 105 appeared) it is strange that this point of view, which entirely contradicts Einstein's special theory of relativity (1905) as well as his general theory of relativity (1916) was still actively supported by Klein's pupils in a book published in 1928. Actually some of the theories of contemporary cosmology return to the state of affairs where the three geometries of constant curvature, elliptic, Euclidean, and hyperbolic again come to the forefront.

152 See Lie's paper "Remarks on Helmholtz's Paper 'On the Facts that lie at the Foundations of Geometry'" (*Bemerkungen zu v. Helmholtz' Arbeit Über die Tatsachen, die der Geometrie zu Grunde liegen*, Leipziger Berichte, Bd. 38, 1886, S. 337–342; reprinted in Lie's collected works).

153 The more limited character of Klein's constructions is revealed in particular by the fact that for every fixed dimension of space he introduces a finite number of geometric systems, not an infinite number as did Riemann.

154 *Phil. Trans. Roy. Soc., London*, **149**, 1859, pp. 61–70; reprinted in Cayley's Collected Works (*The Collected Mathematical Papers of Arthur Cayley*, Vol. 2, 1889, Cambridge, Univ. Press, 1869, pp. 561–592), where Cayley added important remarks indicating the article's connection with Klein's interpretation of non-Euclidean geometry, and expressing his disagreement with Klein.

155 At the time only a few mathematicians were familiar with non-Euclidean geometry. The first "non-Euclidean boom" occurred after the publication of Gauss's correspondence in the late 1860s, and the first public speech in its favor in Germany was given only in 1870 (see Notes 129 and 148).

156 *Mathematische Annalen*, Bd. 6, 1873, S. 112–145; this article also had further additions and clarifications printed in the same journal and all brought together in the first volume of Klein's collected works (F. Klein, *Gesammelte mathematische Abhandlungen*, Bd. 1, Berlin, Springer, 1973—the most recent

edition—where the text of the main article "On the so-called non-Euclidean Geometry" is also published). (Incidentally, for the paper's critics—including Cayley—further explanations proving the mathematical validity of all of Klein's constructions were not convincing: he who does not wish to see will not see.)

[157] An exposition of Klein's ideas on non-Euclidean systems is contained in his book quoted in Note 107. For a more detailed (and more modern) exposition of the same topic see, for example, I.M. Yaglom, B.A. Rosenfeld, E.V. Yasinskaya, "Projective Metrics," *Russian Mathematical Surveys*, Vol. 19, No. 5, 1964, pp. 49–107 and B.A. Rosenfeld, *Non-Euclidean Spaces*, Moscow, Nauka, 1969 (in Russian). A more elementary exposition of the plane Cayley–Klein geometries (and some general information on space geometries, for example, a list of them) is contained in *Supplements A–C* of I.M. Yaglom's book referred to in Note 159.

[158] Hermann Minkowski was born in a small Jewish community in Byelorussia (the place is now part of Lithuania). As a child he showed brilliant and varied aptitudes. In tsarist Russia it was difficult for Jewish children to get a thorough education. This forced the Minkowskis to emigrate to Germany, where Hermann finished secondary school and the University in Königsberg (now Kaliningrad in the USSR). He began his teaching career at the same university. At that time he got to know a fellow student named David Hilbert (whom we have repeatedly mentioned before and who will appear again in these pages). This acquaintance quickly developed into a strong friendship, and the intimate personal and scientific contacts between Minkowski and Hilbert continued until Minkowski's death (see, for example, the book *Hilbert* by C. Reid, quoted in Note 305). In 1887 Minkowski moved to Bonn to further his scientific and teaching career; there he became first an extraordinary then an ordinary professor. However, when Hilbert moved from Königsberg to Göttingen in 1895, Minkowski returned to Königsberg to take his place. This time Minkowski's stay in Königsberg was short: in 1896 he accepted the offer of a professorship at the famous Zürich *Technische Hochschule*, and in 1902 he moved to Göttingen, where he once again worked with his friend Hilbert. He was not to leave Göttingen until the end of his days.

In Zürich one of Minkowski's students was Albert Einstein. Minkowski considered Einstein a rather ordinary student and did not suspect that many of his own best (and certainly best-known) future achievements would be related to Einstein's ideas. The famous Minkowski lecture "Space and Time" (*Raum und Zeit*), containing the geometric interpretation of special relativity theory, was read in 1907 to the Göttingen Scientific Society and first published in *Jahresbericht der Deutschen Math. Vereinigung*, Bd. 18, 1909, S. 75, and in *Physikalische Zeitschrift*, Bd. 10, 1909, S. 105. Later this lecture was repeatedly edited and translated into many languages (usually with the appendix written by Klein's pupil A. Sommerfeld, whom we mentioned earlier). It was in Minkowski's lecture that spaces with metrics of the type $d^2 = (x_1 - x)^2 - (y_1 - y)^2$ (or $d^2 = (x_1 - x)^2 + (y_1 - y)^2 - (z_1 - z)^2$ in 3-dimensional space) appeared for the first time. In Minkowski's lecture this metric was introduced in four-dimensional space with coordinates x, y, z, t, where t is time, and, was of the form

$$d^2 = (x_1 - x)^2 + (y_1 - y)^2 + (z_1 - z)^2 - c^2(t_1 - t)^2.$$

Here c is the velocity of light in a vacuum. The four-dimensional space with this metric is now called the *pseudo-Euclidean Minkowski space* (the two-dimensional version is called the *Minkowskian plane*). On the further development of this idea (used by Minkowski also in his subsequent paper on the theory of relativity) see also Klein's report on the "The Geometric Basis of the Lorentz Group" ("Über die geometrischen Grundlagen der Lorentz Gruppe," *Jahresberichte der deutschen Math. Vereinigung*, Bd. 19, 1910, S. 281). It is also included in Klein's collected works. Incidentally, prior to Minkowski a similar idea on the connection between Einstein's (physical) relativity theory and the geometries of (four-dimensional) pseudo-Euclidean space (differing from Euclidean space by the minus sign before one of the squares of the differences of the coordinates of two points in the formula for the distance between them) was expressed by one of the creators of the special theory of relativity, the outstanding French mathematician and physicist Henri Poincaré (whose name will turn up many times in these pages) in his fundamental paper "On the Dynamics of the Electron" ("Sur la dynamique de l'éléctron," *Rendiconti del Circolo Math. di Palermo*, **21**, 1906, pp. 129–176), which was repeatedly reedited and translated. It may be that it was precisely in this connection that Poincaré, in the second part ("Space") of his famous book on the methodology of natural sciences *Science and Hypothesis* (H. Poincaré, *La Science et l'Hypothèse*, Paris, Flammarion, 1902, which was also reedited and translated many times), points out the existence of four main geometric systems: the geometries of *Euclid*, *Lobachevsky* (hyperbolic), *Riemann* (elliptic), and a nameless "*fourth geometry*", which, despite the brevity of its characterization, is easily seen to be pseudo-Euclidean geometry. Nevertheless, we feel that this geometry is deservedly called "Minkowski geometry", since it was Minkowski's report and his subsequent publications in *Jahresbericht* and *Phys. Zeitschrift* that brought about general interest in the new geometry and in its connection with the physical constructions of Einstein. As to the remarkable paper "On the Dynamics of the Electron," these ideas were only developed in the form of brief theses, and so were not noticed by anyone, while in the book *Science and Hypothesis* there is only one brief paragraph devoted to a single fact of pseudo-Euclidean geometry, namely *the existence of lines that are perpendicular to themselves*. Of course a reader unacquainted with the relevant materials could never decipher this paragraph.

The main scientific achievements of Minkowski are related to the theory of *convex polyhedra* (or, more generally, arbitrary *convex point sets*) and to number theory, where he introduced a new, geometric, interpretation of number-theoretic problems. It is not by chance that one of Minkowski's main books is called *Geometry of Numbers* (*Geometrie der Zahlen*, Leipzig, Teubner, 1896). The main role in Minkowski's number-theoretic papers is played by the connection between the set (ring) of integers and the *integer lattice* (in the plane and in space) of the points with integer Cartesian coordinates, and by a novel metrization of the plane, also due to Minkowski and used in three-dimensional and in multidimensional spaces (Minkowski spaces with convex length indicatrix; see for example Chapters 24 and 48 of H. Busemann, P.J. Kelly, *Projective Geometry and Projective Metrics*, N.Y., Academic Press, 1953, or—for an elementary exposition—Chapter 7 of the article quoted in Note 116: B.A. Rosenfeld and I.M. Jaglom, *Nichteuklidische Geometrie*). Unlike the pseudo-Euclidean geometry of Minkowski, this geometric system (which later played a vital role in the progress of functional analysis) is often called the *Banach–Minkowski geometry* (after the Polish mathematician Stefan Banach (1892–1945), who widely applied the Minkowski construction to infinite-dimensional spaces).

[159] See I.M. Yaglom. *A Simple Non-Euclidean Geometry and Its Physical Basis*, N.Y., Springer, 1979.

[160] See, for example, some of the books quoted in Note 116 (here Baldus-Löbell has virtually the status of a classic and so should probably top the list). A rather elementary exposition of hyperbolic geometry along these lines is contained in the Supplement of the book (quoted in Note 73) I.M. Yaglom, *Geometric Transformations* III. Among more advanced books we mention H.S.M. Coxeter, *Non-Euclidean Geometry*, also listed in Note 116, and the more fundamental L. Redei, *Begründung der euklidischen und nichteuklidischen Geometrien nach F. Klein*, Budapest. Akademiai Kiado, 1965.

[161] Recall that it was only in the late 1860s that mathematical opinion finally accepted the existence of two equally valid geometries—the Euclidean and the hyperbolic. As for elliptic geometry, which had appeared in Riemann's 1854 lecture (the third geometry), the fact that it is just as valid as Euclidean and hyperbolic geometry became generally accepted after Klein's memoir of 1871. Nowadays, the term "non-Euclidean geometries" usually refers only to hyperbolic and elliptic geometry and not to all the Cayley–Klein geometries. Thus almost all of the Klein book referred to in Note 107 is devoted to just these two geometries (and, of course, to Euclidean geometry); this in spite of its stated aim to explain the essence of more general constructions. Only these two non-Euclidean systems appear in almost all the books and articles mentioned in Note 116. Except for the book mentioned in Note 159, devoted largely to plane semi-Euclidean geometry, I know of no book that develops any of the Cayley–Klein geometries other than the Euclidean, hyperbolic, and elliptic ones.

[162] See, for example, the very clear account in B.N. Delonay, *Elementary Proof of the Consistency of Lobachevskian Geometry*, Moscow, Gostekhizdat, 1956 (in Russian), which is accessible to beginners in mathematics.

[163] See the references given in Note 157.

[164] Because of his careless attitude to the classification problem and the determination of the number of geometric systems that he had considered, Klein in his 1871 memoir and in the book mentioned in Note 107 speaks of seven plane "non-Euclidean" geometries, including classical Euclidean geometry. Actually here Klein is counting not the geometries but their so-called absolutes, some of which are associated with more than one geometry. The number of geometries in space given in the book is also incorrect.

[165] It is well known that the European Renaissance, a period of cultural revival noted for its intense interest in the culture of antiquity, including ancient Greek mathematics, and for its attempts to continue the creative efforts of Greek and Roman scholars and artists on a new basis, took place primarily in Italy. Accordingly, it is to Italy that we look in order to find the leading mathematical researchers of the new era, personified by such outstanding figures as Nicolo Tartaglia and Girolamo Cardano (see page 4). The prestige of Italian mathematics in the 17th century was upheld by the excellent school headed by Galileo Galilei (1564–1642), whose members included such outstanding scientists as Evangelista Torricelli (1608–1647), and Bonaventura Cavalieri (1598?–1647). However, in the 18th century and in the first half of the 19th, Italian mathematics went through a period of relative degradation (the Franco-Italian J.-L. Lagrange should, of course, be considered as a representative of the French, rather than the Italian, school). This is one of the reasons why the outstanding results of Paolo Ruffini, published in Italian, were not noticed and appreciated at the time. However, one of the consequences of the national and cultural upheaval characteristic of Italy of the mid 19th century, which concluded in the political unification of the country, was the revival of the great traditions of Italian science. In mathematics, this revival was marked by the emergence of a distinguished school of Italian *geometers* (this is typical of mathematics of 19th century) which reached its peak at the end of the 19th and the beginning of the 20th century. Here we especially note the achievements of Antonio Maria Bordoni (1789–1860), a professor at Pavia University, who headed the group that included representatives of *differential geometry* such as Angeli Gaspare Mainardi (1800–1879), Delfino Codazzi (1824–1873), Francesco Brioschi (1824–1897), and E. Beltrami, as well as the founder of the Italian school of *algebraic geometry* Luigi Cremona (1830–1903). The most important figure in this pleiad of scientists, which played a key role in the creation of the Italian school of geometry was undoubtedly E. Beltrami. Beltrami held professorial

posts in Bologna, Pisa, Rome, Pavia, and again in Rome, where he became a member, and afterwards the president, of the National Academy of Sciences.

This distinguished Italian school of geometry undoubtedly occupied a leading position in *differential geometry*, and, in particular, in *tensor analysis*, which first appeared in connection with geometric (and, to a lesser extent, mechanical) applications (Gregorio Ricci-Curbastro, 1853–1925, Tullio Levi-Civita, 1873–1941, Luigi Bianchi, 1856–1928), in algebraic geometry (L. Cremona, Corrado Segre, 1863–1924), in the foundations of geometry [G. Peano and his pupils, among whom we should name Mario Pieri (see Note 185) and F. Enriques (see Note 53)], and in topology (Enrico Betti, 1823–1892). Unfortunately, after World War I, fascism dealt Italian culture and, in particular, Italian mathematics, a devastating blow.

166 See pages 62–63 and Note 142.

167 E. Beltrami, *Saggio di interpretazione della geometria non-euclidea*, published in the Neapolitan journal of mathematics (*Giornale di Matemat.*, **6**, 1968, pp. 284–312) and also (in the same year) as a separate book (Napoli, Torino e Firenze, 1968), is included in Volume I of Beltrami's *Collected Mathematical Papers* (*Opere matematiche*, Milano, V. I, 1902, pp. 374–405), and was later repeatedly reedited and translated. The continuation of this work is Beltrami's article on the foundations of the theory of spaces of constant curvature ("Teoria fondamentale degli spazzi di curvatura costante," *Annali di Matemat.*, Milano, **2** (2), 1868, pp. 232–255), also included in Volume I of Beltrami's collected papers.

We note that surfaces of rotation of constant negative curvature, including the *pseudosphere* (the name is due to Beltrami), were described in 1839 (i.e., before Beltrami) by a professor at the German University in Dorpat, Russia (now the city of Tartu in Estonia), Ernst Ferdinand Adolf Gottlieb Minding, 1806–1885. Minding also discovered in 1840 that the trigonometric relations in a triangle formed by geodesic (shortest) curves on a surface of constant Gaussian curvature K can be obtained from the formulas of spherical trigonometry by replacement of the radius of the sphere by $\sqrt{K}$, which is purely imaginary in the case of negative curvature K. Minding failed to notice the connection between these results and the hyperbolic geometry of Lobachevsky–Bolyai (about which he apparently knew nothing at the time).

168 The parametric equations of a pseudosphere (a surface of rotation!) can be written in the form

$$r = \sqrt{x^2 + y^2} = 1/\cosh t, \qquad z = t - \tan t,$$

where $\cosh t = \frac{1}{2}(e^t + e^{-t})$ is the hyperbolic cosine of the parameter t (or, equivalently, as $x = \cos \varphi/\cosh t$, $y = \sin \varphi/\cosh t$, $z = t - \tan t$). One obtains this surface by rotating about the z-axis the so-called *tractrix*, defined as the trajectory of a mass point $M(x, z)$ which, at the initial moment of time, is located at the point $(1, 0)$ of the (horizontal!) plane $x0z$ and is joined to the origin by a flexible unstretchable string whose other end is pulled by someone along the z-axis.

169 Hilbert's paper "On Surfaces of Constant Gaussian Curvature" ("Über Flächen von konstanter Krümmung") was published in 1901 in an American journal (New York, *Trans. Amer. Soc.*, **2**, 1901, pp. 86–99); in 1903 Hilbert included this paper in the second edition of his *Foundations of Geometry* (*Grundlagen der Geometrie*) and from then on it invariably appeared in all the new editions and translations of his *Foundations of Geometry*.

170 The map of the hyperbolic plane onto the interior of the (say, unit) disk, which gives us the Beltrami–Klein model, assigns to each point M of the Lobachevskian plane two numbers x and y, where $x^2 + y^2 < 1$—the coordinates of the corresponding point of the (Euclidean) disc of radius 1 centered at the origin $O(0, 0)$ of the Euclidean plane. These numbers, which appeared in Beltrami's memoir *Attempt at an Interpretation of Non-Euclidean Geometry* are now called the *Beltrami coordinates* of the points of the hyperbolic plane. These coordinates were already considered, in essence, by Lobachevsky (see page 56).

171 This model is considered in Poincaré's article "On the Fundamental Hypotheses of Geometry" ("Sur les hypothèses fondamentales de la géomètrie," *Bulletin de la Société Math. de France*, **15**, 1887, pp. 203–216), where it is included in a series of interpretations covering all the main geometric systems—the Euclidean, hyperbolic and elliptic (see Note 173).

172 A *sphere of imaginary radius* exists in complex Euclidean space with coordinates x, y, z and distance $d = d_{MM_1}$ between the points $M(x, y, z)$ and $M_1(x_1, y_1, z_1)$; it is given by the formula

$$d^2 = (x_1 - x)^2 + (y_1 - y)^2 + (z_1 - z)^2$$

provided that the coordinates x, y, z of the points can be arbitrary complex numbers. In particular, the "two-dimensional sphere of imaginary radius i" in complex Euclidean space is given by those points $M(x, y, i\zeta)$ where x, y and ζ are real numbers and $x^2 + y^2 - \zeta^2 = -1$. It is not difficult to see that this "sphere" essentially coincides with the hyperboloid shown in Fig. 15 in real (actually pseudo-Euclidean) space with coordinates x, y, ζ.

173 Poincaré considered a system of geometries which he called *quadratic* because the "domain of action" of the geometry is a *quadric*, a surface in three-dimensional Euclidean (or, more precisely, affine) space with coordinates x, y, z given by the quadratic equation

$$ax^2 + by^2 + cz^2 + 2dxy + 2exz + 2fyz + 2gx + 2hy + 2kz + l = 0.$$

Special cases of such quadrics are: the sphere $x^2 + y^2 + z^2 = 1$ on which *elliptic* geometry is realized; the hyperboloid $x^2 + y^2 - z^2 + 1 = 0$ on which *hyperbolic* geometry holds; the paraboloid $x^2 + y^2 - z = 0$ on which we have *Euclidean* geometry if we take as its "straight lines" its sections by "vertical" planes (i.e., planes parallel to the z axis).

We note also that the connection between *Lobachevskian* geometry and *pseudo-Euclidean* geometry (in a space of dimension one greater; see Chapter 5) and between pseudo-Euclidean geometry and *relativity theory* (compare Note 158) determines a simple non-Euclidean (hyperbolic) interpretation of facts of the special relativity theory, in which points of three-dimensional Lobachevskian space correspond to the velocities of uniform motions in four-dimensional (the fourth dimension being time) Minkowskian space. This relationship between Lobachevskian geometry and relativity theory (very briefly developed in elementary terms in, say, supplement A of the book cited in Note 159) was first noted by Klein's pupil Arnold Sommerfeld, in his paper "On the Sum of Velocities in Relativity Theory" ("Über die Zusammensetzung der Geschwindigkeiten in der Relativtheorie," *Physikalische Zeitschrift*, Bd. 10, No 22, 1909, S. 826–829). It was the subject of much research by the Yugoslav (Croatian) physicist Vladimir Varicák (1865–1942), who devoted to it the book *Die Stellung der Relativtheorie in dreidimensionalen Lobotschefskychen Räumen*, Zagreb, 1924.

Chapter 5

174 Note that on the real line the *only* closed convex figure (in fact the only closed figure that is connected (i.e., consists of only one piece) is the closed interval, which for this reason (since there is nothing to choose from!) is simultaneously the one-dimensional analogue of the triangle, the disc and the parallelogram.

175 Note, for example, the geometric versions of the names of the powers of numbers greater than three ("squared square," "squared cube," etc.) used by ancient and Arab mathematicians, during the Renaissance in Italy, and by the German "Cossists" (i.e., algebraists—the German noun *Coss* stands for the unknown in an equation). These names clearly appealed to some sort of multidimensional intuition (see the review in B.A. Rosenfeld, *A History of Non-Euclidean Geometry*, Moscow, Nauka, 1976, p. 148 ff., where the arguments of the Cossist Michael Stifel (1486–1569) are presented in detail; they explain the admissibility of "multidimensionality" in arithmetic and its inadmissibility in geometry—one cannot "go beyond the limits of the cube as if there were more than three dimensions, since this would be unnatural"). Nicolas Bourbaki, in the book referred to in Note 117, sees the origin of the idea of n-dimensionality in Pierre de Fermat's (1601–1665) analytic geometry. There Fermat states problems leading to *points*, *curves*, and *surfaces* and ends

with the vague conclusion: "this gives rise to *spacial geometric loci* as well as [to loci] that follow."

A relevant and more substantial item is a remark by Immanuel Kant (1724–1804) in his early paper "Thoughts about the True Assessment of Living Forces" ("Gedanken von der wahren Schätzung der lebendigen Kräfte und Beurtheilung der Beweise derer sich Herr von Leibniz und andere Mechaniker in dieser Streitsache bedient haben nebst einigen vorhergehenden Betrachtungen welche die Kraft der Körper überhaupt betreffen") (included in most of his collected works in many languages). Here Kant attempts to explain the fact that the world is three-dimensional by saying that (according to God's will) "substances in our universe interact with each other so that the acting force is inversely proportional to the square of the distance." This is indeed related to the three-dimensionality of real space, as may be established by using the mathematical theory of potential created much later. (Actually Kant might have been aware of the heuristic arguments given by Newton's contemporary and rival Robert Hooke (1635–1703), who stated the universal law of gravitation earlier than Newton and apparently could never understand why, or accept that, Newton was viewed as the true author of this law. Hooke's starting point is that the area of a sphere of radius r is proportional to r^2 and therefore, if we assume (as Hooke and Newton did) that the "gravitational force" of a mass m located at a point M does not decrease with distance and is proportional to m, then at each point of the sphere of radius r with center M this force will be a $(1/r^2)$-th part of the "force of the mass m" at a unit distance from M. This crude reasoning leads us to expect that in n-dimensional (physical!) space gravitational, electric, and other forces due to mass or to charge must decrease with the distance r as $1/r^{n-1}$.) "If the number of dimensions were different," continues Kant, "the forces of attraction would have different properties and dimensions. The science of all such possible space forms would undoubtedly be the most sublime geometry that finite reason could pursue ... If the existence of spaces of other dimensions were possible, then God would most likely have placed them somewhere."

A less profound observation is that, just as plane figures symmetric with respect to a line cannot be superimposed by a motion within the plane, but can be brought into coincidence after being turned over in three-dimensional space, so too figures in space symmetric with respect to a plane (like right-hand and left-hand gloves) can be superimposed by moving them in four-dimensional space. This is explained by A.F. Möbius in his "Barycentric calculus" (1827), with the reservation that "since four-dimensional space cannot be imagined, a real superposition is not possible here".

Finally, an important step in the development of multidimensional considerations was J.L. Lagrange's *Mécanique Analytique* (1787), where the concept of generalized coordinates of a mechanical system (independent variables ξ, ψ, φ, ... whose choice determines the position of the system) was introduced and used throughout. To be sure, Lagrange himself warned that since they are purely algebraic these constructions do not require any geometric considerations and his book did not contain a single diagram. Nevertheless, many readers must have drawn pictures of certain situations considered in the book—realistically in the case of two or three degrees of freedom (i.e., in the case of two or three generalized coordinates)—and must have arrived at the idea of "multidimensionality" when the number of degrees of freedom was greater than three. Moreover, while few of Lagrange's contemporaries are likely to have noticed this, his book contains in essence the notion of the "phase space" of a mechanical system, which is $2n$-dimensional in the case of n degrees of freedom; the coordinates of a point in phase space, characterizing the state of the system, are $2n$ numbers $\xi, \psi, \varphi, \ldots, \dot{\xi}, \dot{\psi}, \dot{\varphi}, \ldots$; here $\xi = \xi(t)$, $\psi = \psi(t)$, ... and $\dot{\xi} = d\xi/dt$, $\dot{\psi} = d\psi/dt$, ... are the rates of change of the parameters of the system and t is time. Phase space, in a somewhat different guise, played an especially important role in W.R. Hamilton's works on mechanics. The curve $\xi = \xi(t)$, $\psi = \psi(t)$, ...; $\dot{\xi} = \dot{\xi}(t)$, $\dot{\psi} = \dot{\psi}(t)$, ... in $2n$-dimensional space, characterizing the evolution of the system in time, is now called the *phase portrait* of the system.

[176] The name of this outstanding scientist, philosopher and author was given to him by a police inspector, to whom someone brought him as a baby abandoned on the steps of the Jean le Rond chapel in Paris. The inspector named the baby Jean le Rond. Out of pity for the boy, he did not hand him over to the Children's Home, to which abandoned babies were usually dispatched and where conditions were quite harsh, but presented him to a peasant woman who agreed to bring him up. However, the child's father soon appeared on the scene. He was a General Detouch, who had been abroad and had not known that such a fate had befallen his illegitimate son. He found the boy and had him transferred to the family of a poor glass blower. The future scientist lived in this family for about 40 years. Subsequently Jean le Rond made the family's surname, Alembert, his own. D'Alembert did not know his mother at all. His father occasionally visited him and, during his lifetime, paid for the child's education. The foster parents, whom the boy loved ardently, were too poor to do so. After his father's death, Jean le Rond received a small pension (his father died when he was 10).

The period between the 16th and the 18th centuries abounded in universalists, but even against the background of his time d'Alembert's breadth of knowledge, interests, and achievements was striking. One of the 18th century's three greatest mathematicians and mechanicians (the other two being Euler and Lagrange), d'Alembert was at the same time a leading authority in philosophy, history, literature, and music. An acclaimed writer, he was saluted by Voltaire in a letter as "the best writer of our age." It is not surprising that d'Alembert was a member of both French academies. In 1754 he was elected to the Académie Française (also known as the Academy of the Immortals, because the number of members always remains the same: only when a member dies is a successor elected), whose members are men of letters and philosophers. In 1765 he became full member of the Royal Paris Academy of

Sciences; this gave him a small allowance to live on. It is notable that d'Alembert became a member of the Berlin (Prussian) Academy of Sciences in 1747 and was elected to the St. Petersburg (Russian) Academy earlier than to the Paris Academy (in 1764). From 1772 on, he was the permanent secretary of the Paris Academy and, in effect, its head. D'Alembert was the permanent secretary of the Académie Française (which did not have a president) from the time he was elected, i.e., from 1754. Characteristically—and this is not such a rare case for a brilliant talent—d'Alembert was not elected to either the Paris Academy of Science or to the Académie Française on the first ballot.

(It may be appropriate to explain the historical reasons for the existence of two academies in France, the Académie Française and the Paris Academy of Sciences. (The latter was subsequently reorganized by Lazare Carnot into *L'Institut*; see Note 65 on page 153.) The Académie Francaise was founded in 1635 by Cardinal Richelieu, who virtually ruled France at the time and regarded himself as a writer and a patron of writers. Naturally, he was included in the initial membership and was the master of the academy during his lifetime. The rivalry we know so well from Dumas's *Three Musketeers* (where, however, it is presented in a grotesque and exaggerated form) between Richelieu and Louis XIII resulted in the almost simultaneous founding of the *Academia Parisiensis*, patronized by the King, whose members were scientists, chiefly physicists and mathematicians. However, the Paris Academy was fully organized and received serious financial support only during the reign of Louis XIV—in 1666, when the farsighted finance minister Colbert became its patron. At that time the outstanding scientist Christian Huygens (1629–1695) was invited from Holland to be its president.)

D'Alembert was one of the first professional scientists in European history. He never lectured and did not occupy any official posts except in the academies. (In contrast we may recall that Newton was a professor at Cambridge University, later Chancellor of the Exchequer and, for a number of years, a member of parliament, while Leibniz was the historiographer of the dukes of Hannover.) Accordingly, d'Alembert's material position was always rather unsatisfactory. This, however, caused him little concern—we pointed out above that he spent most of his life in the family of a poor glass blower. Frederick "the Great" of Prussia invited him to Berlin, intending to make him president of the Berlin Academy (a post once held by Leibniz) and offering him a large salary—but in vain. Even larger sums were promised by Catherine II of Russia, who hoped to assign to d'Alembert the upbringing of the heir, Pavel Petrovich (the future Emperor Pavel I). But d'Alembert explained to the mighty monarchs that he preferred his modest position, since it spared him trouble—he had nothing to lose, yet could still help those who were poorer than he. Moreover, d'Alembert wrote to Frederick: "I owe nothing to the French government from which I can expect many bad things and few good ones, but I have duties to my country; to leave would be highly ungrateful on my part." This attitude to life is also confirmed by the following story about Laplace. The young Laplace came to d'Alembert with letters of recommendation from nobles, but d'Alembert refused to receive him. On the next day Laplace sent d'Alembert his first mathematical papers. D'Alembert immediately received him and inquired, perplexed: "Having such recommendations, he seeks favors from the nobles?" As a result, Laplace soon obtained the post of professor of mathematics in the Military Academy.

D'Alembert stands at the source of such branches of mathematics as the *theory of functions of a complex variable* (underlying which are the so-called Cauchy–Riemann equations, actually first stated by d'Alembert) and the *theory of partial differential equations*, which begins with the equation $\partial^2 u/\partial t^2 = a^2(\partial^2 u/\partial x^2)$ of a vibrating string, first stated and solved by d'Alembert; here $u = u(x, t)$ is the deviation of the string, assumed fixed at its ends, at time t at the point determined by the abscissa x, while $\partial^2 u/\partial t^2$ and $\partial^2 u/\partial x^2$ are the (second) partial derivatives of u. Finally, in mechanics his best-known achievement is the "d'Alembert principle" (the foundation of the *Traité de la dynamique*, Paris, 1743, republished and translated on many occasions), which reduces problems of dynamics to problems of statics.

D'Alembert's work on the famous Encyclopedia, of which he was initially coeditor with Diderot, are particularly impressive. (However, tired of official persecution and hostility, d'Alembert had his name subsequently removed from the title page, although he remained Diderot's close friend to the end, as well as an active and prolific contributor.) D'Alembert wrote almost all the entries on mathematics, the natural sciences and technology, as well as numerous entries on philosophy, history, literature, aesthetics, and ethics. Moreover, the great enterprise opened with a long introduction on the origin and development of the sciences, entirely written by d'Alembert and containing a coherent account of his philosophical views (and one of the first attempts to classify the sciences, within which the author also included the arts). D'Alembert's individual entries in the Encyclopedia contained many profound thoughts, often far ahead of their time. Above (page 54 and Note 114) we have already dealt with the significance of the entry "Geometry"; here we are concerned with the rather unexpected entry "Dimensions." In the entries "Limit" and "Differential" d'Alembert (before A. Cauchy, to whom the achievement is usually attributed) presented the first outline of a theory of limits, and in "Definition," long before Hilbert (20th century!), he set out his view of geometry as a science studying abstractly given notions characterized by their properties but entirely devoid of graphic form. Geometry, wrote d'Alembert, would retain its rigor, although it would sound funny, if we called a triangle what we usually call a circle and vice versa.

[177] *Philos. Magazine*, London, 1843 and *Cambridge Math. Journal*, **4**, 1884; reprinted in Cayley's *Selected Mathematical Papers*, Vols 1–13; Cambridge University Press, 1889–1898; see Vol. 1, pp.55–62.

The discrepancy between the title and the contents of the work (n-dimensional geometry as against $(n-1)$-dimensional geometry) is due to the fact that Cayley considers sets of elements determined by n numbers x_1, x_2,

$\ldots, x_n$, which he views as *projective coordinates* of points in projective space (recall that a point of the projective plane is determined by three homogeneous coordinates, a point of projective space by four coordinates and, correspondingly, a point in n-dimensional projective space by $n + 1$ coordinates; compare page 40 and Note 81).

[178] In addition to the books and articles mentioned in Note 157, it would be appropriate here to mention again the book by G. Buseman and P. Kelly mentioned in Note 116—its title contains the words "projective metrics."

[179] Similarly, Carl Jacobi, whose name will appear repeatedly below, did not use geometric terminology when computing (in 1834) the volume and area of a sphere of radius r in n-dimensional Euclidean space. According to Jacobi, the area in question equals $2\pi^{n/2}r^{n-1}/(n/2 - 1)!$, when n is even and $2\pi^{(n-1)/2}[((n - 1)/2)!]r^{n-1}/(n - 1)!$ when n is odd; one obtains the volume of the corresponding ball from these formulas by multiplying by r/n.

[180] One of the most brilliant figures in European (and American) mathematics in the second half of the 19th century, James Sylvester was in many respects Cayley's direct opposite. In particular Cayley was very cautious in using new terms, while Sylvester called himself "Adam the name giver"; it was from him that the entire terminology of the theory of invariants, including the word "invariant," was derived. Having graduated from Cambridge University in 1837 (his late graduation was due to a serious illness from which he suffered in his student years), Sylvester worked from 1838 as a professor of natural philosophy (i.e., physics) at University College in London.

Unable to get along with colleagues, he left England for America, where he taught mathematics from 1841 to 1845 at the provincial Virginia University. In 1845 he returned to England, where he worked for ten years, first as an insurance agent and later as a lawyer, without abandoning mathematics. From 1855 to 1871 Sylvester was a professor of mathematics at the highly regarded Military Academy in Woolwich. From 1871 to 1876 Sylvester led the life of a private person, not working anywhere officially and living mostly in Paris. In 1876 he was invited to teach at the prestigious Johns Hopkins University in Baltimore; he stayed there for eight years, carrying out very extensive research and teaching activities. For this reason the Americans consider him to be one of the founders of their school of mathematics. In particular, he founded the first specialized mathematics journal in the Western hemisphere, the *American Journal of Mathematics*, which still enjoys a reputation for excellence. In 1884 the seventy-year-old Sylvester returned to his homeland and accepted a professorship at Oxford, which he did not leave until the end of his days.

Sylvester's numerous changes of work were due partly to his caustic character and to his biting sense of humour, which often expressed itself in short sharp verses which offended his colleagues. Moreover, traditionally inclined

English professors were often irritated by the extreme character of Sylvester's pedagogical views (which had a considerable influence on Felix Klein). Sylvester's destructive criticism centered on the teaching of mathematics in British schools "according to Euclid" (more precisely, along the lines of the so-called school version of Euclid's *Elements*—compare Note 96); in contrast, the traditionally inclined Cayley supported the old English mathematics curriculum entirely. Finally, Sylvester's Jewish origins (his family, which in previous generations had no surname, acquired the anglicized family name Sylvester only in James's generation) may have irritated some of Sylvester's reactionary colleagues.

[181] Cayley (whose life, like Newton's, passed between Cambridge and London) was a representative of elitist English intellectual circles; the Jew Sylvester was cosmopolitan. George Salmon represented a third type of British scientist, very different from the other two. During most of his life the Irishman Salmon was at Trinity College in Dublin—a Protestant institution that traditionally brought together the study of mathematics, philology (mainly classical), and theology. A long time before, George Berkeley had graduated from this college; it is there that he acquired the knowledge needed to successfully apply higher mathematics (the calculus) in theological discussions. A few years before Salmon, W.R. Hamilton had graduated from Trinity College. Hamilton, whom we will discuss in more detail below, combined interest in mathematics and in philology. The Irish–Protestant atmosphere at Trinity College was always quite strictly traditional and very conservative. Even Cambridge, itself quite traditional, was viewed at Trinity as following continental (French, German) fashions and neglecting theology and philosophical contemplation. It is understandable that Salmon, who was brought up in this atmosphere, and who remained at Trinity College as a teacher after graduation and never left that institution until his death, was always inclined to religious thought; in fact, he left his mathematics professorship after 25 years to become professor of theology. He was extremely conservative (especially on pedagogical questions) and thus, in this respect, the opposite of Sylvester.

Nevertheless Salmon, like Sylvester, was an outstanding teacher. His textbook of analytic geometry and higher algebra, translated into practically all European languages, played a very important role in disseminating the ideas of English mathematics. In particular, Klein once explained that he had learned of Cayley's ideas, which were to influence him so much, in a German translation of Salmon's textbook.

[182] We cannot deny the reader the pleasure of contemplating the titles of both versions of this outstanding work. They are typical of Grassmann's involved scientific style—one of the reasons why he was underestimated by his contemporaries. The title page of the 1844 edition reads: "The Science of Linear Extension, A New Branch of Mathematics Developed and Explained in Its Applications to Other Branches of Mathematics as well as to Statics,

Mechanics, the Science of Magnetism and Crystallonomy (*sic*!) by Hermann Grassmann, mathematics teacher at the Frederick-Wilhelm School in Stettin" ("Die lineale Ausdehnungslehre, ein neuer Zweig der Mathematik dargestellt und durch Anwendungen auf die übrigen Zweige der Mathematik, wie auch auf Statik, Mechanik, die Lehre von Magnetismus und die Krystallonomie erläutert"). The alternative title was: "The Science of Extensive Magnitude or the Study of Extensions, A New Mathematical Discipline Developed and Clarified by Means of Applications by Hermann Grassmann" ("Die Wissenschaft der extensivenen Grösse oder die Ausdehnungslehre, eine neue mathematische Disziplin dargestellt und durch Anwendungen erläutert"). Further it was indicated that the present work was a first part, containing the science of linear extension only. Here one must bear in mind that the terms "extension" (*Ausdehnung*), and "extensive magnitude" (*extensive Grösse*) in the sense in which Grassmann used them, were Grassmann's inventions and entirely unknown to his readers.

The 1862 version reads: "The Science of Extension, Entirely Revised and Rigorously Developed" ("Die Ausdehnungslehre, vollständig und in strenger Form bearbeitet"). The shorter title shows that its author took into consideration the commercial and scientific failure of the first version of the book. Here we have reproduced the title page—but it should not be trusted entirely. The book was not published by Enslin Publishers in Berlin as indicated on the title page, but in Stettin, by a publishing house owned by Hermann Grassmann's younger brother and associate Siegmund Ludolf Robert Grassmann (1815–1901). The book was sent to the Berlin publishers on a commission basis, but without success: there was no more demand for the second version than for the first. Further, the book appeared in Stettin in 1860 and was partly distributed by mail by the author, although the title page indicates 1861 as the year of publication—a minor commercial trick.

[183] A very detailed biography of Grassmann, by Lie's pupil Friedrich Engel, was included by the latter in a three-volume collection of Grassmann's mathematical and physical works, constituting the second book (second half-volume) of Volume III (see H. Grassmann, *Gessammelte mathematische und physikalische Werke, Bd. III, Thl.* 2: *Grassmanns Leben, geschildert von F. Engel, nebst einem Verzeichnis der von Grassmann veröffentlichen Schriften und einer Übersicht des handschriftlichen Nachlasses*, Leipzig, B.G. Teubner, 1911, XV +400 S). This is only one of the many books about Grassmann available today.

[184] Grassmann, who taught school for many years, planned to write a three-volume *Mathematics Textbook for Secondary Schools* (*Lehrbuch der Mathematik für höhere Lehranstalten*). The three volumes were to contain, respectively, arithmetic, plane geometry, and solid geometry, while trigonometry was to be divided between the second and third volumes. This intention was only realized in part: in 1860 the first volume (Arithmetic) appeared and

in 1864 a second volume (Trigonometry) followed. (The title page of the 1860 book reads, more briefly, *Lehrbuch der Arithmetik* and the title page of the 1864 book is similar. Besides, both books name the Enslin publishing house (Berlin), where the book was sent on a commission basis, and not Robert Grassmann's Stettin publishing house, which actually printed them; the first book gives the year of publication as 1861 rather than 1860, when it actually came out.) These books were not successful and remained unappreciated in their time; but today the volume on arithmetic, which strikes us in places as close to the trends of our computer age, is justly regarded as one of the classical works of the mathematical literature. This book, which can hardly be viewed as a simple school text (we will return to this below), was one of the starting points of so-called "recursive arithmetic", based on *recursive* (or *inductive*) definitions of the natural numbers; in modern notation $a + 1 = a'$ and $a + b' = (a + b)'$, where "$'$" denotes the passage from the given natural number to its successor, $a \cdot 1 = a$ and $a \cdot b' = ab + a$. Of course, today we view the symbol "$'$" as the *operator* which sends n into $n + 1$. The recursive definitions $a + b' = (a + b)'$ and $ab' = ab + a$ can easily be implemented as computer instructions or as algorithmic descriptions of arithmetic operations in the spirit of modern computer-oriented constructivism. Grassmann himself wrote $a + e$ instead of a', where e is the so-called "unit" (*die Einheit*) of his system. Besides, he considered not only the natural numbers but all the integers, so that the definition of addition had to contain two more rules: $a + 0 = a$, $a + b'' = (a + b)''$, where $0 = 1 + (-e)$ or $e + (-e)$; here "$''$" is the symbol indicating that we pass from a to the previous number $a + (-e)$ (of course the notation $'a$ would be more suggestive in this context). It was from Grassmann's constructions that R. Dedekind eventually developed the generally accepted axiomatic definition of the natural numbers (see R. Dedekind, *Was sind und was sollen die Zahlen*?; Braunschweig, Vieweg, 1888, often republished and translated). This book also inspired Peano (we will have more say about this mathematician in Note 185; on his definition of natural number, given in 1889–1891, which in essence coincides with Dedekind's, see for example G. Peano, *Arithmetices principia, nova methodo exposita*, Torino, 1889). Today, recursive arithmetic is a significant chapter of modern logic and of the study of the foundations of mathematics (see, for example R.L. Goodstein, *Recursive Number Theory*, Amsterdam 1957); it is also widely used in teaching (see, for example, S. Federman, *The Number Systems*, Reading (Mass.), Addison-Wesley, 1963, or the computer science oriented book by A.S. Blokh, *Numerical Systems*, Minsk, Vishaishaya Shkola, 1982 (in Russian).

Note that Grassmann did not consider his book as having to do with research, but only as a textbook for the upper classes of secondary school. It is known that he used the book as a text in his classes in the gymnasium and, despite the opinion of his son (Hermann Grassmann Jr, also a mathematician) who took part in publishing his father's collected works and according to whom this teaching experience was successful, the very fact that the older Grassmann used his book when working with schoolchildren bears out Klein's poor opinion of Grassmann as a teacher—the *Lehrbuch der Arithmetik* would have been completely out of place in the classroom.

[185] The generally accepted axioms of an (*n*-dimensional) linear (vector) space used today are based on Grassmann's axioms and were stated by the Italian Guiseppe Peano (1858–1892) in his book *Geometric Calculus, according to Grassmann's Science of Extensive Magnitudes.* Peano's book is prefaced by an exposition of the operations of deductive logic (*Calcolo geometrico secondo l'Ausdehnungslehre di H. Grassmann, preceduto dalle operazioni della logica deduttiva,* Torino, 1888). Peano was the head of the Italian school in the foundations of mathematics and professor at the Turin Military Academy, where Lagrange had once taught. He played an important role in the flourishing of logical and axiomatic studies typical of the turn of the century, and had a significant influence on the creation of computer-oriented mathematics and in the birth of the concepts which gave rise to the French school of Nicolas Bourbaki. Peano himself was concerned with questions of the foundations of arithmetic (see Note 184), geometry (Euclidean and affine), and analysis. Part of the achievements of his school are brought together in his five-volume *Formulaire de Mathématiques,* Torino, 1895–1905. Peano's pupil Mario Pieri (1860–1913) was the author of the first really rigorous axiom system of Euclidean geometry, developed in his book *On Elementary Geometry as a Hypothetical Deductive System* (*Della geometria elementare come sistema ipotetico-deduttivo,* Torino, 1899). This book appeared a few months before D. Hilbert's famous *Grundlagen der Geometrie.* Trying to be as rigorous as possible, Peano, like János Bolyai, invented a special "logical language" containing a minimum of words and a maximum of logical and mathematical symbols. He used this language to write most of his articles—which made them practically unreadable, as were all the articles in the journal *Rivista di Matematica,* which specialized in the foundations and appeared in Turin under Peano's editorship.

[186] Grassmann distinguished between "factual" sciences, which have to do with the real world, and "formal" sciences, whose object is created by human thought. In this connection it is natural to recall Plato's distinction, developed in the *Republic,* between the "visible world" and the "world of Ideas" (within which he includes mathematics). According to Grassmann, there are only two formal sciences: philosophy, which studies "the general" (as it is created and perceived by thought) and mathematics, which studies "the particular" as created by thought (*die Wissenschaft des besonderen Seins, als eines durch des Denken gewordenen*). This particular is called "thought-form" (*Denkform*) by Grassmann. It is with these "forms of thought" that mathematics is concerned. Today it is easy to decipher in these somewhat inflated expressions a very profound understanding of the axiomatic foundations of mathematics in terms of Bourbaki's mathematical structures. However, 19th-century mathematicians felt that, to use a Bourbaki expression,

they were "pas dans leurs assiettes," when they came across terms with general statements far removed from ordinary mathematical formulas and equations. The time when the mathematical climate was determined by mathematicians prone to philosophical generalizations, such as Georg Cantor (1845–1918), G. Peano, D. Hilbert and H. Poincaré, Luitzen Egbertus Jan Brouwer (1881–1966), Hermann Weyl, Bertrand Russell (1872–1970), and Alfred North Whitehead (1861–1947) was yet to come.

[187] Compare Grassmann's statement given in Note 186 with G. Boole's view of mathematics as the science studying operations considered *per se* rather than the different objects to which they can be applied (see Chapter 1 of Boole's *An Investigation of the Laws of Thought*, London, McMillan, 1854, repeatedly republished, in particular, in its author's *Collected Logical Works*, Chicago–London, P. Jourdain, 1916, see Vol. I, p. 3). Compare also what we say about Hankel in Note 189. Note that Boole, like Grassmann, was an amateur mathematician, who had received no formal mathematical education and was far removed from "official" mathematical circles. Thus neither of these scholars expressed what might be called the general mathematical atmosphere of their time.

[188] The vector sum of the linear subspaces U and V is the set of all sums of vectors of these spaces

$$U + V \stackrel{\text{def}}{=} \{a + b \mid a \in U, b \in V\},$$

where "$\stackrel{\text{def}}{=}$" means equal by definition.

[189] The general mathematical views and specific scientific interests of H. Hankel, formed in part under Grassmann's influence and, in part, independently, were very close to the latter's. Compare, for example, Hankel's definition of the essence of mathematics: "a purely intellectual, pure theory of form whose subject matter is not the combination of magnitudes or their representations, numbers, but abstract thought objects (*Gedankendinge*) which may correspond to actual objects or relations, although such a correspondence is not necessary" (*Theorie der complexen Zahlensysteme*, Leipzig, Voss, 1867). Compare this with Grassmann's attitude to mathematics (see Note 186). We shall soon give a more detailed exposition of the contents of Hankel's book, which is also very close to Grassmann's *Ausdehnungslehre*.

[190] Siegmund Ludolf Robert Grassmann (1815–1901) was six years younger than his brother Hermann, but the two brothers were always very close. Robert Grassmann also taught in the Stettin schools (mathematics, physics, philosophy, geology, chemistry, botany, zoology, German, French, Greek, and Latin). However, he gave up teaching in the early 1850s, concentrating his efforts on the publishing of a Stettin newspaper and on his editorial and printing activities. It was in his publishing house that almost all of Hermann's

works appeared, as well as the many books, on the most varied topics, that he wrote himself—political pamphlets on the burning issues of the day (the Franco-Prussian war, Bismarck) anti-Catholic leaflets, textbooks, popular science, advanced monographs in mathematics, physics, chemistry, biology, geology, geography, theology, law, government, ethics, aesthetics, history, philology ... The high point of Robert Grassmann's scientific work was to be a ten-volume treatise *The Edifice of Knowledge* (*Das Gebäude des Wissens*) which was to cover—at least in the author's view—all the existing sciences. Grassmann's printing shop and publishing house brought forth an unbelievably varied assortment of literature, including ancient Hebrew and Aramaic texts (in particular, the Hebrew Talmud which was to be sold abroad) with their complex alphabet and unusual typesetting rules.

In the preface to his arithmetic textbook, Hermann Grassmann indicated that the book arose from his discussions of the topic with his brother Robert—who indeed wrote a book popularizing and explaining the *Ausdehnungslehre.* Undoubtedly, Robert Grassmann was not a scholar of the same stature as his elder brother, and his unbelievable breadth of interests resulted in noticeable superficiality. But, as a popularizer of science and culture, he certainly deserves remembrance and honor, while his philosophical works retain a certain interest even in our day. Thus, for example, in 1981 a thesis in philosophy was defended in Leningrad by G.I. Malykhina on the subject of Robert Grassmann's logical studies.

[191] H. Grassmann and R. Grassmann, *Leitfäden der deutschen Sprache*, Stettin, Druck und Verlag von R. Grassmann, 1876; H. Grassmann, *Deutsche Pflanzennamen*, Stettin, Druck and Verlag von R. Grassmann, 1870. In the foreword to the latter book, its author expresses his gratitude to his brother Robert.

[192] To understand the modern attitude to Grassmann's scientific heritage, see Laurence Young, *Mathematicians and Their Times* (*History of Mathematics and Mathematics of History*) Amsterdam, North Holland, 1981. Its author is one of the members of a popular mathematical family; he writes interestingly and with gusto; but some of his views, due to his unconcealed bias, are controversial and his books unfortunately are not free of factual errors. Young tends to accuse Klein of insufficiently appreciating Grassmann's achievements, and claims that the first to really understand his work were the leading 20th-century French mathematicians Henri Poincaré and Elie Cartan (1868–1951) and the famous Swiss mathematician Georges de Rham (1903–1969).

[193] The Euler formula (which should really be called the Decartes-Euler formula, since Euler did not have a rigorous proof of it, while its statement was known to René Decartes (1596–1650) a century earlier) asserts that the number of vertices N_0 of any convex or, more generally, any simply connected

polyhedron is related to the number N_1 of its edges and to the number N_2 of its faces by the equation $N_0 - N_1 + N_2 = 2$. The two-dimensional analogue of this formula is the relation $N_0 - N_1 = 0$ involving the number N_0 of vertices and the number N_1 of sides of an arbitrary polygon. (On the question of a rigorous proof of the Decartes–Euler formula, see the discussion of various rigorous and nonrigorous ways of proving it in the remarkable book by the well known Hungarian-British logician Imre Lakatos (1903–1974): *Proofs and Refutations*, Cambridge University Press, 1976.)

The Schläfli formula

$$1 - N_0 + N_1 - N_2 + \cdots + (-1)^n N_{n-1} + (-1)^{n+1} = 0$$

where N_k (for $k = 0, 1, 2, 3, \ldots, n - 1$) is the number of k-dimensional faces of an n-dimensional polytope; in particular, N_0 is the number of its vertices. Thus, for example, in four-dimensional space

$$N_0 - N_1 + N_2 - N_3 = 0, \quad \text{i.e., } N_0 + N_2 = N_1 + N_3.$$

[194] In n-dimensional Euclidean space, when $n \geqslant 5$, there exist only three types of regular convex polytopes all of whose faces are identical, regular $(n - 1)$-dimensional polygons, and all of whose polygonal angles are identical regular polygonal angles, i.e., angles congruent to the angle at the vertex of a regular pyramid. These regular polytopes are similar to the ordinary (three-dimensional) regular tetrahedron (the regular simplex), the cube (the regular parallelotope) and the regular octahedron, the polytope dual to the regular parallelotope. When $n = 3$, there are, as is well known, five types of regular polyhedra ("Platonic solids"); besides the three mentioned above, they include the regular dodecahedron (twelve faces) and icosahedron (twenty faces). In four-dimensional space there are six types of regular polytopes; the regular simplex, the analogue of the tetrahedron (with five faces), the cube (a parallelotope with eight faces), the four-dimensional cross (the analogue of the octahedron, with sixteen faces), as well as regular polytopes with 24, 120, and 600 faces. Finally, in one-dimensional space (on the real line) there exists only one regular polytope, the closed interval (check that the Euler–Schläfli formula also holds for it), while in two-dimensional space (in the plane) there are infinitely many of them (a regular polygon with n sides exists for any n). These facts are summarized in the following table:

Dimension of space	Number of types of regular polytopes
$n = 1$	1
$n = 2$	∞
$n = 3$	5
$n = 4$	6
$n \geqslant 5$	3

See, for example, the elementary article by B.A. Rosenfeld and I.M. Yaglom, "Mehrdimensionale Räume," *Enzyklopädie der Elementarmathematik* (*EEM*), Bd. V (Geometrie), Berlin (DDR), Deutscher Verlag der Wissenschaften, 1971, pp. 337–383, or the literature listed in Notes 195 and 197).

[195] A detailed description of this question's history (as well as of the question itself) with substantial biographical data on Schläfli and the story of the publication of his works, is contained in H.S.M. Coxeter's book *Regular Polytopes*, N.Y., Dover, 1973.

[196] A short summary of this memoir appeared in Jordan's Academy of Sciences papers (in C.R.) in 1872, and the complete text in 1875. The memoir and the summary are included in vol. 3 of C. Jordan, *Œuvres*, v. 3, Paris, Gauthier-Villars, 1964.

[197] See the literature on *multidimensional geometry*, e.g., P.H. Schoute, *Mehrdimensionale Geometrie*, Bd. 1, 2, Leipzig, B.G. Teubner, 1902–1905; D.M.Y. Sommerville, *An Introduction to the Geometry of N Dimensions*, N.Y., Dover; 1958; B.A. Rosenfeld, *Multidimensional Spaces*, Moscow, Nauka, 1966 (In Russian).

[198] See, above all, the book by E. Cartan, *Leçons sur les invariants intégraux* (a course of lectures delivered by Cartan at the Paris Faculté des Sciences), Paris, Hermann, 1922, which contains a detailed exposition of Grassmann's "exterior algebra" and Poincaré's "exterior analysis" (the study of the differentiation and integration of Grassmann's exterior products of differentials) and their application to mechanics. Actually, all of Cartan's research work is saturated with the ideas of "exterior algebra and analysis".

[199] In order to relate an "extensive magnitude of the second order" (or bivector, as we would say today) $\sum x_i e_i = \sum x_{ij}[e_i e_j]$ to a parallelogram in some (two-dimensional) plane of n-dimensional space, say, the parallelogram spanned by the vectors $a = \sum x_i e_i$ and $b = \sum y_j e_j$, we must assume that the coordinates x_{ij} of the bivector satisfy certain quadratic "Grassmann conditions" (otherwise known as "simplicity conditions" for the bivector) which single out a so-called Grassmann manifold in the $(n(n-1)/2)$-dimensional "space of bivectors." Of course, the area of a bivector in the ordinary geometric sense can be considered only in the case when the bivector is contained in this manifold. Here there is no need to dwell on these (actually quite simple) questions.

[200] See any text on vector calculus, e.g., the elementary article by W.G. Boltjanski, I.M. Jaglom, "Vektoren und ihre Anwendungen in der Geometrie," *EEM* (see Note 194), Bd IV (Geometries), 1980, pp. 295–390.

[201] In the literature there are many symbols for denoting the inner and outer products of vectors; apparently all attempts to unify them have been abandoned. This is why we indicate two systems of notation here: "·" and "×" as well as parentheses and brackets. These notations, however, are not the only ones used. In this connection, Felix Klein, in Chap. IV (Complex Numbers) of Part I (Arithmetic) of his book *Elementarmathematik vom höheren Standpunkt aus*, Bd. 1, Heidelberg, Springer, 1968, relates how a special commission for the unification of vector notation was created during the 1903 Natural Science Congress in Kassel. The commission members, however, had different opinions on the subject. And since they were all tolerant of each other's views, the only result of their work was the appearance, along with the previous systems of notions, of three new ones! Klein adds that the unified system of units in physics (concerning which, incidentally, not all is smooth today!) was created as the result of powerful pressures due to industry. Since the vector calculus has no such stimuli, there is no hope for a unified notation.

[202] The first geometric interpretation of complex numbers as points of the plane (or more precisely as line segments joining the origin to the given point, an interpretation which is even closer to the vector calculus), with a complete description of the geometric meaning of operations on complex numbers, was given by the Norway-born Danish cartographer and geodesist Caspar Wessel (1745–1818). It is contained in his only mathematical work, which is remarkable for its clarity and substantial contents. Wessel presented this in 1797 to the Danish Academy of Sciences, and published it in Danish in 1799. However, his work (which incidentally included the first attempt to find an appropriate space analogue of complex numbers!) was not noticed by anyone. It was only 100 years later, through the efforts of S. Lie (ever the Norwegian patriot!), that Wessel's work was published in French (in Copenhagen, 1897) and came to be known. In 1806 the geometric interpretation of complex numbers was discovered again by the Swiss-born French mathematician Jean Robert Argand (1768–1822) in his anonymously published, and also unnoticed, pamphlet *Essai sur une manière de représenter les quantités imaginaires dans les constructions géometriques*, Paris, 1806. (For a more detailed account see A. Dahan-Dalmédico, J. Peiffer, *Routes et dédales*, Paris, Etudes Vivantes, 1982, Ch. 7.) However, in 1813–1814 Joseph Diaz Gergonne, the editor of the most popular French mathematical journal *Annales des mathématiques pures and appliquées* (we mentioned Gergonne in connection with his works on projective geometry) published Argand's pamphlet in his journal, where it was finally noticed. And yet the geometric interpretation of complex numbers was introduced into general use only by Gauss.

[203] Two modifications of complex numbers were proposed by William Kingdon Clifford—namely, *dual numbers* $x + \varepsilon y$, where $\varepsilon^2 = 0$ (mentioned above), and *double numbers* $x + ey$, where $e^2 = 1$. Both types of numbers may

be given geometric meanings. Dual and double numbers (under different names and notations) were introduced by Clifford in connection with their geometric applications in his paper "Preliminary Sketch of Biquaternions," *Proc. Lond. Math Soc.*, 1873, pp. 381–395, reproduced in the *Mathematical papers of W. K. Clifford*, pp. 181–200. For the geometric interpretation of these numbers (in particular, double numbers, which can be represented as straight lines of the ordinary Euclidean plane), see I.M. Yaglom, *Complex Numbers in Geometry*, N.Y., Academic Press, 1968; also see Supplement *C* in the Yaglom book referred to in Note 159.

Dual numbers, originating in Plücker's work and, in part, in Hamilton's early work on geometric optics, and appearing in *line-element geometry* (as well as in *non-Euclidean line-element geometry*), were used by Plücker's successor in the chair of mathematics at Bonn university (and later professor at Greifswald university) Eduard Study (1862–1922) and by the Khazan geometer Alexander Petrovich Kotelnikov (1865–1944). In this connection see the conclusion of Part I of W. Blaschke, *Vorlesungen über Differentialgeometrie und die geometrische Grundlagen Einstein's Relativitätstheorie*, Bd. I: *Elementare Differentialgeometrie*, Berlin, Springer, 1930.

[204] Here, in certain cases, it is necessary to enlarge the original definition of a Clifford algebra, as Clifford did when passing from ordinary complex numbers (with unit i, where $i^2 = -1$) to double numbers (with unit e, where $e^2 = +1$). Thus in "generalized Clifford numbers" (also called *alternions* and discovered by the famous English physicist Paul Adrien Maurice Dirac (1902–1985) in the course of his work on quantum mechanics) the squares of the principal units e_i may equal -1 or $+1$. It is clear that for $n = 1$ the Clifford numbers with one unit e satisfying $e^2 = -1$ coincide with the *complex numbers*; if $e^2 = 1$, we get Clifford's *double numbers*. For $n = 1$ the Grassmann numbers, where (as always in his case) the square of the only unit is zero, yield Clifford's *dual numbers*. For the Clifford algebra in the case $n = 2$, see page 89.

All the systems of numbers considered above can be introduced in a uniform manner. Thus consider a Euclidean n-space with basis $\mathbf{e}_1, \mathbf{e}_2, \ldots, \mathbf{e}_n$ and inner product

$$(\mathbf{x}, \mathbf{y}) = F(\mathbf{x}, \mathbf{y}) = (x_1\mathbf{e}_1 + \cdots + x_n\mathbf{e}_n, y_1\mathbf{e}_1 + \cdots + y_n\mathbf{e}_n) = \sum a_{ij}x_iy_j$$

with symmetric bilinear form F. Define the product $\mathbf{xy}$ of vector $\mathbf{x}$ and $\mathbf{y}$ by putting

$$\mathbf{xy} + \mathbf{yx} = (\mathbf{x}, \mathbf{y}) \quad (= F(\mathbf{x}, \mathbf{y}))$$

and using the distributive and associative laws. If the quadratic form $F(\mathbf{x}, \mathbf{x}) = F$ is written in the canonical form $F = \pm X_1^2 \pm X_2^2 \pm \cdots \pm X_k^2$, $k \leqslant n$, then we obtain an alternion algebra (singular if $k \leqslant n$) which corresponds to the conditions $\mathbf{E}_i^2 = \pm 1$ for $i \leqslant k$ and $\mathbf{E}_j^2 = 0$ for $j > k$ (here $\mathbf{E}_1, \mathbf{E}_2, \ldots, \mathbf{E}_n$ is a canonical basis of the space for the form F and, at the same time, a system of generators for our "algebra of numbers"). The nonsingular case, the negative-definite case and the null case of the form F correspond, respectively, to the *Dirac numbers*, the *Clifford numbers* and the *Grassmann numbers*. Note that $F = (x_1\mathbf{E}_1 + x_2\mathbf{E}_2 + \cdots + x_n\mathbf{E}_n)^2$, so that any quadratic form is the square of a linear form whose coefficients $\mathbf{E}_1, \mathbf{E}_2, \ldots, \mathbf{E}_n$ are "noncommutative numbers" of a new kind. Dirac arrived at his numbers from similar considerations: he viewed the Laplace operator $\Delta = (\partial^2/\partial x^2) + (\partial^2/\partial y^2) + (\partial^2/\partial z^2)$ as the square of a linear operator,

$$\Delta = \left(L\frac{\partial}{\partial x} + M\frac{\partial}{\partial y} + N\frac{\partial}{\partial z}\right)^2 \qquad (*)$$

and, although any mathematician could easily show that the representation (*) is impossible, Dirac was led to accept it on the basis of physical considerations.

[205] Here is a very clear formulation of what "symbolic algebra" is about by another one of the leaders of the Cambridge formalist group, George Peacock (1791–1858): symbolic algebra is "the science of symbols and their combinations, constructed according to their own rules, which may be applied to arithmetic and other sciences by means of an interpretation" (see pp. 194–195 in Peacock's *Report on the recent progress and present state of certain branches of analysis*, Rept. of the British Assoc. for the Adv. of Sci. for 1833, London, 1834).

[206] Systems of numbers $u = x_0 + x_1e + x_2e^2 + \cdots + x_{n-1}e^{n-1}$, where x_0, $x_1, \ldots, x_{n-1}$ are real numbers and the formal sums u and $v = y_0 + y_1e + y_2e^2 + \cdots + y_{n-1}e^{n-1}$ are added and multiplied in the ordinary way, using the relation $e^{ij} = e^{i+j}$, are now called *cyclic* numbers if $e^n = +1$, *anticyclic* numbers if $e^n = -1$, and *plural* numbers if $e^n = 0$. Thus in modern terminology C. Graves's triplets are *3rd-order cyclic numbers*. The terms "cyclic" and "anticyclic" are due to the fact that the algebra of (say) cyclic numbers is the same as the algebra of so-called "cyclic matrices," whose rows are obtained from each other by cyclic permutations, i.e., have the form $(x_0, x_1, x_2, \ldots, x_{n-1})$, $(x_1, x_2, \ldots, x_{n-1}, x_0)$, $(x_2, x_3, \ldots, x_{n-1}, x_1, x_0)$, $\ldots$, $(x_{n-1}, x_0, x_1, \ldots, x_{n-2})$. It is easy to show that the algebras of cyclic and anticyclic numbers can be written as the direct sums of a certain number of copies of the field of complex numbers and of at most two copies of the field of real numbers. The most interesting geometric applications are those of *plural numbers*, which generalize the dual numbers of Clifford–Study–Kotelnikov, but we will not discuss this here.

[207] C. Graves orthogonally projects the point of ordinary Euclidean space (x, y, z) corresponding to the triplet $u = x + ye + ze^2$ on the line l: $x = y = z$ and on the plane $\pi \perp l$ with equation $x + y + z = 0$. Then the multiplication of two triplets reduces to the multiplication of their projections on l as real numbers on the l-axis and of their projections as complex numbers in the plane π. (Thus, "geometrically," the algebra of triplets is represented as the "direct sum" of the real line l and the complex plane π; see Note 206.)

[208] Compare, for example, a modern textbook on mechanics that clearly presents the mathematical aspect of "Hamiltonian formalism", V.I. Arnold's *Mathematical Methods in Classical Mechanics* (N.Y., Springer, 1978).

[209] Hamilton himself liked to recall that he spent nearly ten years vainly trying to construct such a system of numbers with three units—he called them *triplets*, copying the terminology of de Morgan and Graves. In one of his later letters to his son, he recalls how each morning, coming down to breakfast, his son would ask him: "Well, father, have you learned to multiply and divide triplets?", and he would sadly reply: "No, I still only know how to add and subtract them."

[210] In the letter quoted in the previous note, Hamilton recalls how the idea of giving up triplets and going on directly to quaternions (numbers with four units) came to him, together with the understanding that commutativity of multiplication must be sacrificed and together with the main formulas of "quaternion algebra." He was walking with his wife along the Royal Canal to a session of the Royal Irish Academy, where he was to preside; his wife was telling him something, but he did not hear the words. The solution of the problem which had occupied him for such a long time came to him in a flash; crossing the bridge over the canal, Hamilton wrote out the main formulas on the soft stone of the bridge's railing, using the tip of his penknife. The Moscow shipbuilder and mathematician, fleet admiral and member of the Academy of Science, leading authority on celestial mechanics, and Russian translator of Newton's *Principia*, Alexei Nikolaevich Krylov (1863–1945), retelling the incident, usually claimed that the Dublin municipal authorities periodically freshen up Hamilton's formulas on the bridge railing, so that they can still be seen there today, and that Hamilton was not going to a Royal Academy session but coming home from a party, where he had not neglected the (alcoholic) beverages. This version adds fresh color to the historical anecdote —and after all, Hamilton's story was told many years *post factum*, and its reliability may be doubted. It should be mentioned that Krylov (like Hamilton, a drinking man) was prone to exaggerate (as a sailor should) his partiality to alcohol (see his expressive memoirs—*My Recollections*, Leningrad, "Sudostroyenie", 1979; in Russian). Of course Krylov never saw the Hamilton formula on the bridge railing—his story is apocryphal.

[211] Hamilton viewed the vector $\mathbf{v} = a\mathbf{i} + b\mathbf{j} + c\mathbf{k}$ as a translation operator sending the point $A(x, y, z)$ into the point $B(x + a, y + b, z + c)$. He uses the terms *vehend* for A and *vectum* for B. Hamilton considers the triplet of terms vehend–vectum–vector as similar to diminuend–difference–subtrahend and dividend–divisor–quotient. However, only the term "vector" survived in mathematics.

[212] For a time the *vector calculus*, so useful to physicists and engineers today, existed only in the form of a "quaternion calculus." In particular, it was

in quaternion form that James Clerk Maxwell (1831–1879) wrote his famous *Treatise on Electricity and Magnetism.* Thus the fundamental Maxwell equations of electromagnetic field theory, familiar to us in vector form, were first written by their author not in vector but in quaternion terms. This was possible because Hamilton, in his study of quaternions, laid the foundations not only of vector algebra, but of *vector analysis* as well: he considered the "symbolic vector" (or "purely vector quaternion")

$$\nabla = i\frac{\partial}{\partial x} + j\frac{\partial}{\partial y} + k\frac{\partial}{\partial z},$$

which he called "nabla" after the Biblical instrument "nebela," a kind of triangularly shaped harp; here i, j, k are "quaternion units", and $\partial/\partial x$, $\partial/\partial y$, $\partial/\partial z$, are partial-derivative operators. Further, Hamilton considered the formal products $s\nabla$, $S(\nabla v)$, $V(\nabla v)$, where $s = s(x, y, z)$ is a scalar ("purely scalar quaternion"), actually a *scalar field* changing from point to point, while $v = a(x, y, z)i + b(x, y, z)j + c(x, y, z)k$ is a vector ("purely vector quaternion"), i.e., a *vector field.* Hamiton denoted his "symbolic quaternion" by the sign $\lhd$, obtained by rotating the Greek Δ; this operator acquired its modern form ∇ in the book *An Elementary Treatise on Quaternions* (Cambridge, 1873) by the British physicist Peter Guthrie Tait (1831–1901), better known for his physics textbook, written jointly with William Thompson, Lord Kelvin (1824–1907). It is in this *Treatise* that the word "nabla", apparently coined by Hamilton, was first used for the symbol ∇. P.G. Tait played an important part in further discussions of quaternions, where he supported Hamilton's conceptions absolutely. Tait was Hamilton's close friend; at the latter's request, he delayed his own book's publication in order that it appear after Hamilton's book *Elements of Quaternions*, Dublin, 1866 (actually, when the two books came out, Hamilton was no longer living). On the other hand, Tait was a friend of Maxwell, with whom he studied in Edinburgh and then at Cambridge. Apparently it was from Tait that Maxwell learned of Hamilton's creation. (Of course, both Maxwell and Tait took an examination in quaternion theory at Cambridge—at the time a degree was unthinkable without it; Tait's knowledge of the subject went far beyond the examination requirements, and Maxwell was able to demonstrate his own perfect mastery.)

The vector calculus did not acquire its quaternion-free modern form in the works of mathematicians, but in *Elements of Vector Analysis*, New Haven, 1881–1884, by the outstanding American physicist Josiah Willard Gibbs (1839–1903), who worked all his life at Yale University, which he helped make world-famous, and in *Electromagnetic Theory*, London 1903, by the English engineer and electrician Oliver Heaviside (1850–1925), creator of the so-called "symbolic calculus," member of the Royal Society, who, for almost all his career, led the life of a private person. Both these authors deleted the minus sign in Hamilton's formula for the scalar product of vectors.

[213] In particular, Hankel first stated the so-called *permanence principle*, which must be taken into account when we extend algebraic (e.g., numerical)

systems: operations on elements of the new system must be defined so that their application to the original elements (now part of the new system) gives the same result as before. (When we pass to the new system, we must extend our knowledge, not learn anew!) Thus the operations on complex numbers applied to (real) numbers $x + 0i(=x)$ give the same result as operations applied to x's viewed as real numbers; quaternion operations on numbers of the form $s + xi + 0j + 0k$ do not differ from their analogues on complex numbers $s + xi$, etc.

[214] The importance of the (now somewhat neglected) topic of (hyper) complex numbers at the turn of the century is attested in the very large article on complex numbers (*Complexe Zahlen*) in Klein's *Encyclopedia of Mathematical Sciences*; cf. Chapter 8. This article was written by Eduard Study (see Note 203). A French version was contributed to the enlarged French edition of this Encyclopedia by the great French mathematician Elie Cartan, mentioned in Note 192; see E. Study, E. Cartan, *Nombres complexes*, Encycloped. Sciences Math., Edition Française, Paris, Gauthier-Villars, 1908, article I, 5.

It is clear that Grassmann's number system (see page 82) and Clifford numbers (see page 84) of the nth order with principal units $\mathbf{e}_1, \mathbf{e}_2, \ldots, \mathbf{e}_n$ are actually hypercomplex numbers with 2^n complex units; these units may all be written in the form $\mathbf{e}_{i_1}\mathbf{e}_{i_2}\cdot\ldots\cdot\mathbf{e}_{i_k} = \mathbf{e}_{i_1 i_2 \ldots i_k}$, where $0 \leqslant k \leqslant n$ (for $k = 0$ our unit e does not contain any of the factors $\mathbf{e}_1, \ldots, \mathbf{e}_n$ and can simply be identified with the number 1) and $i_1 < i_2 < \cdots < i_k$. The two systems of numbers differ only in the "multiplication tables" for the units $\mathbf{e}_{i_1 \ldots i_k}$.

[215] Thus the two "quotients" $\mathbf{t}_1(=\mathbf{u}\mathbf{v}^{-1})$ and $\mathbf{t}_2(=\mathbf{v}^{-1}\mathbf{u})$ of the vectors $\mathbf{u}$ and $\mathbf{v} \neq 0$ considered by Hamilton (i.e., of two "purely vector quaternions") are equal to $(|\mathbf{u}|/|\mathbf{v}|)\cdot(\cos\varphi \pm \mathbf{w}\sin\varphi)$, where $|\mathbf{u}|$ and $|\mathbf{v}|$ are the lengths of the vectors $\mathbf{u}$ and $\mathbf{v}$, while $\mathbf{w}$ is the unit vector perpendicular both to $\mathbf{u}$ and to $\mathbf{v}$ (or the zero vector $\mathbf{0}$, if $\mathbf{u}$ and $\mathbf{v}$ are in the same line) and φ is the angle between $\mathbf{u}$ and $\mathbf{v}$. Therefore the "Hamilton quotient" of two vectors is not a vector, but a "general" quaternion; for collinear vectors (i.e., vectors contained in the same line) this quotient is unique and is a scalar (the real number t such that $\mathbf{u} = t\mathbf{v}$). In any system of hypercomplex numbers, for any $\mathbf{u}$, the expressions $\mathbf{u}\mathbf{u}^{-1}$ and $\mathbf{u}^{-1}\mathbf{u}$ (the quotients of $\mathbf{u}$ by itself) are equal to each other and to a fixed element $\mathbf{e}$, which plays the role of the *identity element* in our system, i.e., for any $\mathbf{v}$ we have $\mathbf{ev} = \mathbf{ve} = \mathbf{v}$.

[216] We have already mentioned the efforts expended by Hamilton (and by the "Hamiltonians" and "quaternionists" who followed him) to develop the general *theory of analytic functions of a quaternion variable.* Its authors expected that it would have applications as wide and fruitful as those of the theory of analytic functions of a complex variable (created by A. Cauchy, B. Riemann, and K. Weierstrass) to classical analysis and differential equations. Alas!—these expectations were to be disappointed. S.M. Lie's closest pupil

and collaborator Georg Scheffers extended the research of the "quaternionists" to the theory of functions of an arbitrary associative hypercomplex variable, similar to the Cauchy–Riemann–Weierstrass theory of functions of a complex variable. However, he achieved success only in the case of *commutative* multiplication; in this case he found "analyticity conditions" generalizing the classical Cauchy–Riemann conditions for functions of an (ordinary) complex variable. The Cauchy–Riemann conditions, which appear in every book on complex analysis or on the theory of analytic functions, are the following: a function $w = u(x, y) + iv(x, y)$ of a complex variable $z = x + iy$ is analytic if and only if

$$\frac{\partial u}{\partial x} = \frac{\partial v}{\partial y}, \qquad \frac{\partial u}{\partial y} = -\frac{\partial v}{\partial x}.$$

In the case of a *double* variable $z = x + ey$, where $e^2 = +1$, and a *dual* variable $z = x + \varepsilon y$, where $\varepsilon^2 = 0$, the "Scheffers conditions" are

$$\frac{\partial u}{\partial x} = \frac{\partial v}{\partial y}, \quad \frac{\partial u}{\partial y} = \frac{\partial v}{\partial x} \qquad \text{and} \qquad \frac{\partial u}{\partial x} = \frac{\partial v}{\partial y}, \quad \frac{\partial v}{\partial x} = 0.$$

But Scheffer's beautiful constructions (*Verallgemeinerung der Grundlagen der gewöhnlichen komplexen Funktionen*, Sitzungsberichte Sächs Ges. Wiss, Math.-phys. Klasse, Bd. 45, 1893, pp. 828–842) have never been used in other branches of mathematics and its applications, and now appear to be a typical "mathematical plaything."

[217] Lie's pupil at Leipzig University, the geometer Friedrich Heinrich Schur (who should not be confused with the great algebraist Issaï Schur, 1875–1941), always followed in the footsteps of his teacher Lie and, in part, in those of the latter's friend Felix Klein. Schur's work on hypercomplex numbers was very highly rated at the time, but it is now less well known than his characterization of *Riemannian spaces of constant curvature*, which can be viewed as bearing on the *Helmholtz–Lie problem*. (On the grandiose development of this topic, mentioned in passing in Note 144, see T.A. Wolf, *Spaces of Constant Curvature*, 1972; the 1982 Russian translation of this book contains a Supplement by Yu.D. Burago covering some of the latest developments in the field.) More popular still is Schur's axiomatic presentation of Euclidean geometry in his book *Grundlagen der Geometrie*, Leipzig–Berlin, Springer, 1909. This is a reassessment of Hilbert's book of the same name in the spirit of Klein's "Erlangen program" (see Chapter 7): Schur's book is based on axioms of Euclidean plane and solid geometry involving the corresponding isometry groups.

[218] T.E. Molin was born in Riga (Latvia). He graduated from Dorpat (now Tartu—Estonia) University and taught at Dorpat and later in Tomsk (Siberia). Thus most of his life was spent in Russia, where he was called Fiodor Eduardovich; recall that Riga and Dorpat, when Molin lived there, were part

of the Russian Empire. Nevertheless Molin is undoubtedly a representative of the German mathematical school, not because he was German (which is not important in his case), but because he was educated in Leipzig, wrote his first paper there and had Sophus Lie for a teacher, and in Dorpat, which was a purely German university in spirit. (Note that Friedrich Schur, as we mentioned in Note 217, also studied under Lie at Leipzig; for a time he was also a professor at Dorpat University.)

[219] Of course, it would be more logical to write the hypercomplex numbers (5.7) in the form $u = x_0e_0 + x_1e_1 + \cdots + x_ne_n$, where the "complex identity element" e_0 satisfies $e_0e_i = e_ie_0 = e_i$ for all $i = 1, 2, \ldots, n$. This allows us to identify e_0 with the number 1.

[220] This result was obtained by G. Frobenius in his fundamental paper "Über lineare Substitutionen and bilineare Formen," *Crelle Journ.*, **84**, 1878, pp. 1–63; C.S. Peirce's publication was "Upon the logic of mathematics," *Proc. Amer. Acad. of Arts and Sci.*, **7**, 1865–1868, pp. 402–412; see also C.S. Peirce, "On the algebras in which division is unambiguous," *Amer. Journ. of Math.*, **4**, 1881, pp. 225–229. An elementary exposition of this result (and other results mentioned in this book) can be found in a book that is accessible to a wide class of readers: *Hypercomplex Numbers* by I.L. Kantor and A.S. Solodovnikov (Moscow, Nauka, 1973, in Russian, but an English translation is in preparation), as well as in I.V. Arnold's book *Theoretical Arithmetic*, Moscow, Uchpedgiz, 1939 (but his Russian book is hardly accessible to the English-reading public). (The Moscow mathematician and teacher Igor Vladimirovich Arnold (1900–1948), should not be confused with his son, Vladimir Igorievich Arnold (b. 1937), mentioned in Note 208.)

[221] Another method for introducing octaves is based on the elegant operation of "doubling" systems of hypercomplex numbers. This operation, applied to real numbers, yields the complex numbers; applied to complex numbers, it yields the quaternions; and applied to quaternions it yield the octaves. The exposition in the book by Kantor and Solodovnikov mentioned in the previous note is based on this method. In this connection see also the historical article by J. Gueridon and J. Dieudonné, "L'Algèbre depuis 1840," in the book *Abrégé d'histoire des mathématiques (1700–1900)*, sous la direction de J. Dieudonné, v. I, Paris, Hermann, 1978, pp. 91–127, especially pp. 106–111.

[222] The existence of a well-defined division operation in a system of hypercomplex numbers is related to the absence of so-called *divisors of zero*, i.e., numbers $\mathbf{u} \neq \mathbf{0}$ such that there exist numbers $\mathbf{v} \neq \mathbf{0}$ satisfying $\mathbf{uv} = \mathbf{0}$. For example, in the systems of double numbers and dual numbers, numbers of the respective forms $x(1 \pm e)$ and $x\varepsilon$ are divisors of zero. Incidentally, for certain "extensions" of the set of existing systems of hypercomplex numbers (double and dual numbers, four types of quaternions, six types of octaves), it is possible to define a division operation which yields an "ideal" number in the case when

the divisor is a divisor of zero; for the sets of double and dual numbers this procedure is described in detail in I.M. Yaglom's book *Complex Numbers in Geometry*, mentioned in Note 203.

223 It is precisely in this form that the Frobenius theorem (in "generalized form") is proved in the Kantor–Solodovnikov book mentioned in Note 220.

224 Another generalization, akin to alternativity, of the notion of associativity of hypercomplex systems is its Jordan property (see page 100.), first introduced by the outstanding German theoretical physicist Ernst Pascual Wilhelm Jordan (b. 1902); this property has come to the forefront in the work of many physicists and mathematicians in the last decade.

225 It is typical that, in contrast, Lawrence Young, in his recent book mentioned in Note 192, rates Cayley's discovery of the octaves highly enough, but stresses that it is Cayley's only achievement which has retained its importance in our day. (Young, who is often very harsh and subjective in his estimates, writes that Cayley was the author of 900 mathematical papers which, except for the ones on octaves, have entirely lost their interest. He indicates that that the 13 volumes of Cayley's *Collected Mathematical Papers* (Cambridge University Press, 1889–1898, Vols. 1–13) cannot be compared to the mere 60 pages of mathematical notes that the unfortunate Evariste Galois was able to write in his short life. Without going further into this rather uncalled-for comparison (the two persons were too far apart in time, in their lives, in temperament and in scientific style), I would nevertheless like to venture an opinion as to why Young underestimates Cayley's contributions to mathematics so very much. Many of the ideas that Cayley introduced into mathematics (multidimensional spaces, matrices, Cayley tables for group multiplication, etc.) are now so familiar that they seem obvious to us and are viewed as "mathematical folklore," unrelated to the person who first introduced them.)

226 Hurwitz's original and very beautiful proof is contained in his article "Über die Komposition der quadratischen Formen mit beliebig vielen Variablen," *Göttinger Nachrichten*, 1898, pp. 300–316. See also the book by Kantor and Solodovnikov mentioned in Note 220.

227 On the "generalized" Hurwitz theorem for associative number systems (not for octaves!) see A.A. Albert, "Quadratic forms permitting composition," *Ann. of Math.*, **43**, 1942, pp. 161–177. The case of "generalized octaves" has been studied by the Moscow geometer David Borisovich Persits (b. 1941).

Chapter 6

228 Compare with what is said in the foreword to P. Alexandroff's book *Introduction to the Theory of Groups* (see Note 32).

[229] See the fundamental article by C. Jordan, "Mémoire sur les groupes de mouvements," *Ann. math. pures et appl.*, ser. 2, **2**, 1868–1868, pp. 167–215, 322–345, also reproduced in Jordan's *Œuvres*.

[230] If, in the formulas (6.2), we restrict ourselves to values $\Delta > 0$, then we obtain transformations which may be called *direct* affine transformations: they preserve the orientation of the basis formed by two noncollinear vectors $\mathbf{e}_1$, $\mathbf{e}_2$. (The orientation of such a basis is positive if the smallest rotation that sends the direction of $\mathbf{e}_1$ to that of $\mathbf{e}_2$ is counterclockwise and negative in the opposite case.) If in the formulas (6.2) we require that $\Delta = \pm 1$, then we obtain the class of so-called *equiaffine* transformations, which are area-preserving. If in (6.2) we require $\Delta = +1$ then we obtain the *direct equiaffine* transformations. Finally, transformations (6.2) for which $\Delta < 0$ are called *opposite* affine transformations.

[231] The classification of crystals according to their symmetry properties has its origins in ancient times. It was noticed that crystals can have axes of symmetry of orders 2, 3, 4, and 6, but cannot have axis of symmetry of order 5—although such axes often appear in living things (e.g., in sea-stars and in many flowers)—or of orders $\geqslant 7$. The mathematical theory of crystals, however, is entirely a product of the 19th century. In particular, a list of all possible crystallographic groups appeared in the (independent) research works of the Russian crystallographer Efgraf Stepanovich Fedorov (1853–1919) in 1891, of the German mathematician Arthur Moritz Schönflies (1853–1928), also in 1891 but later than Fedorov, and of the English crystallographer William Barlow (1845–1934) in 1894. The number of crystallographic groups (in space) turned out to be 230. (None of the three researchers mentioned above obtained this exact result—all their papers contained (easily filled) gaps, so that each obtained less than 230 groups, and the complete list was obtained by comparing the three results.)

The number of plane crystallographic groups is 17. All of them have been (in effect!) known to and used by ancient designers, almost from Cro-Magnon times; in particular, Arab architects in medieval Spain certainly knew all of them.

In the 20th century the list of "Shubnikov" or "black-and-white" symmetry groups was established. In these groups, elements of the same shape but of different colors (black and white, say), or possessing electric charges of different sign, are distinguished. The number of such groups is 122 in the plane, and 1651 in space. (Alexei Vassilievich Shubnikov, 1887–1970, was a Russian crystallographer.)

At present, the problems of finding all possible groups of plane color symmetries when the elements can have more than two colors, and all symmetry groups in four-dimensional and in multidimensional spaces (Fedorov and Shubnikov groups, not to mention the "colored" groups) remain, to the best of my knowledge, open. (For a beautiful illustration of a plane color ornament, see the picture "Reptiles" in *The World of M.C. Escher*, N.Y., Abrams, 1971, Color plate II.)

The literature on geometric crystallography and crystallographic groups (including their role in art and in nature) is too vast to be listed here. I shall mention only Chapter 4 ("Two-dimensional crystallography") in H.S.M. Coxeter's book *Introduction to Geometry* (see Note 72); Chapter II of the well-known book by D. Hilbert and S. Cohn-Vossen, *Geometry and the Imagination*, N.Y., Chelsea, 1952; Chapters 10, 11, 17 of E. Martin's *Transformation Geometry*, N.Y. Springer, 1982; the album book C. McGillavary, *Symmetry Aspect of M.C. Escher's Periodic Drawings*, Utrecht, 1976; and classical books by D'Arcy W. Thompson, *On Growth and Form*, London, Cambridge University Press, 1952 and (especially) H. Weyl, *Symmetry*, Princeton University Press, 1952. Among books a bit further away from our topic, but rich in content, see B. Grünbaum and G.C. Shepard, *Tilings and Patterns*, N.Y. Freeman, 1987 and *Patterns of Symmetry* (ed. M. Senechal and G. Fleck), Amherst (Mass.), University of Massachusetts Press, 1977; A.V. Shubnikov, V.A. Koptsik, *Symmetry in Science and Art*, N.Y., Plenum Press, 1974. Finally, miscellaneous books on the subject include G.I. Bradley, A.P. Gracknell, *The Mathematical Theory of Symmetry in Solids*, Oxford, Clarendon Press, 1971; M.J. Bürger, *Elementary Crystallography*, N.Y., 1956; J.J. Burkhart, *Die Bewegungsgruppen der Kristallographie*, Basel, Birkhäuser, 1966; M.A. Jaswon, *An Introduction to Mathematical Crystallography*, London, 1965.

[232] In the present exposition, intended for the beginner, we do not dwell on the difference between *Lie groups* and *continuous groups* (the former are determined by equations involving smooth (i.e., differentiable) functions). Nor do we consider the history of *Hilbert's 5th problem*, which is concerned with the connection between these two notions. See, for example, "The Mathematical Developments Arising from Hilbert's Problems" (*Proc. of Symposia in Pure Math.*, Vol. XXVIII), Providence (R.I.), American Math Soc., 1976, pp. 12–14 (the statement of Hilbert's problem) and pp. 142–146 (brief comments by C.T. Yang about this problem). The initial situation, as it was at the end of the 1930s, is described in a nonelementary, but beautifully written, book by L.S. Pontrjagin, *Topological Groups*, Princeton University Press, 1939. The exposition in the German edition of Hilbert's problems (with comments by Russian mathematicians), *Die Hilbertschen Probleme*, Leipzig, Akademische Verlagsgesellschaft Geest und Portig, 1979, pp. 43–47 (Hilbert's text), pp. 126–144 (comments by E.G. Skljarenko; one should keep in mind that, unfortunately, the proofs of A.N. Kolmogorov's results [3], mentioned in these comments, were never published by the latter.

Perhaps the simplest exposition of the proof of the solution of Hilbert's 5th problem (not meant for the beginner, however), is contained in Part II of I. Kaplansky's small book *Lie Algebras and Locally Compact Groups*, Chicago, the University of Chicago Press, 1972.

[233] I hope that the reader realizes the tentative character of this rather arbitrary division of German algebraists into a Leipzig and a Berlin group. As an undoubted pupil and follower of Lie, Study studied in Leipzig as well as in Jena, Strasbourg, Munich, and taught in Leipzig for three years; he then taught at a number of universities, including Johns Hopkins in Baltimore. Study's longest stays were at the universities of Bonn and Greifswald. On the other hand, Frobenius, who came from Berlin, was also under the strong influence of Lie, with whom he had constant friendly contacts. Such clarifications might also be made about other persons mentioned here, but this is hardly necessary.

[234] Or the "exterior product" due to Grassmann, which, as we know, is almost the same.

[235] Mathematicians call the expression δ_i^j, which equals zero for $i \neq j$ and 1 for $i = j$, the *Kronecker symbol*, after the Berlin mathematician Leopold Kronecker (1823–1891) who introduced this notation and studied its properties. Kronecker was the direct opposite of Weierstrass as well as his rival.

[236] Cayley defined the multiplication of matrices in the way it is done today in all linear algebra textbooks: if the matrix $A = (a_{ij})$ corresponds to the linear transformation from the variables x_i to the new variables x_i' (i.e.,

$$x_i' = a_{i1}x_1 + a_{i2}x_2 + \cdots + a_{in}x_n = \sum_{j=1}^{n} a_{ij}x_j,$$

where $i, j = 1, 2, \ldots, n$, while the matrix $B = (b_{ij})$ in the same sense determines the transformation from the variables x_i' to the variables x_i'' (i.e., $x_i'' = \sum_{j=1}^{n} b_{ij}x_j'$), then the passage from the variables x_i to the variables x_i'' can be carried out directly by means of one matrix $C = (c_{ij})$ (i.e., $x_i'' = \sum_{j=1}^{n} c_{ij}x_j$), where $C = B \cdot A$. The Peirces were the first to notice that these definitions (known for $n = 2$ and 3 long before Cayley) were those of a certain (associative) system of (hyper) complex numbers (a matrix algebra in the sense which they assigned to the word "algebra").

[237] It is clear that all the elements (5.1) of the Grassmann algebra whose coefficient x_0 is zero, in particular all the "principal units" $e_1, e_2, \ldots, e_n$ which generate this algebra are examples of *nilpotent* elements (for any e_i, where $i = 1, 2, \ldots, n$, we even have $e_i^2 = 0$). Examples of *idempotent* elements are provided by Clifford's double numbers $e_1 = (1 + e)/2$ and $e_2 = (1 - e)/2$, where $e^2 = +1$ (see Chapter 5; check that $e_1^2 = e_1$ and $e_2^2 = e_2$; note that, on the other hand, Clifford's "dual unit" ε is a nilpotent element of the corresponding algebra). Other examples of idempotent elements are the Clifford numbers $(1 \pm e_{123})/2$ and $(1 \pm e_{1234})/2$, but not $(1 \pm e_{12})/2$ or $(1 \pm e_{12345})/2$ (check this!); here $e_{i_1 i_2 \cdots i_k}$ denotes the "Clifford product" $e_{i_1}e_{i_2}\ldots e_{i_k}$ of the principal units $e_{i_1}, e_{i_2}, \ldots, e_{i_k}$.

[238] In the Russian scientific literature (and perhaps not only there) the expression "almost associative algebras" is often used to include alternative algebras (see relations (5.10) and (5.10a)), Jordan and Lie algebras.

[239] If $u * v = uv + vu$, where we now denote the main (associative!) product of elements from our algebra without putting a dot between them, then $(u^2)_* = u * u = 2u^2$; therefore the left-hand side of (6.8) equals

$$[2u^2v + 2vu^2] * u = 2(u^2v + vu^2)u + u(2u^2v + 2vu^2)$$
$$= 2(u^3v + u^2vu + uvu^2 + vu^3).$$

The right-hand side of (6.8) is equal to the same expression:

$$2u^2 * (vu + uv) = 2u^2(vu + uv) + (vu + uv)(2u^2)$$
$$= 2(u^3v + u^2vu + uvu^2 + vu^3)$$

(since the sum of elements in our algebra is always commutative!).

If we put $u \circ v = uv - vu$, then

$$(u \circ v) \circ w = (uv - vu) \circ w = (uv - vu)w - w(uv - vu)$$
$$= uvw - vuw - wuv + wvu,$$

which obviously implies (6.9):

$$(u \circ v) \circ w + (v \circ w) \circ u + (w \circ u) \circ v = 0$$

(check this!).

[240] Carl Gustav Jacob Jacobi (see the chapter on him in E.T. Bell's book *Men of Mathematics*), one of the leading mathematicians of the 19th century, made important contributions to almost all branches of mathematics and mathematical mechanics. (His elder brother, Moritz Hermann (called Boris Semionovich in Russia) Jacobi (1801–1872), creator of galvanoplastics and author of numerous papers on the practical use of electricity, is now ranked among the greats of Russian physical science; he was a member of the St. Petersburg Academy of Sciences, spent most of his life in Russia and even acquired Russian citizenship. During the life of Carl Jacobi, his elder brother's renown far surpassed his own, but today that relation is reversed.) The brothers came from a rich Jewish banking family; Carl lost his fortune as the result of unsuccessful financial operations and, at the end of his life, had to earn a living from mathematics. He was widely educated; In particular, he was a connoisseur of classical philology, and this had an important influence on his attitude to mathematics, where he was prone to see the aesthetic side (compare with what we said about F. Klein and F. Enriques in Chapter 2). His vast scientific background also influenced his creative work. Jacobi touched upon almost all the branches of "pure" mathematics, but also did work in applied mathematics and astronomy, as well as fundamental research in mechanics where, incidentally, the identity now bearing Jacobi's name first arose in

connection with properties of differential operators. For most of his life Jacobi worked in Königsberg (now Kaliningrad). It was to his scientific, pedagogical and organizational abilities (as well as to the work of the famous astronomer and mathematician Friedrich Wilhelm Bessel, 1784–1846) that the physico-mathematical department at Königsberg University owed its high reputation, maintained for many years, until F. Klein succeeded in "seducing" the leading Königsberg mathematicians to come to Göttingen (see Chapter 8). Jacobi's colossal workload at Königsberg finally exhausted him and, mostly for this reason, he moved to Berlin, where he no longer tried to retain the same level of productivity. Typically, he did not even try to make scientific contacts with Riemann, who listened to his lectures in Berlin, leaving this initiative to his friend Dirichlet. Jacobi died of smallpox in Berlin at the age of 47 ("died in Blattern," as one can read in the Russian translation of Klein's *Vorlesungen über die Entwicklung der Mathematik* ...; the German word *Blattern*, however, means smallpox, and nouns in that language are always capitalized, whether they be names of cities or of diseases).

[241] The literature on the theory of Lie groups and algebras (the two topics are usually studied simultaneously; their relationship is discussed below) is too vast to be reviewed here. We cannot even make a comprehensive list of the most important works in the field (Lie's contributions will be discussed separately). The "main" textbook in Lie algebras (but not Lie groups!) is generally held to be N. Jacobson, *Lie Algebras*, N.Y.–London, Interscience Publishers, 1962; a shorter introduction to the topic is contained in Part 1 of Irving Kaplansky's short book mentioned in Note 232. Among the large treatises we note the 4-volume exposition in Nicolas Bourbaki's *Eléments de Mathématique, Groupes et algèbres de Lie*, Ch. I, II–III, IV–VI, VII–VIII, Paris, Hermann, 1971, 1972, 1968, 1975 and C. Chevalley, *Théorie des groupes de Lie*, Vol. I–III, Paris, Hermann 1946, 1951, 1955. These books appeared in the legendary series "Publications de l'Institut Mathématique de l'Université de Nancago," named after the (imaginary!) city of Nancago, where the imaginary mathematician N. Bourbaki was professor. Nancago = Nancy + Chicago: it was in these two cities that two of the founders of the Bourbaki group first worked—Jean Aléxandre Dieudonné (b. 1906) and André Weil (b. 1906); Claude Chevalley (b. 1909) was also one of the founders of the group. Two shorter expositions, are J.-P. Serre, *Lie Algebras and Lie groups*, N.Y., Benjamin, 1965 and Chapter XIX of J. Dieudonné, *Eléments d'analyse*, Vol. IV, Paris, Gauthier-Villars, 1971, pp. 119–213. Among the classic books and articles by Elie Cartan note the very clearly written article *La géomètrie des groupes de transformations*, Journ. Math. pures et appl., **6**, 1927, pp. 1–119, which also appears in his Collected works (*Œuvres complètes* en 6 volumes, Paris, Gauthier-Villars, 1952–1955) and his textbook-monograph, *La théorie des groupes finis et continus et la géomètrie différentielle traitées par la méthode du repère mobile*, Paris, Gauthier-Villars, 1937.

The simplest expositions of Lie groups and algebras are those intended not for mathematicians but for "users of mathematics", almost all of whom consider the topic important. As an example, we refer the reader to the book of the well-known Israeli physicist H.J. Lipkin with the catchy title *Lie Groups for Pedestrians*, (Amsterdam, North-Holland, 1966).

It is also typical that recently, in Moscow, when a philology student interested in mathematics asked one of the most authoritative Moscow mathematicians where he should begin studying in order to use mathematics in linguistics, the answer was immediate: "Study the theory of Lie groups."

[242] See any exposition (as elementary as you wish!) of the elements of vector calculus, e.g., the article mentioned in Note 200: W.G. Boltjanski, I.M. Jaglom, *Vektoren und ihre Anwendungen in der Geometrie*. I should also like to cite the carefully written textbook of Ya.S. Dubnov, *The Foundations of the Vector Calculus*, Vol. I, Moscow–Leningrad, Gostechizdat, 1950, but this book is hardly accessible to the English-reading public. The Jacobi identity immediately follows from the easily proved relation $[\mathbf{a}[\mathbf{b},\mathbf{c}]] = (\mathbf{b},\mathbf{a})\mathbf{c} - (\mathbf{c},\mathbf{a})\mathbf{b}$.

[243] The *nondegeneracy* requirement for the scalar product is the following: for any vector $\mathbf{a} \neq \mathbf{0}$, there exists a vector $\mathbf{b}$ such that $(\mathbf{a},\mathbf{b}) \neq 0$; it is easy to see that $(\mathbf{0},\mathbf{b}) = 0$ for all $\mathbf{b}$. Concerning the *positive definiteness* requirement, see below.

[244] In the literature, a vector space supplied with a scalar product $(\mathbf{a},\mathbf{b})$ satisfying, besides the properties listed above, the requirement of *positivity* or *positive definiteness*, is called a *Euclidean space*. Actually, the term Euclidean space" has come to denote a vector space with a scalar product satisfying only the three properties listed in the main text.

In the literature, spaces with positive scalar product are sometimes called *proper Euclidean*; with nondegenerate but not necessarily positive product, *pseudo-Euclidean*; with possibly degenerate scalar product, *semi-Euclidean* (cf. Chapter 4).

[245] For this reason an important place in the theory of Lie algebras (and groups) is occupied by the problem of classifying particular types of such algebras (groups) satisfying certain supplementary conditions, e.g., the classification problem for simple (or semi-simple) Lie groups, which will be discussed below.

The fact that the classification problem for Euclidean spaces is much simpler than that for Lie algebras is easy to explain: n-dimensional Euclidean space is a vector space with coordinates $x_1, x_2, \ldots, x_n$ and a metric determined by the scalar product of vectors $\mathbf{a}(x_1, x_2, \ldots, x_n)$ and $\mathbf{b}(y_1, y_2, \ldots, y_n)$ in accord with the formula (valid for arbitrary—not necessarily Cartesian—coordinates)

$$(\mathbf{a},\mathbf{b}) = g_{11}x_1y_1 + g_{12}x_1y_2 + \cdots + g_{nn}x_ny_n = \sum_{i,j=1}^{n} g_{ij}x_iy_j,$$

where it is natural to assume $g_{ij} = g_{ji}$ for all $i, j = 1, 2, \ldots, n$ and $i \neq j$; additional nondegeneracy or positive definiteness requirements may be imposed on the quadratic form (g_{ij}) appearing in this formula. On the other hand, a Lie algebra is determined by the "Lie product"

$$[\mathbf{a}, \mathbf{b}] = c_{11}^1 x_1 y_1 \mathbf{e}_1 + c_{11}^2 x_1 y_1 \mathbf{e}_2 + \cdots + c_{nn}^n x_n y_n \mathbf{e}_n = \sum_{i,j,k=1}^{n} c_{ij}^k x_i y_j \mathbf{e}_k$$

of vectors $\mathbf{a}(x_1, x_2, \ldots, x_n)$ and $\mathbf{b}(y_1, y_2, \ldots, y_n)$, where $\mathbf{e}_1, \mathbf{e}_2, \ldots, \mathbf{e}_n$ are the basis unit vectors, i.e., it is determined by the structural constants c_{ij}^k satisfying the conditions $-c_{ij}^k = c_{ji}^k$ for all $i, j, k = 1, 2, \ldots, n$ (the antisymmetry or anticommutativity requirement for $[\mathbf{a}, \mathbf{b}]$) and the rather complicated condition

$$\sum_{r=1}^{n} (c_{ir}^s c_{jk}^r + c_{jr}^s c_{ki}^r + c_{kr}^s c_{ij}^r) = 0, \qquad i, j, k, s = 1, 2, \ldots, n, \qquad (*)$$

which is equivalent to the *Jacobi identity*. Thus in one case we must classify *second-order tensors* g_{ij} (characterized by two indices i and j; concerning the notion of tensor see, for example, I. Gelfand, *Lectures on Linear Algebra*, N.Y., Interscience Publishers, 1961, or any elementary text on the tensor calculus), while in the other case we must classify *third-order tensors* c_{ij}^k (antisymmetric with respect to i and j and satisfying relation (*)), depending on three indices i, j and k, which makes the problem incomparably more difficult. (For comparison, we note that while the classification problem for Grassmann's *bivectors* ("extensive magnitudes of the second order") in spaces of arbitrary dimensions (which reduces to classifying antisymmetric second-order tensors ε_{ij} satisfying $\varepsilon_{ji} = -\varepsilon_{ij}$) presents no difficulty, the classification problem for *trivectors* (i.e., Grassmann's "extensive magnitudes of the third order," or third-order tensors ε_{ijk}, antisymmetric with respect to any two lower indices (i.e., such that $\varepsilon_{ijk} = -\varepsilon_{jik} = -\varepsilon_{ikj} = \ldots$) is still a long way from solution, although recently there have been some advances, associated, in particular, with the name of Ernest Borisovich Vinberg (b. 1937). It is clear that, in n-dimensional space, for $n < 3$, there exist no nonzero trivectors, while in 3-dimensional and 4-dimensional space there is only one type of trivector; the classification of trivectors in 5-dimensional space is simple enough. Trivectors in 6- and 7-dimensional space were classified by the outstanding Dutch specialist in linear algebra and tensor calculus Jan Arnoldus Schouten (1883–1973). When Grigory Borisovich Gurevich (1898–1980) was defending his doctoral thesis (Moscow, 1935) on the classification of trivectors in 8-dimensional space, Schouten wrote in his review of this work that finally there was hope that the trivector classification problem might be solved in the reviewer's lifetime. However, G.B. Gurevich, who spent most of his life trying to solve the problem, was unable to resolve the classification problem of trivectors in 9-dimensional space; the solution was eventually given by Vinberg. As for the general problem of listing all possible types of trivectors, we still have no approaches to its solution today; even the 10-dimensional case remains unsolved. This being so, the prospects for solving the much more difficult general classification problem of Lie algebras (i.e., "Lie tensors" c_{ij}^k) are, to put it mildly, not very good. As for the classification theorems obtained by Sophus Lie himself, e.g., the list of possible types of *geometric transformation groups of the line* and *of the plane* (depending, of course, on a finite number of parameters), they are a huge and beautiful monument to the difficulty of the general problem, rather than an approach to its solution!)

[246] The choice of the angular velocity vector $\mathbf{l}$ determines not only the one-parameter subgroup $\mathfrak{v}$ of the group $\mathfrak{B}$ but also the "canonical parameter" (time) corresponding to this subgroup. This parameter "enumerates" the transformations $\beta \in \mathfrak{v}$ (rotations about a fixed axis) by means of real numbers subject to the condition

$$\beta(t_1), \beta(t_2) \in \mathfrak{v} \Rightarrow \beta(t_1) \cdot \beta(t_2) = \beta(t_1 + t_2).$$

A similar construction plays an essential role in the general assignment of a Lie algebra to a Lie group.

[247] Of course, just as the multiplication of complex numbers $z(r, \varphi) \cdot z(a, \alpha) = z(r', \varphi')$ (written in polar coordinates) generates the group of rotations and similitudes of the plane (the transformation with parameters (a, α) sends the point $M(r, \varphi)$ into the point $M'(r', \varphi')$, where $r' = ar$, $\varphi' = \varphi + \alpha$; see Fig. 19(b), so too any algebra (system of hypercomplex numbers) with multiplication

$$(x_1 e_1 + x_2 e_2 + \cdots + x_n e_n)(a_1 e_1 + a_2 e_2 + \cdots + a_n e_n)$$
$$= x_1' e_1 + x_2' e_2 + \cdots + x_n' e_n$$

determines a continuous (Lie) group which acts on the "group space" whose points are given by the coordinates $x_1, x_2, \ldots, x_n$. The choice of parameters $a_1, a_2, \ldots, a_n$ determines a transformation of our group. This transformation sends the point $M(x_1, x_2, \ldots, x_n)$ into the point $M'(x_1', x_2', \ldots, x_n')$. In the above-mentioned case of ordinary complex numbers, the role of the group space is played by the complex plane itself. This rather obvious observation was made by H. Poincaré in a brief article published in 1884; N. Bourbaki (see his text on the history of mathematics mentioned at the beginning of Chapter 3) notes the strong impression that this article made on Lie and his followers (among whom Bourbaki includes, in this connection, Study, Scheffers, F. Schur, Molin, and E. Cartan) who were particularly interested in the connection between Lie groups and algebras and in the classification of both. In particular, the transformations determined by the elements of the "group space" were to play a large part in E. Cartan's considerations (see, for example, his long article mentioned in Note 241).

[248] Lie did not doubt the possibility of extending any "local group" (a small neighborhood of its identity element ε) to a "complete" continuous group, but,

apparently (showing that this chronologically near period is psychologically distant from us!), never tried to prove this fact. This result was obtained by one of the founders of *modern* Lie group theory, Elie Cartan, repeatedly mentioned above.

[249] Let $\mathfrak{p}$ and $\tilde{l}$ be subspaces of a Lie algebra $\mathscr{L}$. Denote the subspace of $\mathscr{L}$ generated by all elements of the form $[k, l]$, where $k \in \mathfrak{p}$, $l \in \tilde{l}$ by $[\mathfrak{p}, \tilde{l}]$. In that case $\mathfrak{p}$ is a subalgebra of $\mathscr{L}$ if $[\mathfrak{p}, \mathfrak{p}] \subset \mathfrak{p}$, and $\mathfrak{p}$ is an ideal of $\mathscr{L}$ if $[\mathfrak{p}, \mathscr{L}] \subset \mathfrak{p}$. The Jacobi identity implies that for any three subspaces $\mathfrak{a}$, $\mathfrak{b}$ and $\mathfrak{c}$ of the Lie algebra $\mathscr{L}$ we always have $[[\mathfrak{a}, \mathfrak{b}], \mathfrak{c}] \subset [[\mathfrak{b}, \mathfrak{c}], \mathfrak{a}] + [[\mathfrak{c}, \mathfrak{a}], \mathfrak{b}]$, where the "+" sign stands for the vector sum of subspaces; therefore if $\mathfrak{i}$ and $\mathfrak{j}$ are ideals of the Lie algebra $\mathscr{L}$ then so is $[\mathfrak{i}, \mathfrak{j}]$. Since the Lie algebra $\mathscr{L} = l^{(0)}$ itself is trivially an ideal of $\mathscr{L}$, we obtain a decreasing (more precisely, nonincreasing) sequence of ideals:

$$\mathscr{L} = l^{(0)} \supset l^{(1)} \supset l^{(2)} \supset \cdots \supset l^{(k)} \supset \cdots,$$

where $l^{(1)} = [l^{(0)}, l^{(0)}]$, $l^{(2)} = [l^{(1)}, l^{(1)}]$, $l^{(3)} = [l^{(2)}, l^{(2)}], \ldots$.

The Lie algebra $\mathscr{L}$ is called *sovable* if there exists a (natural) number k such that $l^{(k)} = o$, where $o = \{0\}$ is the trivial ideal consisting of only one element—the 0 element of the algebra (vector space) $\mathscr{L}$. Another definition of a solvable algebra (equivalent to the first one, as can easily be checked) is the following: $\mathscr{L}$ is solvable if and only if there exists a sequence of subalgebras

$$\mathscr{L} = \mathfrak{w}_0 \supset \mathfrak{w}_1 \supset \mathfrak{w}_2 \supset \cdots \supset \mathfrak{w} = o,$$

such that the dimension of each $\mathfrak{w}_i$ is less than that of the previous one precisely by 1 (here $\dim \mathfrak{w}_i = n - i$, where "dim" stands for "dimension" and n is the dimension of the vector space $\mathscr{L}$; here, as above, $o = \{0\}$) and each subalgebra in the sequence is an ideal of the previous subalgebra ($[\mathfrak{w}_i, \mathfrak{w}_{i-1}] \subset \mathfrak{w}_i$ for all $i = 1, 2, \ldots, n$).

The Lie algebra $\mathscr{L}$ is called *semisimple* if it contains no solvable ideal different from the trivial one $o = \{0\}$. The correspondence between Lie algebras and Lie groups allows us to limit ourselves to the definition of solvable and semisimple Lie algebras, since Lie groups are called solvable and semisimple if such are their corresponding Lie algebras.

[250] The Cartan theorem mentioned in Note 248 established a correspondence between Lie algebras and simply connected ("not containing any holes") Lie groups. In the present elementary exposition, we have neither the possibility nor the necessity of dwelling on the possible types of Lie groups that are not simply connected. Nor shall we consider such groups as the group (6.2) of affine (or linear) transformations of the plane, which consists of two disconnected parts, corresponding to the values $\Delta > 0$ ("direct" affine transformations; they are the ones that constitute the simply-connected group appearing in the Cartan construction) and $\Delta < 0$ ("opposite" affine transformations, which do not constitute a group, since the product of two opposite transformations is a direct affine transformation).

[251] The classification problem for *semisimple* Lie groups and algebras is of no independent interest, since it can be shown that every semisimple Lie algebra is the "direct" (or "vector"—see Note 188) sum of simple Lie algebras.

[252] Wilhelm Karl Joseph Killing studied at the universities of Munster and Berlin. He then taught at gymnasia in Brilon and Braunsberg, and later at Munster university. Thus he had no connection with the universities and cities where Sophus Lie worked, so that in this sense he was not a pupil of Lie's. However, the deep scientific (and later personal) relationship between Killing and Lie, as well as Killing's dependence in his entire research work on the ideas and problems formulated by Lie, allows us to say that Killing was indeed Lie's pupil—perhaps by his loyalty and the scope of his talent, the leading one. Nevertheless, this did not stop Lie, who was never very easy to get along with, from expressing unjust remarks about Killing.

[253] W. Killing, "Die Zusammensetzung der stetigen endlichen Transformationsgruppen I–IV," *Math. Annalen*, **31**, 1888, pp. 259–290; **33**, 1889, pp. 11–48; **34**, 1889, pp. 57–122; **36**, 1890, pp. 161–189.

[254] E. Cartan, *Sur la structure des groupes de transformations finis et continus* (Thèse), Paris, Nony, 1894; 2nd edition: Paris, Vuibert, 1933. This also appears in Cartan's *Collected Works*, cited in Note 241.

[255] What we say about the greater meaningfulness of the van der Waerden-Dynkin construction as against Cartan's thesis should not be taken literally. Of course, both van der Waerden's clear and elegant geometric constructions (see Note 256), mainly originating in H. Weyl's fundamental research (see the latter's "Theorie der Darstellung kontinuierlicher halbeinfacher Gruppen durch lineare Transformationen I–III und Nachtrag," *Math. Zeitschift*, **23**, 1925, pp. 271–309; **24**, 1926, pp. 328–376; pp. 377–395; pp. 789–791, which also appears in H. Weyl's Collected Works (*Gesammelte Abhandlungen*, Bd. 1–4, Heidelberg, Springer, 1968)) and the "Dynkin diagrams" (see Note 257), inspired in turn by van der Waerden's ideas, are typical of the modern mathematical style, with its unexpected points of contact of topics which at first glance seem far apart. All the same, Cartan's *Thèse* remains one of the "mathematical classics."

[256] B.L. van der Waerden, "Die Klassifizierung der einfachen Lie'schen Gruppen," *Math. Zeitschrift*, **37**, 1933, S. 446–462.

[257] Also see the nonelementary but clearly and beautifully written article by E.B. Dynkin, "The structure of semisimple Lie algebras," *Uspehi mat. nauk*, 2, vyp. **4**(20), 1947, pp. 59–127 (in Russian) or the briefer exposition of the result concerning us here: E.B. Dynkin, "Classification of simple Lie groups," *Mat. sbornik*, **18**(60), 1946, pp. 347–452 (in Russian). Note that Dynkin's classifying construction reduces to finding a system of plane graphs corre-

sponding to simple Lie groups; their number (more precisely, the number of series of plane graphs, because to each "series" of Lie groups there corresponds a "series" of graphs) turns out to be very small. As an illustration of deep general properties of the science of mathematics, reflecting a certain simplicity and harmony of our world (here our point of view differs little from that of the Pythagoreans of the 6–5th centuries B.C. and, more generally, from that of the mathematicians of antiquity), we note that Dynkin diagrams, which first appeared in a rather narrow mathematical problem, have since turned up in various topics far removed from the original one. Thus the diagrams classifying simple Lie groups also happen to classify the singularities of smooth maps, caustics, wave fronts, etc. (See V.I. Arnold's book, meant for a wide reading public, *Catastrophe Theory*, N.Y., Springer Verlag, 1984, or the detailed monograph V.I. Arnold, A.N. Varchenko, S.M. Husein-Zade, *Singularities of Smooth Maps*, Part I, *Classification of Critical Points, Caustics and Wave Fronts*, Part II., *Algebraic-topological Aspect*, Moscow, Nauka, 1982, 1983 (in Russian; an English translation is in preparation). In this connection, the following quotation from Arnold's first-named book above is appropriate: "There is something mysterious in the theory of singularities, as in all mathematics: it is the extraordinary coincidences and relationships appearing in topics and theories which seem far removed from each other at first glance." Moreover, some Moscow mathematicians and natural scientists have recently discovered connections between Dynkin diagrams and classification problems arising in the study of social and natural phenomena. If their expectations were to come true, this would confirm the profound character of the Dynkin diagrams, which uncover—as mathematics should—certain deeply hidden laws of the Universe ("world harmony", as the Pythagoreans thought and said).

[258] *Vectors* in 2-dimensional, 3-dimensional and n-dimensional affine (vector) and Euclidean spaces are often also called *points*. This identification is connected with the possibility of assigning to each point A of the plane the vector $\mathbf{a} = \overline{OA}$, say; here 0 is the "origin of vectors" (which corresponds to the zero vector 0, appearing in the axioms of vector space). For example, J. Dieudonné does this in the book cited in Note 64, which he views as a preliminary sketch of a school textbook. On the other hand, the special role played in vector algebra by the zero vector predetermines the presence of a distinguished point 0 in this construction of geometry. For this reason the plane or space, viewed as a set of vectors identified with points, is sometimes called the *centroaffine* plane (or space) or the *centro-Euclidean* plane (or space); the centro-Euclidean plane is sometimes described as the "punctured" plane, i.e., the plane from which a distinguished point has been removed. If we do not distinguish any point of the plane, then we are led to a description of geometry like that in Herman Weyl's famous textbook on relativity theory, *Raum, Zeit, Materie*, Berlin, Springer, 1923. Thus the basic elements of the system under consideration include both the set $\mathscr{V} = \{\mathbf{a}, \mathbf{b}, \mathbf{c}, \ldots, \mathbf{0}\}$ of vectors

and the set $\mathscr{T} = \{A, B, C, \ldots\}$ of points; the connection between vectors and points is effected by assigning to each pair of points $A, B \in \mathscr{T}$ a unique vector $\mathbf{a} \in \mathscr{V}$ denoted by $\overline{AB}$; this operation must satisfy two axioms:

$$A_1: \forall A \in \mathscr{T}, \mathbf{a} \in \mathscr{V}\ \exists!\ B \in \mathscr{T} \mid \overline{AB} = \mathbf{a};$$

$$A_2: \forall A, B, C \in \mathscr{T} \mid \overline{AB} + \overline{BC} + \overline{CA} = \mathbf{0}.$$

259 Consult the fundamental (but not easily accessible) article by H. Freudenthal, "Oktaven, Ausnahmegruppen und Oktavengeometrie," Mimeographed, Utrecht, 1951, or, for example, J. Tits, "Le plan projectif des octaves et les groupes de Lie exceptionels" *Bull. Acad. Roy. Belg. Sci.*, **39**, 1953, pp. 300–329.

260 The current great interest in the extremely difficult and, it would seem, rather narrow topic of *finite simple groups* is a curious outgrowth of the explosive increase of attention, typical of contemporary "pure mathematics," being paid to finite (or discrete) mathematics as opposed to the mathematics of the continuous. On this shift of interest away from continuous mathematics (which includes the calculus and Sophus Lie's theory of continuous groups) see Notes 42 and 45, as well as I.M. Yaglom's "Elementary Geometry, Then and Now," in the book *Geometric Vein* (The Coxeter Festschift), edited by C. Davis, B. Grünbaum, F.A. Sherk, N.Y., Springer, 1981, pp. 258–269. The theory of finite simple groups is worth looking at in more detail. Already Galois established that the *alternating group* A_n, i.e., the group of so-called *even* permutations of n elements, is simple for any $n \neq 4$ (see Chapter 1 above). On the other hand, since "geometries" (vector spaces) can be constructed not only over the field of real numbers, the field of complex numbers, the noncommutative field of quaternions and the algebra of octaves, but also over finite (Galois) fields (see page 20), it is possible to carry over the "classical" simple Lie groups and their geometric interpretation to "finite geometries," containing only a finite number of points. Thus we come to new meaningful examples of finite simple groups, known as *simple groups of Lie type*, more often called (and this is only fair) *Chevalley groups*, after the French mathematician Claude Chevalley, whom we had the occasion to mention previously, and who was the first to investigate such groups in depth. Finally, besides these two large classes of finite simple groups (Galois's alternating groups; Chevalley's groups of Lie type) there also exists a finite set of simple groups not included in any classification. These groups are now called *sporadic* simple groups. There are 26 of them; their list begins with the group M_{11} of order (number of elements) 7920 and concludes with the so-called *Baby Monster* (of order $2^{41} \cdot 3^{13} \cdot 5^6 \cdot 7^2 \cdot 11 \cdot 13 \cdot 17 \cdot 19 \cdot 23 \cdot 31 \cdot 47 \approx 10^{34}$) and *The Monster* (or *Big Monster*) of order $2^{46} \cdot 3^{20} \cdot 5^9 \cdot 7^6 \cdot 11^2 \cdot 13^2 \cdot 17 \cdot 19 \cdot 23 \cdot 29 \cdot 31 \cdot 41 \cdot 47 \cdot 59 \cdot 71$ ($\approx 10^{54}$). Try to imagine a Cayley table for these groups! It seems that at present the proof of the existence of all these groups may be viewed as completed. It was elaborated during a long period by many researchers in

several countries with the essential aid of computers; see the already partially obsolete but very expressive review by M. Ashbacher, *The Finite Simple Groups and their Classification*, New Haven, Conn., Yale University Press, 1980, which can be supplemented by its author's later interview in the Los Angeles *Times* of October 24, 1980. See also D. Gorenstein's (advanced) book *Finite Simple Groups*, N.Y. Plenum Press, 1982, or his more accessible article "The Enormous Theorem," *Scientific American*, **6**, pp. 104–115.

Actually, the words "proof" and "complete" in this context have a somewhat nonstandard (and perhaps not very clear) meaning. The stupendous amount of work carried out convinces us that the list of 26 sporadic simple groups is correct, but in what sense are we, the heirs of Pythagoras, Plato, Aristotle, Gauss, Weierstrass, Russell, and Hilbert, to understand this work to be a "proof"? The first (and yet far from the most complicated!) of the questions that arise here concerns the use of computers. Can we say without any misgivings, for example, that the famous four-colour problems has truly been solved if the arguments advanced for its solution required thousands of hours of computer time and parts of the proofs carried out by computer have never been checked by anyone? (See, in this connection, the article by K. Appel and W. Haken, "The Solution of the Four-Color Problem," *Scientific American*, October 1977, pp. 108–121, or K. Appel, W. Haken, "The Four-Color Problem" in the book *Mathematics Today* (*Twelve Informal Essays*), L.A. Steen, editor, N.Y., Springer Verlag, 1979, pp. 153–188.) Concerning the difficulties which arise here, the well-known American specialist in combinatorial problems Daniel Cohen stated as early as 1978 that he could simplify the Appel–Haken proof. This, incidentally, was not obtained by Kenneth Appel and Wolfgang Haken alone, but by a large group, including many mathematicians, programmers, and computer scientists—among the latter John Koch deserves to be singled out—and the powerful IBM360 computer at Illinois University. The simplified proof projected by Daniel Cohen was meant to be "verifiable by hand", i.e., the computer parts of the proof were to be short enough to be checked by human mathematicians; Cohen appropriately called the book under preparation *Human Solution of the Four-Color Problem.* However, this book, to my knowledge, has not appeared in print, so that there is no definite answer to the question of whether the four-colour problem is "solved" or "unsolved."

But in the case of sporadic simple groups the situation is much more complicated than in the case of the four-color problem. The arguments leading to the conclusion that the list of all such groups is complete have never been written down. This is not because there seem to be doubts about the uniqueness of one or two of the 26 groups (or because there were such doubts recently), but because it is impossible to write down, read, proofread and verify the thousands upon thousands of pages the text would require. Specialists like Aschbacher feel that we are just beginning the difficult work of elaborating certain links of the entire chain of arguments. This elaboration, in which many researchers from different countries will take part, will probably stretch out for several years, after which succeeding generations will have at their disposal "a number of fat, carefully prepared volumes," containing an alleged solution of the problem.

Now if finding the sporadic finite simple groups requires such titanic efforts, is it really worthwhile? Is the result worth the effort? Here we return to the general considerations about modern mathematics touched upon in Note 257, which alone justify the very existence of mathematical science. In this connection see the book of P.J. Davis and R. Hersh, *The Mathematical Experience*, mentioned in Note 109, as well as the following articles by outstanding physicists, Nobel prize winners: E.P. Wigner, "The Unreasonable Effectiveness of Mathematics in the Natural Sciences," *Comm. in Pure Appl. Math.*, **13**, 1960, pp. 1–14; C.N. Yang, *Einstein and the Physics of the Second Half of the Twentieth Century* (the text of Yang's report at the II Marcel Grossman seminar devoted to Einstein's 100th anniversary is included in Yang's *Selected Papers*, 1945–1980, San Francisco, Freeman, 1983), or the less declarative and more concrete text by C.N. Yang, "Fibre Bundles and the Physics of the Magnetic Monopole," in *The Chern Symposium 1979*, N.Y., Springer, 1980, pp. 247–253). Finite simple groups of the Lie type or Chevalley groups may also seem to be an elegant plaything, not worthy of serious attention or of great efforts—but the discovery of these groups was not the conclusion but the starting point of large series of interconnected investigations in algebra, analysis, number theory, geometry, algebraic geometry, etc. (See, for example, the review by R. Steinberg, *Lectures on Chevalley Groups*, New Haven, Conn., Yale University Press, 1967.) Here we should perhaps also mention the wild idea about the relationship between the order of the first of the simple groups of type E_8 (the group E_8, the last and most complicated of the singular Lie–Killing–Cartan simple groups, has been the focus of attention of mathematicians and physicists for a number of reasons) and the number of protons in the Universe. This idea is discussed at the end of Section 9.8 of H.S.M. Coxeter and W.O.J. Moser, *Generators and Relations for Discrete Groups*, Heidelberg, Springer, 1972. Similarly, the theory of sporadic groups may turn out to be related to important branches of science having general mathematical significance. The very difficulty of this theory attests, to some extent, to its depth. A fervent—but perhaps as yet insufficiently founded—article by J.H. Conway, one of the leading *dramatis personae*, discusses the possible consequences for mathematics of the deciphering of the Big Monster's structure (*The Math. Intelligencer*, 1980, 2, #4).

[261] See, for example, C. Jordan, *Cours d'Analyse de l'École Polytéchnique*, Vol. I–III, Paris, Guathier-Villars, 1909–1915; E. Picard, *Traité d'Analyse de la Faculté des Sciences de Paris*, Vol. I–III, Paris, Gauthier-Villars, 1891–1896 (Lie's theory is presented in the last volumes of these famous calculus courses).

[262] A typical example of the different approaches to the solution of differential equations in the first and second halves of the 20th century, reflecting

the changes of attitude towards the "mathematics of infinity" and "discrete mathematics," which we discussed above (e.g., in Note 260), is the change in the relation between *differential equations*, viewed since Newton's day as the principal mathematical language describing the laws of nature, and *difference equations*, which are their "discrete analogue." Before World War II, difference equations were thought to be rather primitive models of differential equations, instruments of the engineer and the natural scientist. Thus, for example, the theory of linear difference equations with constant coefficients, very similar to the theory of differential equations with constant coefficients, was usually studied in universities *after* differential equations and viewed as a "toy"—an arithmetical model of linear differential equations; it attracted students precisely because of its resemblance to a "real" mathematical theory. Whenever a difference equation arose in applications, mathematicians would usually approximate it by a similar differential equation and estimate the solutions of the former by means of the solutions of the latter. In our time the converse occurs more often: in order to prepare a differential equation for computer solution (solution on a machine which is discrete (numerical) in principle) one replaces the differential equation by the difference equation which approximates it. This approach is reflected in all modern textbooks on differential equations. The "Newtonian" idea that all nature's laws are described by differential equations has been criticized in an entirely different way—though from a similar standpoint, *typical of our computer* age—in a brilliant book by B.B. Mandelbrot, *The Fractal Geometry of Nature*, San Francisco, Freeman, 1982. In somewhat simplified and exaggerated form, Mandelbrot's point of view may be described as follows. *All* the functions which describe the phenomena of the natural and social sciences are continuous but nowhere differentiable (i.e., functions which "change direction" at each point); Newton's and Leibniz's differentiable functions are nothing more than an idealized approximation to the real state of affairs.

[263] Two relevant books are P.J. Olver, *Applications of Lie Groups to Differential Equations*, N.Y., Springer Verlag, 1986, and L.V. Ovsiannikov, *Group Analysis of Differential Equations*, Moscow, Nauka, 1978 (in Russian).

[264] Edmond Laguerre, an outstanding French mathematician, was extremely versatile. He worked in the theory of functions of a complex variable, in classical mathematical analysis, and in geometry. A graduate of (and later, for many years, a teacher at) the Paris École Polytéchnique, which we have often mentioned here, Laguerre was a typical product of this institution. In particular, the school has very difficult examinations (Galois failed the examinations for *entrance*!) and special instructors to prepare students for them; Laguerre served in this role for several years. Undoubtedly these rigorous examinations influenced Laguerre greatly, diversified his scientific interests, and developed his taste for the difficult problems which he produced all his life and regularly published in French mathematical journals.

Laguerre devoted a series of articles to his circle geometry, in which the role of isometries in the plane is played by "line-element circle transformations." In these articles, in particular, he describes the family of all "Laguerre cycles", i.e., curves with "Laguerre self-transformations" (cf. what we say below about Klein's and Lie's *W*-curves). All of these articles appear in Volume 2 of Laguerre's works (*Œuvres*, Vol. 1–2, Paris, Gauthier-Villars, 1898–1905). Laguerre's own approach to his transformations differs from the one developed here and in Note 265. As to a general assessment of Laguerre as a mathematician, see H. Poincaré's introductory article in Vol. 1 of Laguerre's works.

[265] The Möbius inversion (or pointwise inversion) described in Chapter 3 (note that it was first discovered by Apollonius from Perga (*c.* 262–190 B.C.) and then rediscovered after Möbius, by William Thompson, Lord Kelvin (1824–1907) in connection with certain problems in electrostatics and is often attributed to him) is often described as follows. (This construction is due to J. Steiner.) The *power of a point A with respect to a circle* S with center Q and of radius r, denoted by $\mathrm{po}(A, S)$, is defined as the square of the (real or purely imaginary) length of the segment $t(A, S)$ of the tangent to S from A, i.e., $\mathrm{po}(A, S) = t^2 = d^2 - r^2$, where $d = AQ$. We can also write $\mathrm{po}(A, S) = \overline{AB_1} \cdot \overline{AB_2}$, where B_1 and B_2 are the intersections of an arbitrary line a passing through A with the circle S. The family of circles $\mathscr{A} = \{S \mid \mathrm{po}(Q, S) = k\}$ is called a *bundle* of circles with radical center Q and power k. Now let $S \in \mathscr{A}$ and $S \ni A$. All such circles S, contain, besides the point A, another point A' (in the limiting case A' may coincide with A; then all the circles are tangent to each other at A). The transformation $A \mapsto A'$ is called the *Möbius inversion* with center Q and power k. Similarly, we can define $\mathrm{po}(a, S) = (r - d)/(r + d)$, where $S = S(Q, r)$ is a circle, a is an oriented straight line and d is the positive or nonpositive distance from Q to a (or $\mathrm{po}(a, S) = \tan^2(\angle(a, S)/2)$, where $\angle(a, S)$ is the real or imaginary angle between a and S; or $\mathrm{po}(a, S) = \tan(\angle(a, b_1)/2) \cdot \tan(\angle(a, b_2)/2)$, where b_1 and b_2 are the two tangents drawn from an (arbitrary!) point $A \in a$ to S and $\mathrm{po}(S_1, S_2) = d^2 - (r_1 - r_2)^2$, where $S_1 = S_1(Q_1, r_1)$ and $S_2 = S_2(Q_2, r_2)$ are two arbitrary oriented circles and $d = Q_1Q_2$; or $\mathrm{po}(S_1, S_2) = (t(S_1, S_2))^2$, where $t(S_1, S_2)$ is the (real or imaginary) *tangential distance* between S_1 and S_2 i.e., the length of the segment of the common tangent between them). Now let $\mathscr{B} = \{S \mid \mathrm{po}(q, S) = k\}$ be the *net* of circles with axis q and power k and $\mathscr{C} = \{S \mid \mathrm{po}(S, \Sigma) = k\}$ the *bunch* of circles with central circle Σ and power k. The family of all circles S, where $S \in \mathscr{B}$ and $S\tau a$ (here a is a fixed oriented line and τ means "tangent to"), is tangent to another line a' (in the limiting case a' may coincide with a). The family of all circles $\mathscr{s}$, $\mathscr{s} \in \mathscr{C}$ and $\mathscr{s}\tau S$, where S is a fixed oriented circle, is tangent to a second circle S' (which may coincide with S in the limiting case). The transformation $a \mapsto a'$ on the set of all oriented straight lines is the *Laguerre inversion*; such inversions generate the Laguerre transformations of the plane—line-element transformations of the plane. It would be appropriate to call the transformation $S \mapsto S'$ a *Lie inversion*, for such inversions generate the family of *Lie*

tangential circle transformations. For the details see I.M. Yaglom, "On the Circle Transformations of Möbius, Laguerre and Lie," in *The Geometric Vein* (see Note 260), pp. 345–353. All these notions and constructions can be carried over to 3-dimensional and *n*-dimensional geometry.

An elementary introduction to the three types of circle geometries is contained in I.M. Yaglom, "Geometrie der Kreise," in *Enzyklopädie der Elementarmathematik* Bd. IV (Geometrie), (see Note 72), pp. 457–526. A detailed theory of all these geometries is contained in W. Blaschke, *Vorlesungen über Differentialgeometrie*, Vol. III, *Differentialgeometrie der Kreise und Kugeln*, Berlin, Springer, 1929. [On Wilhelm Blaschke (1885–1962), one of the leading geometers of the 20th century see, for example, my afterword to the Russian translation of Blaschke's book *Kreis und Kugel*, published by Nauka in Moscow in 1967, pp. 201–227 (in Russian).] In less detail, "circle transformations" and "circle geometries" are treated in the "Higher Geometry" textbooks for German universities by Klein and Bieberbach, which were very popular in the first third of this century.

In view of the general trend of modern mathematics pointed out in Notes 42, 45, 260 and 262, exponents of circle geometry, like exponents of projective geometry (see Note 75), try to preserve its scientific significance in the context of a general decrease of interest in geometry by shifting to finite circle geometries (circle geometries over finite (Galois) fields which lead to "planes" containing only a finite number of points; these planes can also be characterized in "purely geometric" terms, by means of appropriate axioms), and by stressing the relation between circle geometry and algebra. In this connection, see the elementary book I.M. Yaglom, *Complex Numbers in Geometry* (see Note 203), where circle geometries appear; better still, consult the more recent monograph by W. Benz, *Vorlesungen über Geometrie der Algebren*, Heidelberg, Springer, 1973. (Walter Benz now directs the famous Hamburg University Mathematics Institute founded by W. Blaschke, where he heads the modern school of circle geometry; this deals, in particular, with *finite* circle geometries.)

[266] See, for example, V.I. Arnold's book mentioned in Note 208.

[267] The term "*W*-curves", in Klein's and Lie's work, derives from the German word *der Wurf*, whose mathematical meaning is hard to explain to the modern reader. The literal translation is "the number of points for a throw of dice." Von Staudt uses the term in the sense of the *cross ratio* of four collinear points *A*, *B*, *C*, and *D* understood in the purely projective sense (without using distance, which is a notion of metric or Euclidean geometry) or in the sense of the *simple ratio* of three collinear points (a notion appearing in affine geometry). It was precisely his desire to stress the independence of this notion from nonprojective concepts that led von Staudt to coin the new term *der Wurf*. An elegant exposition of von Staudt's "Wurf calculus" in a language close to the modern is given in T.W. Young's short, beautifully written book *Projective Geometry*, 1983 (The Carus mathematical monographs series). The

key role in von Staudt's "Wurf calculus" is played by the extensive group of projective self-transformations of the straight line; it was this fact that led Klein and Lie to call *all* curves possessing projective self-transformations "Wurf curves" or, briefly, "*W*-curves."

[268] For a simple solution to the problem of finding all the curves with "self-similitudes" see, for example, I.M. Yaglom and V.G. Ashkinuze, *Ideas and Methods of Affine and Projective Geometry*, Part I, *Affine Geometry*, Moscow, Uchpedgiz, 1962 (in Russian), problem 234a and its solution.

The topic of *W*-curves is studied, for example, in W. Blaschke, *Vorlesungen über Differentialgeometrie*, Bd. II—*Affine Differentialgeometrie*, Berlin, Springer, 1923. Also, compare problem 234b and its solution in the Yaglom–Ashkinuze book mentioned above, where all the curves with a group of affine self-transformations are listed. The general *W*-curves can be described as "projective modifications" of the affine *W*-curves discussed in the Yaglom–Ashkinuze book.

Chapter 7

[269] Another general approach to geometry, which does not include projective or circle geometry, but does include in the list of geometries not only Euclidean, hyperbolic, and elliptic geometries but also certain "curved" spaces, is sketched in Riemann's 1854 lecture. But as we pointed out earlier, this metric approach (i.e., based on the notion of distance) was noticed and appreciated only much later.

[270] Actually, the idea that a geometric figure is an arbitrary set of points is too general for geometry: under this definition the study of all figures becomes the subject of *set theory* and *topology*, but certainly not of geometry. In order to make the word "figure" meaningful from the geometric point of view, it is necessary to restrict the family of admissible point sets (compare, for example, I.M. Yaglom and V.G. Boltyanski, *Convex Figures*, N.Y., Holt, Rinehart and Winston, 1961, Appendix II, "On the concepts of convex and nonconvex figures").

[271] Thus, for example, the *meter* was once defined as a forty-millionth part of the Paris meridian. But this definition is no longer the accepted one. The meter was later defined as the distance between two parallel marks on a platinum etalon at 0°C, kept at the Breteuil pavillion at Sèvres (France); now it is defined in terms of wavelengths corresponding to a certain point of the spectrum. The choice of this point is based on ideas and observations of physical chemistry (more precisely, spectral analysis), and not of geometry. (Perhaps a more evocative illustration of the fact that the definition of units of length is not the business of "pure mathematics" is the definition of the *yard*,

still the officially accepted measure of length in the English system of measures a few years ago: the yard was originally defined as the distance from the tip of King Henry I's nose to the end of middle finger of his extended right arm.)

272 Note that unlike units of length, angular units are defined in purely geometric terms. Thus a right angle is half of a straight angle; a degree is one three-hundred-sixtieth part of a full angle; the radian is the central angle corresponding to a circular arc whose length equals the circle's radius. Compare how lengths and angles appear in the statement of the well-known theorem that if the angles C and B of triangle ABC are respectively equal to $90°$ and $30°$, then $AB : AC = 2 : 1$.

273 In the geometry where figures (in particular, triangles) are determined "up to similitude" (i.e., similar figures are viewed as identical or "equal"), the triangle (in this sense) is determined by only *two* independent elements or conditions instead of three. Thus one "typical construction problem" of ordinary school geometry is, say, the problem of constructing a triangle ABC given its sides AB, BC, and its angle B, and another (well known!) problem is that of constructing $\triangle ABC$ given its median BM, bisector BN, and altitude BP. In the "geometry of similar figures" we can mention construction problems such as that of finding ("constructing") a triangle ABC given the ratio $AB : BC$ of two of its sides and the angle B (recall that the notion of length has no meaning in this geometry, while the notion of the ratio of lengths does!), and the (well known) problem of constructing $\triangle ABC$ given that the four angles into which the angle B is divided by the median BM, the bisector BN, and the altitude BP are equal. Note that the assumption in this problem can be stated in the form of *two* relations, e.g., $\angle ABC = \angle MBN$ and $\angle NBP = \angle PBC$.

274 The difference between these two geometries is nicely illustrated by the following fact. In the "geometry of isometry" the only "homogeneous" curves (curves all of whose points are equivalent—a consequence of the existence of a transitive group of self-transformations) are the straight lines and the circles, whereas in the "geometry of similitude" there are other such curves—the logarithmic spirals (see Chapter 6, in particular Fig. 26).

275 The passage from the set $\mathfrak{M}$ to the "set of classes" is described as passing to the *quotient set* of $\mathfrak{M}$ with respect to the equivalence relation "$\sim$". The quotient set is denoted by $\mathfrak{M}/\sim$. Nowadays these notions appear in most school mathematics courses.

276 The group $\mathfrak{I}$ of direct isometries can be described analytically (i.e., by using coordinates) by the equations (6.1) which relate the (rectangular Cartesian) coordinates x, y of the given point $M(x, y)$ and the coordinates x', y' of its image $M'(x', y')$. The group $\mathfrak{A}$ of affine transformations of the plane is

described analytically by equations (6.2) (where now x, y and x', y' are arbitrary linear or affine coordinates, since rectangular Cartesian coordinates are meaningless in affine geometry). Similarly, the group $\mathfrak{P}$ of projective transformations can be described as the group of transformations

$$M(x_0 : x_1 : x_2) \mapsto M'(x'_0 : x'_1 : x'_2)$$
$$\equiv M'(a_{00}x_0 + a_{01}x_1 + a_{02}x_2 : a_{10}x_0 + a_{11}x_1 + a_{12}x_2 : a_{20}x_0 + a_{21}x_1 + a_{22}x_2)$$

where $x_0 : x_1 : x_2$ and $x'_0 : x'_1 : x'_2$ are projective coordinates of the given point and its image (see Chapter 3) and $\Delta = |a_{ij}| \neq 0$; here Δ is the third-order determinant with entries a_{ij}, $i, j = 0, 1, 2$. The group $\mathfrak{A}$ can also be defined as the set of all one-to-one maps of the ordinary or affine plane Π_0 sending each line a into a line a'; the group $\mathfrak{P}$ as the set of maps of the projective plane Π with the same property (see, for example, the book by I.M. Yaglom, *Geometric Transformations III*, mentioned in Note 73).

[277] We can also suppose that $\mathfrak{P} \supset \mathfrak{A}$, where $\mathfrak{P}$ is the group of projective transformations of the plane; then we must assume that the "domains of action" of the projective and affine transformations are the same, i.e., we must extend the affine (or Euclidean) plane to the projective plane by adding to the former the "line at infinity", consisting of "points at infinity." In this sense each theorem and notion of projective geometry acquires a meaning in affine (Euclidean) geometry. Thus the very important projective notion of *cross ratio* $(A, B; C, D)$ of four points on a line can be described in affine or Euclidean geometry as the "ratio of simple ratios": $(AC/BC)/(AD/BC)$. Then the problem, so brilliantly solved by von Staudt, consists in showing that a cross ratio can be described in the "language of projective geometry," without using the "simple ratios" AC/BC and AD/BD, which are meaningless in this geometry.

Also note that our statement, according to which $\mathfrak{G}_1 \supset \mathfrak{G}_2$ implies that each theorem in the geometry Γ_1 holds in the geometry Γ_2, does not include the cases when the group $\mathfrak{G}_1$ appears explicitly in the statement of the theorem. For example, the theorem in the "geometry of similitude" asserting that the logarithmic spiral is a "homogeneous" curve, i.e., has self-transformations sending any given point on it into any other given point on it, is false in Euclidean geometry. In the same way Lie's and Klein's *W*-curves include all the "affine homogeneous curves" but are not limited to them, etc.

[278] Plücker is the creator, so to speak, of "analytic rectilinear geometry." The main objects of "differential rectilinear geometry" (which can be developed in projective as well as in Euclidean space) include one-parameter families of lines or *ruled* surfaces (e.g., the so-called *demiquadrics*, i.e., sets of all straight lines intersecting three fixed straight lines in space; the term "demiquadric" is explained by the fact that in this way we obtain one of the two families of straight lines filling up a "quadric surface", determined by a quadratic equation in space coordinates x, y, z) or *developable surfaces*, formed

by all the tangents to a (smooth) curve in space. Other examples are two-parameter families of straight lines, the so-called *congruences* of lines, e.g., normal *congruences* generated by all the normals to a fixed surface (which proved so interesting to the early masters of geometric optics, e.g., to W.R. Hamilton), or three-parameter sets of lines, the so-called *complexes* of lines. Here all the families of lines must be determined by "smooth" (differentiable) functions of one or several parameters $p_{12} = p_{12}(t)$ or $p_{12}(u, v)$, etc., where $p_{12}, \ldots$ are the Plücker coordinates of the line (see Note 89).

[279] When we speak of isometries of the plane, we always have in mind the group of direct isometries (6.1), thus excluding from the class of isometries fixing a point ξ all the reflections in lines passing through ξ. In Euclidean space, it is more convenient to assume straight lines to be supplied with an orientation, so that the self-transformation group of such a line reduces to translations along the line (reflections in points of the line are excluded).

[280] As an example of an investigation where this general outline works, we note S.S. Chern's article "On Integral Geometry in Klein Spaces," *Annals of Math.*, **43**, 1942, pp. 178–189, also included in S.S. Chern, *Selected Papers*, N.Y., Springer, 1978. Beginning with the mid-thirties, Wilhelm Blaschke (whom we have mentioned previously, e.g., in Note 265) began to develop a new branch of geometry, which he called *integral geometry*, apparently hoping that this new direction of geometric research would soon aquire an importance comparable to that of classical differential geometry. [Incidentally, these expectations did not come true, so that the well-known Moscow mathematician Izrael Moisseievich Gel'fand (b. 1913) even proposed to steal the promising title "integral geometry" from Blaschke's work (since it had not lived up to expectations), and now the term is used more often in Gel'fand's sense. (Cf. W. Blaschke, *Vorlesungen über Integralgeometrie*, Berlin (DDR), Deutscher Verlag der Wissenschaften, 1955; Luis A. Santaló *Integral Geometry and Geometric Probability*, Reading (Mass.), Addison-Wesley (Encyclopedia of Mathematics and Its Applications (ed. Gian-Carlo Rota, Vol. 1) and, on the other hand, I.M. Gel'fand, M.I. Graev, N.Ya. Vilenkin, *Generalized Functions* v. 5, *Integral Geometry and Representation Theory*, N.Y., Academic Press, 1966).] Blaschke's main idea was to compare the measures of elements of different nature coexisting within one geometrical system, for example, the measures of point sets and sets of lines in plane Euclidean geometry (such as the measure of a "pointwise" curve and that of the set of all lines intersecting it) in order to obtain meaningful geometric conclusions. But in 1942, Blaschke's most outstanding pupil, Shiing-Shen Chern (b. 1911), who came from China, studied under Blaschke in Hamburg, taught in China, and later became an American citizen, "closed" the entire direction of Blaschke's research by considering the most general situation, making the study of its particular cases unnecessary (see the article cited above). That is, Chern considered an arbitrary homogeneous Klein space and two distinct generating elements in it, i.e., the group $\mathfrak{G}$ and two distinct subgroups $\mathfrak{g}$ and $\mathfrak{h}$, and

established the connection between the "cosets of $\mathfrak{G}$ with respect to $\mathfrak{g}$" and the "cosets of $\mathfrak{G}$ with respect to $\mathfrak{h}$". He thus carried out in the general case what Blaschke and his pupils, headed by the Argentinian L.A. Santalo, had been doing for specific homogeneous Klein geometries and specific objects in these geometries (compare: L.A. Santalo, *Integral Geometry and Geometric Probability* (Encyclopedia of Mathematics and Its Applications, Vol. 2), Reading, Mass., Addison-Wesley, 1976).

281 It is easy to check (the reader will profit by actually doing it!) that if $\mathfrak{G}$ is the group (6.1) of direct isometries, while $\mathfrak{g}$ is the subgroup of all rotations about the origin:

$$x' = \cos\alpha \cdot x + \sin\alpha \cdot y, \qquad y' = -\sin\alpha \cdot x + \cos\alpha \cdot y$$

then the cosets in the three-dimensional group space (a, b, α) (more precisely, in the layer between the identified planes $\alpha = 0$ and $\alpha = 2\pi$ of this space) can be represented by "vertical sticks" (we assume the α-axis vertical) $a = a_0 (= \text{const})$, $b = b_0 (= \text{const})$. The group $\mathfrak{G}$ interchanges these sticks just as it does the points of the plane $\alpha = 0$, i.e., this group coincides with the ordinary isometry group of the plane. If $\mathfrak{h}$ is the group $x' = x + a$, $y' = y$ of translations along the x-axis, then the cosets are represented by the lines of our layer parallel to the (a, b)-plane: $\alpha = \alpha_0$ $(= \text{const})$, $b = \tan\alpha_0 \cdot a + l$. The group $\mathfrak{G}$ interchanges these lines just as the group of plane isometries of the (a, b)-plane interchanges the lines which are the projections on the plane $\alpha = 0$ of the lines representing the cosets described above, so that we actually obtain line-element Euclidean geometry.

282 Both the geometry of three-dimensional space with the isometry group (3) (sometimes called three-dimensional *semi-Euclidean geometry*) and the "geometry of the Lorentz group" (known as Minkowski *pseudo-Euclidean* geometry, cf. Chapter 4) are "projective metrics" (or non-Euclidean Cayley–Klein geometries; cf. Chapter 4). A more detailed exposition of the connections which here arise between non-Euclidean geometry and mechanics, see the book (written for a wide public): I.M. Yaglom, *A Simple Non-Euclidean Geometry and its Physical Basis*, mentioned in Note 159.

Chapter 8

283 See the letter written by Klein's widow to Young's mother in Young's *Mathematicians and Their Times*.

284 In particular, Lie's lengthy study of "infinite" continuous groups, i.e., continuous groups consisting of transformations that cannot be made to depend on a finite set of parameters, interested Elie Cartan. In Cartan's view, Lie's long memoir fell so short of the standards of rigor prevailing in Cartan's time, that he regarded it as a stimulus for the imagination rather than a

predecessor's research work to be continued. [Infinite continuous groups include, for example, the group (considered locally, i.e., in the neighborhood of one point) of conformal transformations (transformations that preserve angles) of the Euclidean plane x, y determined by the formulas $x = u(x, y)$, $y = v(x, y)$, where u and v are arbitrary functions of two variables satisfying the Cauchy–Riemann equations $\partial u/\partial x = \partial v/\partial y$, $\partial u/\partial y = -\partial v/\partial x$. On such groups see S. Lie, "Untersuchungen über unendliche kontinuierliche Gruppen," *Berichte Sachs. Geselschaft*, **21**, 1895, S. 43–150; included in *Gessammelte Abhandlungen*, **6**, S. 396–493.]

[285] An example (better known to Russian than to English-speaking readers) of a book featuring the "solitary mathematician,"—and depicting Sophus Lie as the mathematician *par excellence*, is the novel *Pussycat Letayev*. This is the first book of the unfinished tetralogy *Moscow* by one of the most interesting Russian writers of the beginning of the 20th century, Andrei Bely (his pen-name; his real name is Boris Nicolaevich Bugayev, 1880–1934), poet, novelist, author of memoirs and works on the theory of literature. The image of Nicolai (Pussycat) Letayev's father, the world famous professor Letayev (in the novel he corresponds with Hermite and is highly regarded by Poincaré, though insufficiently appreciated by Weierstrass), passes through the entire novel; he is aloof from all earthly concerns and engulfed in science, especially in Lie's works. In particular, the author repeats the hero's childhood recollection of his father slinking through the hall of the flat to the bathroom with a candle in his hand (the action takes place in the 19th century) and a volume of Lie under his arm. A rather original personality, Andrei Bely received a mathematical education from Moscow University, where his father Nikolai Bugayev (1837–1903) taught. Hardly remembered today, Nikolai Bugayev was regarded at the time as a leading Russian expert in number theory. He was undoubtedly the model for professor Letayev in the novel. Apparently Bely's mathematical education helped him write his fundamental works on mathematical methods in poetics, works recognized only many years after the author's death and then continued by the famous Moscow mathematician Andre Kolmogorov (born in 1903). Curiously enough, some of the calculations of Bely the mathematician were used by Bely the poet in his verse.

[286] The lecture was first published as *Vergleichende Betrachtungen über neuere geometrische Vorschungen*, Programm zu Eintritt in die philosophische Facultät und den Senat der Universität zu Erlangen, Erlangen, Deichert, 1872. The *Erlangen program* has now been translated into practically all the European languages. It has been published many times in German; for example, it was included in Volume 1 of F. Klein, *Gesammelte mathematische Abhandlungen*, Bd. I, 1921, S. 460–497. In this connection see the concluding part of the present chapter; Klein's *Collected Mathematical Works* were reissued in 1973 (Springer, Heidelberg). In his collected works Klein added to his articles brief—but most valuable—historical and scientific comments. In particular, à propos the "Erlangen program," he mentions that Lie came to Göttingen

at his invitation on September 1, 1872, and together they edited Lie's forthcoming articles (which were close to the Erlangen program's main ideas) as well as the text of Klein's Erlangen lecture. Klein recalls that Lie was immediately very enthusiastic about the main idea of the lecture, and that this encouraged him greatly. On Lie's advice Klein replaced his initial formulation of "different geometric methods, generated by various transformation groups" (this formulation remained in the second part (*zweiter Aufsatz*) of the article "Über die sogenannte nicht-euklidische Geometrie", published later than the "Erlangen program" but written earlier) by the words "different geometries, generated by various transformation groups." Thus this phrase, first stated in October 1872 in Klein's Erlangen lecture and published soon afterwards, the phrase which we identify with "Klein's Erlangen program", was in form (though not in substance) due to Lie rather than to Klein.

[287] Both lectures—Riemann's and Klein's—outlined the possibility of a broad generalization of Euclidean geometry, including hyperbolic and elliptic geometry as particular cases. However, Euclidean, hyperbolic and elliptic geometries are the only schemes which are both Riemann spaces and Klein spaces. [In the subsequent development of Riemann's and Klein's geometry these constructions were widened so that the situation generalizing these two cases would include both Riemann space and Klein space. Here is how this was done. Riemann space R can be imagined as (generally speaking) a "curved" manifold, each point of which has a neighborhood "that looks like Euclidean space." In other words, to each point M of a Riemann space R we can assign a Euclidean space "tangent to R at the point M." In modern terminology this assignment is called the *tangent bundle* of a differentiable (smooth) manifold, whose base is the manifold R itself and whose fiber is the Euclidean space tangent to R at the given point. Neighboring tangent spaces are related (in a certain sense) because if $M_1, M_2 \in R$ are close to each other, then they belong to the same "almost-Euclidean" domain in R. Similarly, if we assign to each point M of a curved (but differentiable) manifold T a flat Klein space (of the same dimension) of an *a priori* chosen type, e.g., an affine space (or a projective, or a conformal space), then we obtain a so-called affine (or projective, or conformal) connection space. In terms of tangent bundles we can say that we have a bundle whose "fibers" are affine, or projective, or conformal spaces (the latter space being the space whose "isometry group" is Möbius's group of circle transformations).

These general concepts are due largely to Hermann Weyl and to Elie Cartan both pupils of Hilbert and Klein. These studies began as attempts to generalize Einstein's general relativity theory by using the notion of a Riemann space; the first such attempts were due to Weyl and the outstanding English physicist Arthur Stanley Eddington (1882–1944). The modern form of the notion of a space with a connection is due to the outstanding French geometer Charles Ehresmann (see, for example *Les connexions infinitésimales dans un espace fibré différentiable*, Col. de topologie, Bruxelles, 1950, pp. 29–55). The relationship between a "general space with a connection" and a homogeneous (uncurved) Klein space and a Riemann space may be conveyed by the following (purely symbolic!) "equality of ratios"

Klein space : Euclidean space

= general space with a connection : Riemann space.

288 See, for example: F. Klein, *Vorlesungen über höhere Geometrie*, Heidelberg, Springer, 1968; L. Bieberbach, *Einleitung in die höhere Geometrie*, Leipzig, Teubner, 1933.

289 See the section on algebra in F. Klein, *Elementarmathematik vom höheren Standpunkt aus*, Bd. I, Heidelberg, Springer, 1968. Klein also deals in some detail with this subject in the book mentioned repeatedly above: *Vorlesungen über die Entwicklung der Mathematik im 19 Jahrhundert* (in the concluding chapter of the first part). Klein's book mentioned in the main text has been translated into English as F. Klein, *Lectures on the Icosahedron and the Solution of Equations of the Fifth Degree*, N.Y., Dover, 1956.

290 From the current viewpoint, Klein's oversight seems inexplicable. Given his deep understanding of non-Euclidean geometries, we find it baffling that he overlooked the simple connection between his own (projective) model and Poincaré's (conformal or Möbius-type) model of plane non-Euclidean geometry. In fact, there are many ways of expressing this connection.

For example, if we project Poincaré's circle model of radius 2 located in a plane π from the north pole of a sphere Σ of radius 1 tangent to π at the center O of the model onto the lower hemisphere of Σ, and then project the resulting image back perpendicularly down on π, then we obtain Klein's model. On the other hand, if we apply to the Lobachevskian plane, given by its Klein model (a circle K with center O), a contraction with ratio 1/2 (i.e., a homothety with center O and coefficient 1/2, assigning to each point A inside the circle K the point A' of the interval OA such that $d_{OA'} = d_{OA}/2$, where d denotes the "hyperbolic distance" between points), then we obtain the Poincaré model. In this connection see Section of Chap. X of Klein's book quoted in Note 107 or the Appendix to Chap. II of I.M. Yaglom's book *Geometric Transformations*, Gostekhizdat, 1956 (in Russian). Nevertheless, in the context of the 1870s, it was hardly obvious that there could be any connection between such distant topics as the theory of automorphic functions of a complex variable on the one hand and non-Euclidean geometry on the other.

291 At the turn of the century Henri Poincaré, one of the greatest French mathematicians, contributed, often decisively, to the founding of a number of new branches of mathematics (for example, *topology*) and to the development of existing branches. Poincaré's scientific interests were distinguished by great breadth, including, besides mathematics, physics (where he should be regarded

as one of the founders—on a par with Albert Einstein—of the special theory of relativity), mechanics, and astronomy. Poincaré's outstanding literary talent enabled him to influence through his articles and textbooks even those fields of science where he had no particular achievements to his credit, for example the theory of probability. This same talent secured his membership both in the French Academy of Sciences (L'Institut), as well as (a very rare case indeed for a scientist) in the famous (literary) *Académie Francaise*. See the English translations of the works on science and philosophy which motivated Poincaré's election to the academy: *Science and Hypothesis, Science and Method*, and *The Value of Science*, also published in a single volume called *The Foundations of Science*, Science Press, 1964. Poincaré, a pacifist, also coined the famous phrase, popular throughout the world before World War I, about his cousin and friend Raymond (later president of France) "Poincaré la guerre": Henri did not share his cousin's political views.

The literature on Poincaré is very extensive. Part of it is listed in the concluding section of the first volume of Klein's *Vorlesungen über die Entwicklung der Mathematik im 19 Jahrhundert*, which is dedicated to Poincaré Klein intended to devote a special chapter in the book to Poincaré, and one to Sophus Lie, but simply did not have time. The book was never finished, and the second volume was published posthumously. Here we limit ourselves to citing the article by Jean Gaston Darboux ("Éloge historique d'Henri Poincaré") in the supplement to the second volume of H. Poincaré, *Oeuvres*, v. 1–11, Paris, Gauthier-Villars, 1916–1956; the chapter devoted to Poincaré (*The Last Universalist*) in the oft-referred to book by E.T. Bell, *Men of Mathematics* and the books: Toulouse, *Henri Poincaré*, Paris, 1910; T. Dantzig, *Henri Poincaré*, N.Y.–London; and A. Bellivier, *Henri Poincaré ou la vocation souveraine*, Paris, 1956.

[292] Poincaré's model of Riemann's elliptic geometry is constructed in much the same way (see, for example, the literature in Notes 288 and 293).

[293] This model is interpreted in ample (perhaps even more than ample) detail in, for example, the D. Pedoe, *A Course of Geometry for Colleges and Universities* (Ch. VI), G. Ewald, *Geometry: An Introduction* (Chapters 6 and 7) (see Note 72), as well as in I.M. Yaglom, *Complex Numbers in Geometry* (see Note 203), Sections 11 and 17; I.M. Yaglom, *Geometric Transformations II*, Supplement to Chapter II (in Russian; see Note 290).

[294] See, for example, the brief Section 3 of Chapter XI in Klein's book mentioned in Note 107.

[295] See *Bulletin de L'Institut Général de Psychologie*, Paris, 1908, N 3 (8 année); also reprinted in the book mentioned in Note 291: Poincare's *Science and Method*.

[296] See, for example: H. Schwerdtfeger, *Geometry of Complex Numbers*, N.Y., Dover, 1979, or I.M. Yaglom, *Complex Numbers in Geometry*, and *A Simple Non-Euclidean Geometry and Its Physical Basis*, mentioned above. The connection between linear fractional transformations of a complex variable and Möbius transformations is dealt with in many other books on geometry and on the theory of functions of a complex variable.

[297] Klein once complained that while in his youth a knowledge of the whole range of questions dealing with automorphic functions, Abelian integrals, and basic algebraic geometry was regarded as absolutely necessary for every mathematician, he lived to see a period when young researchers showed no interest in this entire set of ideas. But in our time the subject is experiencing a revival (compare this with what we said about Lie's theory of differential equations)—international conferences are being held on the theory of Kleinian groups and many relevant monographs are being published (see, for example, I. Kra, *Automorphic Forms and Kleinian Groups*, Reading, Mass., Benjamin, 1972.

[In modern literature the term "Kleinian groups" is encountered much more frequently than the term "Fuchsian groups" (referring to Immanuel Lazarus Fuchs, 1833–1902, German mathematician and Weierstrass's pupil) introduced by Poincaré. This, however, does not mean that we now see Poincaré as having been "defeated" in the rivalry with Klein: Poincaré's works on these questions are also widely used by modern mathematicians.]

[298] See C. Reid, *Courant in Göttingen and New York*, N.Y., Springer, 1976 pp. 230–232, on the origin of the famous book R. Courant and H. Robbins, *What Is Mathematics?*, London, Oxford University Press, 1948; 1st edition 1942. The book was largely written by Herbert Robbins, now a leading authority in mathematical statistics and probability theory, but then only beginning his work in mathematics as a topologist, while the famous Richard Courant was the driving force and leader of the whole enterprise. Courant wanted to leave only one name on the title page, but changed his position immediately when Robbins said "We don't live in Germany."

[299] We note that only the second of the two Klein–Fricke books has two coequal authors on the title page; only Klein was named as the author of the first book, although the title page states that it was *ausgearbeitet und vervollständigt von R. Fricke.*

[300] Thus *Vorlesungen über höhere Geometrie* was compiled by Wilhelm Blaschke, who made much use of Klein's mimeographed lecture notes. (Officially Blaschke was only considered to be the author of Chapters 77–81, which were written without reference to Klein's lectures.) *Vorlesungen über die Entwicklung der Mathematik* were prepared for print by Courant, Otto Neugebauer (b. 1899) and others. Blaschke, Courant, and Neugebauer are all leading scholars.

[301] In Chapter 2 we have already had occasion to compare Klein with the Moscow physicist L.D. Landau. Klein's traits pointed out here also relate him to Landau (and the Moscow mathematician I.M. Gel'fand). Both Landau and Gel'fand absorb very quickly information communicated to them orally, and always prefer this method of learning about new developments in science to reading books and papers. Both also "talk about" works with their coauthors but never write them themselves—which, however, does not diminish in the least their status as authors. The opposite psychological type among Moscow mathematicians, perhaps similar to Sophus Lie, is represented by A.N. Kolmogorov, who absorbs oral information relatively poorly (always transforming it considerably in his mind) and is the sole author even of works appearing as joint publications—the work can be discussed with coauthors but as a rule Kolmogorov writes it alone.

[302] The situation was not saved by a later attempt to translate Klein's encyclopedia into French; this attempt was abandoned early, with only a small part of the work done. It was intended that every entry be the work of an appropriate French mathematician, who would not so much translate the entry as bring it up to date. Thus Cartan was translator and, in this sense, "coauthor" of Study's entry on (hyper) complex numbers.

[303] Perhaps it was the danger that the mathematical sciences would lose their unity and splinter into isolated islands of knowledge that in some way (still unclear to us) contributed to the appearance in the 20th century of a number of universal talents (such as Gauss and Riemann had been in the 19th century). These were David Hilbert, Henri Poincaré, Hermann Weyl, John (in Hungary Janos, in Germany Johann) von Neumann (1903–1957), and Andrei Kolmogorov (b. 1903). The same circumstances were conducive to the appearance of the great mathematician Nicolas Bourbaki whose "figure" marked the transition from individual to group creative effort. Significantly, the well-known Bourbaki article "L'Architecture de mathématique" (in the collected articles *Les grands courants de la pensée mathématique*, F. Le Lionnais, Cahiers du Sud, 1948, p. 35–47, or their English translation in the American Math. Monthly, **57**, 1950, pp. 221–232 or in F. Le Lionnais, *Great Currents of Mathematical Thought*, N.Y., Dover, 1971, pp. 23–36), which may be called the manifesto of the Bourbaki group, begins with a section entitled "La Mathématique ou les Mathématiques?" with the authors favoring mathematics in the singular rather than in the plural.

[304] Compare the words of one of the founders of the Bourbaki group, Jean Dieudonné, in his paper "The Work of Nicolas Bourbaki" *Amer. Math. Monthly*, **77**, 1970, pp. 134–145, which points out, in particular, the differences between Bourbaki's *Elements of Mathematics* and the *Encyclopedia of Mathematical Sciences*, and the difficulties which caused the failure of Klein's enterprise. The differences included, among other things, the fact that the Bourbaki group was much more tightly knit than the group of the Encyclopedia's

contributors. Klein's all-inclusiveness, singled out by Dieudonné as one of the encyclopedia's shortcomings, can be illustrated by the fact that at first Klein even ordered an entry on elementary geometry (a review of the geometry of triangle, the geometry of the circle, and other doubtful "sciences"). However Klein ultimately rejected this entry (see the paper mentioned in Note 260, "Elementary Geometry, Then and Now").

[305] A superb description of his life and research is contained in the article by H. Weyl, "David Hilbert and His Mathematical Work," *Bulletin of the American Mathematical Society*, **50**, 1944, p. 612–654, reproduced in H. Weyl, *Gesammelte Abhandlungen*, Bd. 4, Berlin–Heidelberg–New York, J. Springer, 1968, S. 130–172; also, see C. Reid's books *Hilbert*, Berlin–Heidelberg–New York, J. Springer, 1970, and *Courant in Göttingen and New York* referred to in Note 298. The experience of Göttingen University was undoubtedly taken into account in the founding of other scientific centers in particular, the mathematical and physical centers) so abundant in our age. Examples include the famous *Institute for Advanced Study* in Princeton, where Albert Einstein, Hermann Weyl, John von Neumann, and many other leaders of 20th-century science worked; the French *Institut des Hautes Etudes Scientifique* at Bures-sur-Yvette near Paris; numerous scientific towns in the Soviet Union, like the one near Novosibirsk, incorporating Novosibirsk University and serving all of the USSR east of the Urals; and so on. Finally, New York University's *Courant Institute of Mathematical Sciences*, founded by Courant on the basis of his experience as director of the Göttingen Mathematical Institute, was a direct offspring of the latter.

[306] D. Hilbert (see the literature listed in Note 305) made major (often decisive) contributions to nearly all branches of modern mathematics—geometry, algebra, mathematical analysis. He is justly regarded as the founder of functional analysis. He contributed to logic and to theoretical physics; for his contribution to the theory of relativity see J. Mehra, *Einstein, Hilbert and Theory of Gravitation*, Holland and USA, Reidel, 1974—but this book probably exaggerates Hilbert's achievements. Hilbert's amazing universality was best expressed in his famous report *Mathematical Problems*, delivered at the International Mathematics Congress in Paris in 1900, in which he stated 23 mathematical problems that the 19th century bequeathed to the 20th. The report (see the anotated English edition pointed out in Note 232) was a program of sorts for the subsequent development of mathematics.

Born in Königsberg, Hilbert graduated from a gymnasium there and was first a student and then a professor at Königsberg university. The well-known algebraist and outstanding teacher Heinrich Weber (1842–1913) was his tutor. Felix Klein gladly acknowledged Weber's immense pedagogical talent and leading role in the rise of modern mathematics. It was highly characteristic of Weber that he approached mathematics as a unified science all of whose branches were closely linked. He adopted that idea from Riemann, whom he revered. Weber published a number of Riemann's works after the latter's death, including a lecture course on the theory of partial differential equations which was popular for a long time as a "Riemann–Weber" book. However, the book was apparently a long way from Riemann's original course. It was in Königsberg that Hilbert's close friendship with two other pupils of Weber began. They were the great geometer and number theorist Minkowski and the outstanding analyst Adolf Hurwitz (1859–1919). That friendship also contributed to the enlargement of Hilbert's mathematical interests.

[307] See the book referred to in Note 298, *Courant in Göttingen and New York*.

[308] The "pure theoretician" Born agreed to head the physics institute in Göttingen under the condition that his old friend from student years James Franck (1882–1964) would direct experimental work there. The symbiosis of Born the theoretician and Frank the experimenter proved highly successful. Franck subsequently won the Nobel Prize in physics. (For Göttingen as a physics center, see, for example, the fascinating book by Robert Jungk, *Brighter Then a Thousand Suns*, N.Y., Harcourt Brace Jovanovich, 1958.)

[309] On the latter see, for example, P. Goodchild, *J. Robert Oppenheimer*, Boston, Houghton Mifflin, 1981, and *Brighter Than a Thousand Suns* (Note 308).

[310] In particular, Klein regularly delivered lectures and lecture courses for teachers of German secondary schools. These resulted in his book *Elementarmathematik vom höheren Standpunkt aus*, Bd. I–III (Heidelberg, Springer, 1968), and a few smaller publications, mostly mimeographed in Göttingen.

[311] Wilhelm Frederick Ostwald was undoubtedly a first-rate scientist; in 1909 he received the Nobel Prize in chemistry. At the same time Ostwald was an original philosopher—for example, the well-known division of all scientists into classicists and romanticists, as well as revealing thoughts on the nature of research, are due to him. Ostwald must also be credited with popularizing the classical legacy in science; he founded, and was the first editor of, the well-known series *Ostwald's Klassiker der exakten Wissenschaften*, the importance of which in popularizing classical works of science is difficult to overestimate. His personality was complex. In his time he was a bitter enemy of the atomic theory of matter. He proclaimed a philosophy known as "energetism," which declared energy the basic element of the real world. His extremely sharp and largely unreasonably criticism of the eminent Austrian physicist Ludwig Boltzmann (1844–1906), a greater scientist of greater stature than himself (this is obvious today, but was not clear at all to his contemporaries), was perhaps responsible for Boltzmann's tragic suicide. Ostwald's chauvinism, traces of which can be seen in his book *Grosse Männer*, is also hardly to his credit.

[312] It is necessary to point out, however, that the leading 20th-century French mathematician Jacques Solomon Hadamard (1865–1963), who was no doubt personally acquainted with Klein, accused the latter (in his well-known book on the psychology of mathematics) of nationalism and chauvinism. (See the enlarged and revised French edition: J. Hadamard, *Essai sur la psychologie de l'invention dans le domaine mathématique*, Paris, Librairie Scientifique Albert Blanchard, 1960.) Hadamard refers to one of Klein's publications of 1893, in which the latter says that "it seems that a strong intuition of space was intrinsic to Teutonic science, while the purely logical, critical spirit is better developed among the French and Jewish races". Hadamard adds that since Klein undoubtedly valued intuition above logic, it follows that he believed in the superiority of the Teutonic race over the French and the Jews. Hadamard's bitterness in his judgment of the German Klein doubtless reflected the terrible time when the book was being written—it was based on a course of lectures delivered by Hadamard in New York in 1943, following his exile from France. Of course, Klein considered himself a German, perhaps even a Prussian (Laurence Young, who knew him, wrote that Klein combined the best traits of Prussians). It was pleasant for Klein to point out the merits and achievements of German science, just as it was pleasant for him to believe that hyperbolic geometry was discovered by Gauss alone and passed from him to Lobachevsky and Bolyai (cf. Chapter 4). However, just as Klein later completely rejected the idea of Gauss's influence on Bolyai and Lobachevsky (in the printed version of *Vorlesungen über nicht-euklidische Geometrie*), thereby acknowledging that this idea was incorrect, so too there is not a single word in his *Vorlesungen über die Entwicklung der Mathematik* that can be construed as supporting the (surely incorrect) statement of 1893—on the contrary, the book's import lies in the concept of joint work by scientists of all races and nationalities, contributing perhaps in different ways, but with equal merit, to the construction of the mathematical edifice. Klein highly valued the collaboration of members of different ethnic groups in this work (French, German, English, Irish, and Jewish). To be sure, it is not difficult to point out certain national schools (Russian, Italian) or trends in science (probability theory) which are not given sufficient attention in that book, but Klein can hardly be blamed; he wrote about those branches of mathematics which he knew best or with which he was directly involved. As to the St. Petersburg school in the theory of numbers and probability theory or, say, Italian research in tensor analysis or differential geometry, these were beyond his ken. The idea that Klein was a chauvinist and preferred "mathematicians–physicists" to "mathematicians–logicians" is also contradicted by C. Reid's account (in the book about Hilbert mentioned above) of the following episode. After Minkowski's death, Klein insisted that the post thus vacated at Göttingen should be offered not to Oskar Perron (1880–1975)—a German, a mathematician of the intuitive–geometric frame of mind, who was supported by most of the staff of the mathematics institute—but to Edmund Landau (1877–1938), a Jew and a mathematician of the purely logical type. (Landau

was perhaps even *overly* logical; for example he would always refuse to discuss the "general idea" of a proof with colleagues, insisting that the proof be written out in full.)

In connection with the different types of thinking allegedly intrinsic to different ethnic groups, I would like to recall a story told to me by my friend Lev Kaluzhnin (b. 1914), now a professor at Kiev University (the Ukraine), who studied under Issai Schur in Berlin and, after Schur's exile from Germany, under Emil Artin (1898–1962) in Hamburg. Kaluzhnin attended the notorious survey lecture (*Arische und Jüdische Mathematik*) presented at Berlin University by the well-known mathematician Ludwig Bieberbach (1886–1980), an excellent teacher but a fervent Nazi. As a Jew, Schur refused to attend. After the presentation, Kaluzhnin visited Schur, who asked him to relate the contents of the survey. After listening to Kaluzhnin, Schur said thoughtfully: "I quite agree with Bieberbach when he asserts that there exist mathematicians of completely different, in some ways opposite, psychological quality: for instance, Weierstrass could never understand Riemann, while Leopold Kronecker (1823–1891) could not comprehend the founder of set theory Georg Cantor (1845–1918). But it is hard for me to understand what racial differences have to do with it: both Weierstrass and Riemann were Germans, while Kronecker and Cantor were Jews." It is to Weierstrass's credit that he always supported Riemann, with whom he had little in common, although he believed that it would be disastrous for mathematics if Riemann's style became widespread. On the other hand, Kronecker attacked Cantor viciously—and in consequence Cantor lived out his days in a psychiatric clinic. Weierstrass had supported Cantor despite the fact that he too found Cantor's views quite alien. It would, however, be quite wrong to seek the sources of these differences in racial or religious differences; Kronecker was of the Mosaic faith, Cantor was a Lutheran and Weierstrass a Catholic. The mathematical disagreements were purely individual differences between individual scientists.

Printed in Dunstable, United Kingdom

91220853R00151